Lecture Notes in Mechanical Engineering

AF787805

Lecture Notes in Mechanical Engineering (LNME) publishes the latest developments in Mechanical Engineering—quickly, informally and with high quality. Original research reported in proceedings and post-proceedings represents the core of LNME. Also considered for publication are monographs, contributed volumes and lecture notes of exceptionally high quality and interest. Volumes published in LNME embrace all aspects, subfields and new challenges of mechanical engineering. Topics in the series include:

- Engineering Design
- Machinery and Machine Elements
- Mechanical Structures and Stress Analysis
- Automotive Engineering
- Engine Technology
- Aerospace Technology and Astronautics
- Nanotechnology and Microengineering
- Control, Robotics, Mechatronics
- MEMS
- Theoretical and Applied Mechanics
- Dynamical Systems, Control
- Fluid Mechanics
- Engineering Thermodynamics, Heat and Mass Transfer
- Manufacturing
- Precision Engineering, Instrumentation, Measurement
- Materials Engineering
- Tribology and Surface Technology

More information about this series at http://www.springer.com/series/11236

Alexander D. Pogrebnjak ·
Valentine Novosad
Editors

Advances in Thin Films, Nanostructured Materials, and Coatings

Selected Papers from the 2018 International Conference on "Nanomaterials: Applications & Properties"

Springer

Editors
Alexander D. Pogrebnjak
Department of Nanoelectronics
Sumy State University
Sumy, Ukraine

Valentine Novosad
Materials Science Division
Argonne National Laboratory
Lemont, IL, USA

Department of Nanoelectronics
Sumy State University
Sumy, Ukraine

ISSN 2195-4356 ISSN 2195-4364 (electronic)
Lecture Notes in Mechanical Engineering
ISBN 978-981-13-6132-6 ISBN 978-981-13-6133-3 (eBook)
https://doi.org/10.1007/978-981-13-6133-3

Library of Congress Control Number: 2018967728

© Springer Nature Singapore Pte Ltd. 2019, corrected publication 2023
This work is subject to copyright. All rights are reserved by the Publisher, whether the whole or part of the material is concerned, specifically the rights of translation, reprinting, reuse of illustrations, recitation, broadcasting, reproduction on microfilms or in any other physical way, and transmission or information storage and retrieval, electronic adaptation, computer software, or by similar or dissimilar methodology now known or hereafter developed.
The use of general descriptive names, registered names, trademarks, service marks, etc. in this publication does not imply, even in the absence of a specific statement, that such names are exempt from the relevant protective laws and regulations and therefore free for general use.
The publisher, the authors and the editors are safe to assume that the advice and information in this book are believed to be true and accurate at the date of publication. Neither the publisher nor the authors or the editors give a warranty, express or implied, with respect to the material contained herein or for any errors or omissions that may have been made. The publisher remains neutral with regard to jurisdictional claims in published maps and institutional affiliations.

This Springer imprint is published by the registered company Springer Nature Singapore Pte Ltd.
The registered company address is: 152 Beach Road, #21-01/04 Gateway East, Singapore 189721, Singapore

Preface

Recent progress in the science and technology behind thin films is unprecedented. While thin film materials possess widespread applications, such as in high-tech electronic devices or wear-resistant coatings, the availability of modern characterization and modeling tools for academic researchers enables a drastic improvement of our fundamental understanding of the physics and chemistry of these materials. Importantly, it is possible to target the end-user-oriented properties of various thin film systems by fine-tuning and controlling the synthesis process.

This book discusses several interesting aspects of this highly dynamic field of research from the personal perspectives of the participants of a special focus section on "Thin Films and Coatings" organized during the 2018 International Conference "Nanomaterials: Applications and Properties 2018", which took place September 9–14, 2018, in Zatoka, Ukraine. The book starts with a research article by I. I. Beilis and R. L. Boxman from Tel Aviv University, both of whom are experts in vacuum-arc evaporation and its various technological applications. Next, a series of papers by Pogrebnjak et al. from Sumy State University are presented; these explore the application of vacuum-arc evaporation to a fabrication of multilayer and multicomponent nitride-based protective coatings with enhanced physical–mechanical and tribological properties (including high hardness, resistance to corrosion and wear, and resistance to oxidation under the influence of high temperatures). Peculiarities of the fabrication process for the porous structures of oxide and metallic materials, using electrochemical deposition, are also discussed in detail in several research papers that deal with electrochemically grown thin films. Films and coatings involved in the polymerization of materials, as well as the deposition of thick coatings using high-speed plasma fluxes, are studied in the following articles. Several more papers, devoted to the electro-spark treatment of metallic materials with further nanostructuring, can also be found in the present publication. Considered examples of biomedical applications include ZnO-doped nanoparticles, and thermally treated biocompatible compounds, based on Ti-Ta-Nb or Ti-Ta-Nb-Zr alloys. Finally, the practical aspects of ion implantation to form nanoclusters in semiconducting materials are considered.

We believe that this series of papers could be useful for graduate students and various professionals interested in the synthesis-processing-structure-properties relationship between a variety of thin film systems and in understanding recent advances in the chemical and physical methods used for thin film deposition, including technologically relevant ion- and plasma-assisted processes.

Sumy, Ukraine — Alexander D. Pogrebnjak
Lemont, USA/Sumy, Ukraine — Valentine Novosad

Contents

Abstract

This book highlights the most recent advances in the chemical and physical methods used for thin film deposition and surface engineering, including ion- and plasma-assisted processes. This publication focuses on understanding the synthesis-processing-structure-properties relationship between a variety of thin film systems. The topics of interest include progress in thin film synthesis, new materials in thin film form (such as nitrides, oxides, intermetallic compounds, ultra-hard, wear-, oxidation-resistant, and multifunctional coatings; superconducting, magnetic, semiconducting, and dielectric films; electrochemical and electroless depositions) thin film characterization and instrumentation, and industrial applications.

Thin Film Deposition by Plasma Beam of a Vacuum Arc with Refractory Anodes

I. I. Beilis and R. L. Boxman

Abstract Thin film deposition using hot anode vacuum arcs developed in the last decade is described. Two configurations were used: (i) with an open gap—the hot refractory anode vacuum arc (HRAVA) and (ii) with a closed gap—the vacuum arc with black body assembly (VABBA). In both configurations, the anode was heated by the arc with current I = 145–340 A, and a relatively dense plasma plume of cathode material (Cu, Ti, Cr, Al, Sn, Mo, Nb), was formed by re-evaporation of cathode material from the hot (2000–2500 K) anode, which was fabricated from graphite, Mo, Ta, or W. A steady state mode was reached when the anode was sufficiently hot and a plasma plume expanded, either radially (HRAVA) or directly from the front hot anode surface. As an example, the deposition rate measured in 300 A HRAVAs at distances of 80 mm from the arc axis, to be 3.6; 1.4 and 1.8 μm/min for Cu, Cr and Ti cathodes respectively. Interconnector trenches (100 nm wide ×300 nm deep) on microelectronic wafers were filled using a Cu HRAVA at a rate of 0.5 μm/min.

Keywords Thin film · Vacuum arc · Refractory anode · Interconnector trenches · Deposition rate

1 Introduction

Clean metallic plasma, i.e., without macroparticles (MPs), is required for several applications [1, 2], including metallic film deposition [3, 4], ion implantation, and filling high-aspect ratio trenches with metal on VSLI wafers [5–9]. Clean plasma has been produced in conventional cathodic vacuum arcs by using magnetic filters to separate the plasma jet from the MPs, and in hot-anodic vacuum arcs by anode evaporation [5, 10], however, the equipment is cumbersome, and the process is inefficient.

I. I. Beilis (✉) · R. L. Boxman
Tel Aviv University, POB 39040, Tel Aviv 69978, Israel
e-mail: beilis@eng.tau.ac.il

© Springer Nature Singapore Pte Ltd. 2019
A. D. Pogrebnjak and V. Novosad (eds.), *Advances in Thin Films, Nanostructured Materials, and Coatings*, Lecture Notes in Mechanical Engineering,
https://doi.org/10.1007/978-981-13-6133-3_1

MPs were separated from the arc plasma using a hot refractory anode vacuum arc (HRAVA). This is a discharge mode that uses a nonconsumable anode and consumable cathode [11]. In the HRAVA, metallic plasma is produced by re-evaporation of the cathode material from a hot cylindrical anode. The radially expanding HRAVA plasma is sufficiently hot and dense to be used for film deposition on a substrate placed at some distance from the arc axis around the electrode gap, and the films have much less MP content than obtained from the conventional cathodic arc deposition.

Significantly reduced MPs plasma was generated in a vacuum arc, which enclosed the interelectrode plasma in a mostly closed hot chamber comprised of a hot refractory anode, cathode, and an insulator [12]. This configuration is named a vacuum arc with a black body assembly (VABBA). Plasma flux exited the hot chamber through small holes in the anode. In this paper are reviewed the main important results obtained using the relatively new plasma source based on mentioned vacuum arcs with refractory hot anodes.

2 Conventional Vacuum Arc

2.1 Cathode Spot, Jet, the Effective Voltage

In first, let us introduce the usual conventional arc with ring anode (Fig. 1). The arc current I is supported by cathode spots where the electron emission, cathode heating, and intense local evaporation occurs. The dense plasma and MPs were generated. There also formed a supersonic, fully ionized, high kinetic energy (20–150 eV) plasma jet used for vacuum arc deposition by separating the MPs. The energy Q dissipated in cathode was characterized by effective voltage Ucef = Q/I. The measurements show Ucef = 8 V [13, 14], 6.2 [15].

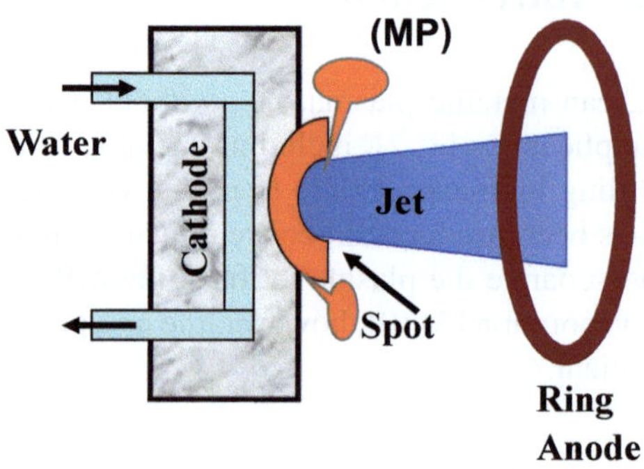

Fig. 1 Conventional arc

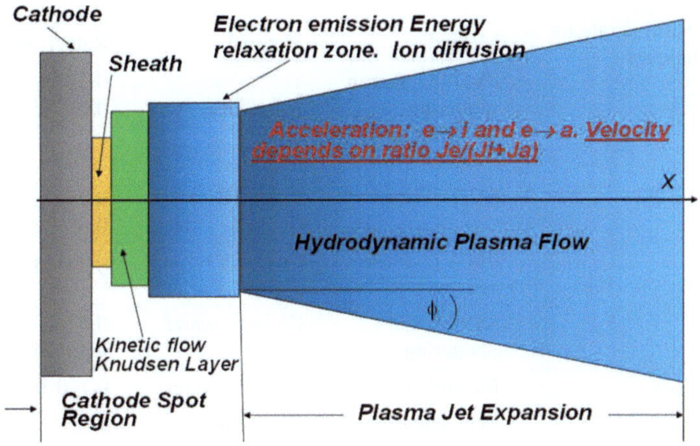

Fig. 2 Schematic presentation of main regions in the cathode plasma and physical phenomena supporting the cathodic arc [16]

2.2 *Model of Cathode Spot and Plasma Jet Generation [16]*

A model was formulated which represents cathode spot phenomena appearing in different physical zones that are limited by boundaries (Fig. 2). The coordinate system origin was placed at the spot center, and the x-direction coincided with the direction of the jet flow axis. The first boundary is the cathode surface, which passes through the coordinate origin. The cathode is heated and emits electrons and neutral atoms due to vaporization and also macroparticles (MP's). The second zone is the space charge layer (sheath ballistic zone). The next zone is at a distance of the Knudsen layer length for the heavy particles (where a returned flux of heavy particle toward the cathode is formed and as result a net of cathode erosion mass is produced) and plasma electrons from the cathode surface. The last zone represents a self-spot region and is located at a distance equal to the emitted electron beam relaxation length from the cathode surface (Fig. 2). In this region the plasma is heated and begins to be accelerated, producing a cathode plasma jet in the expansion region.

3 Plasma Cleaning. Macroparticle Separation

Let us consider the main deposition techniques used at present: magnetron sputtering, chemical vapor deposition (CVD), electrolysis and PACVD-plasma assistance. Table 1 presents the rate of deposition, See details and references in [11].

Vacuum arc deposition is represented by magnetic filtering of plasma jet ejected from the cathodic vacuum arc through a ring anode (Fig. 3). The plasma flux was

Table 1 Deposition rate for different techniques

Method	Material	Rate, μm/min
PVD-magnetron sputtering IPVD-ionized PVD	Ti Al on SiO_2	0.017 0.01
CVD	Cu on TiN Cu on Si wafer	0.05–0.2 0.05–0.1
PACVD- Plasma assisted	Cu on SiO_2, Al	0.004–0.02
SSMD (Self Sputtering)	Cu on Si wafer	0.02–0.4
Electrolysis	Cu on Si wafer	0.1–0.16
Magnetic filtered vacuum arc deposition	Cu on glass	0.1

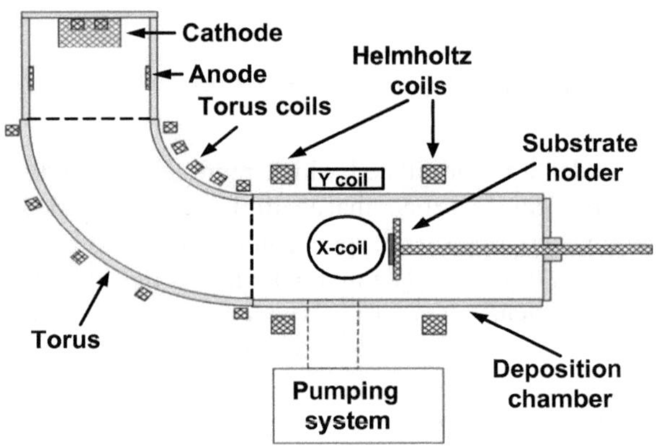

Fig. 3 System of magnetic filtered vacuum arc deposition

filtered from the generated MP's by a magnetic field in a quarter-torus duct. Its rate of deposition is also indicated in Table 1.

3.1 Disadvantages

- Usually low efficiency of material utilization (see Table 1)
- Complex

In two mentioned methods of plasma generating (HRAVA and VABBA) the MP's were not separated but rather converted into the metallic plasma. The MPs are converted using a vacuum arc with a hot refractory anode. This vacuum arc was developed in Tel Aviv University in last two decades [11, 12].

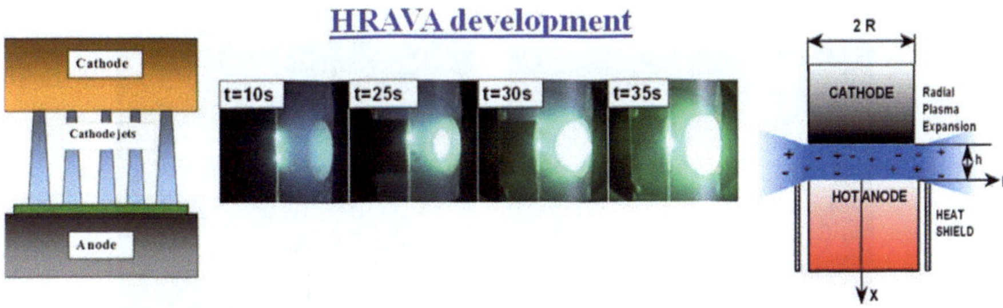

Fig. 4 HRAVA mechanism and principle of working

3.2 *Hot Refractory Anode Vacuum Arc (HRAVA)*

Figure 4 shows the initial HRAVA stage (left) at which the jet plasma is condensed on the cold anode. During the anode heating by the arc the deposit re-evaporates from a hot non-consumable anode and a plasma plume is developed (middle). The radially expanding plasma formed at the hot anode stage is shown on the right side of Fig. 4. The plasma is sufficiently dense which used to deposit films on substrates placed at some distance from the arc axis around the electrode gap. These films had by three orders less MP content in comparison to that in conventional cathodic arc content deposition.

3.3 *Vacuum Arc Black Body Assambly (VABBA)*

In the VABBA, the cathode material is emitted into a closed chamber formed by the end surface of a water-cooled cylindrical cathode and a cup-shaped refractory anode that is heated by the arc (Fig. 5). The material is eroded from the cathode spots as plasma and MPs. They impinge on the hot anode and are re-evaporated from it, forming a dense high-pressure plasma within the chamber. The VABBA electrode assembly acts toward the plasma and MPs somewhat analogously to how a blackbody cavity acts toward photons, i.e., not permitting their escape, except through small holes. In the case of the VABBA, all of the MPs are trapped, while some of the impinging plasma escapes from one or more holes, and can then be used for applications.

Below are presented the results including the anode and cathode effective voltages, the anode temperature, the plasma electron temperature, and density as well as the deposition rates obtained by the vacuum arc with refractory hot anodes.

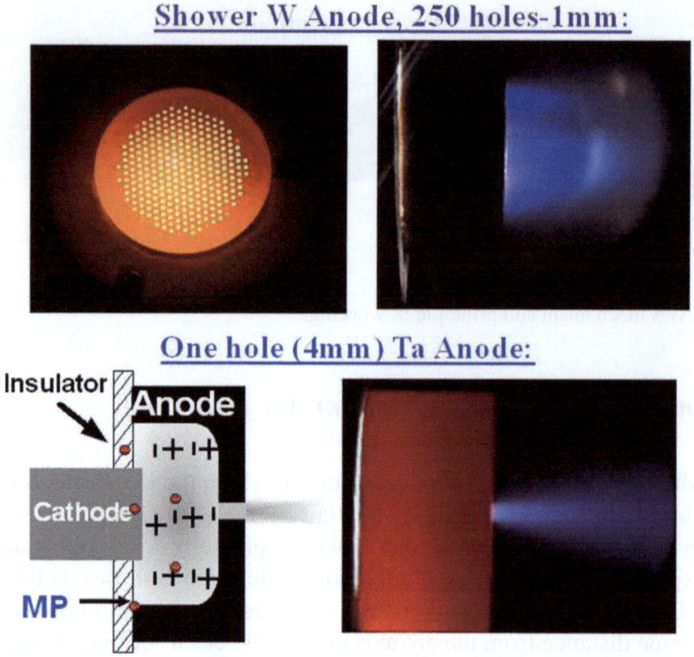

Fig. 5 VABBA plasma plumes using shower and one-hole anodes

3.4 *Experimental Methodology and Conditions*

The methodology includes calorimetric test, optical microscopy and SEM observations, plasma parameters measuring by electrical probe and profilometry of the deposit films. The electrode materials, geometry, gap distance, and other arc parameters are summarized in Table 2.

3.5 *Setup and Measurements*

1. The experimental system includes a vacuum chamber, electrode and diagnostic units that can be seen in Fig. 6.
2. The water cooled cathode, thermocouple, and ion probe measurements schematically are presented in Fig. 7. The metallic film deposition at hot anode arc stage was provided in a system using a shutter schematically presented Fig. 8.

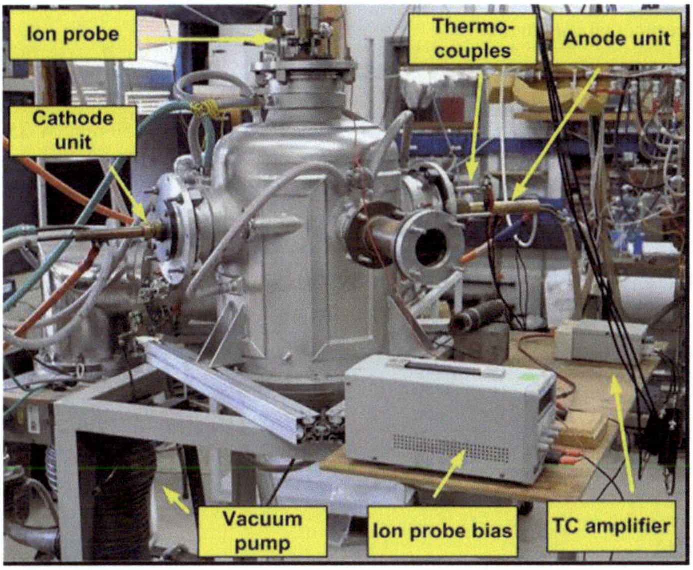

Fig. 6 Experimental system

Fig. 7 Calorimetric and ion current measurements

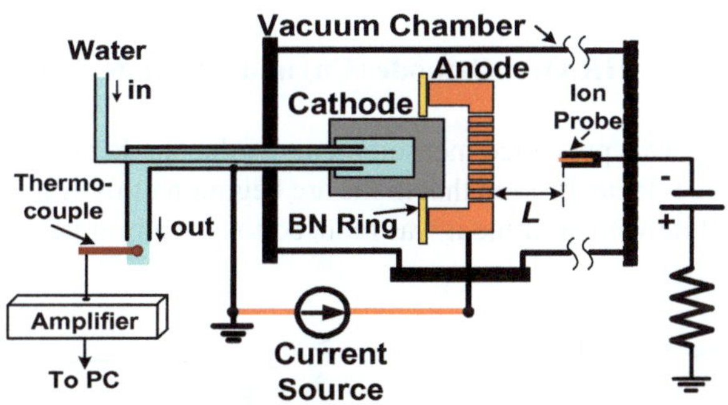

Fig. 8 Deposition using a shutter

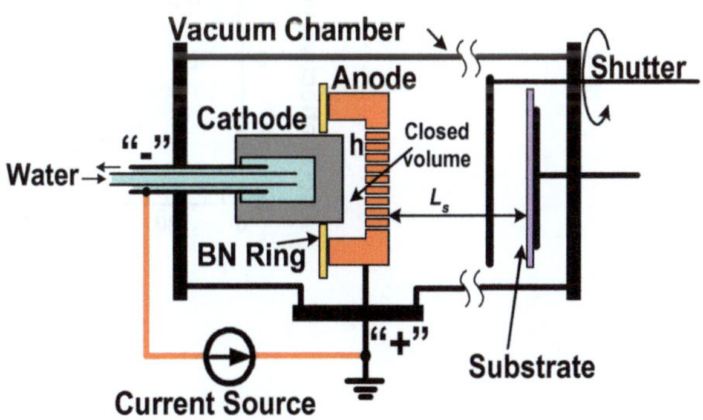

Table 2 Experimental conditions

Cathode materials Geometry	Gr, Cu, Cr, Ti, W Length: 40 mm; Diameter: D = 30 mm
VABBA-anode	Length: 22 mm; D = 50 mm, W-250 holes (D = 1 mm) Ta-1 hole, D = 4 mm
HRAVA anode A-C distance Anode thickness	W, Mo h = 5–10 mm 5, 10, 20, 30 mm
Arc current, I	150, 175, 200, 225, 250 A
Arc time, t	~180 s
W-Thermocouple T-Thermocouple	K-type (chromel-alumel) W/5%Re-W/26%Re
Water flow rate, F	0.193, 0.29 L/s
Ion probe	W rod 1.5 mm diameter

4 Results

4.1 Effective Voltages

4.1.1 HRAVA: Cathode (Cu) and Graphite Anode [17]

Figure 9 presented measurements of the anode effective voltage dependence on arc time. It can be seen that in the arc beginning this voltage is relatively large, 11–12 V, while with arc time it was reduced to a steady state of 6 V.

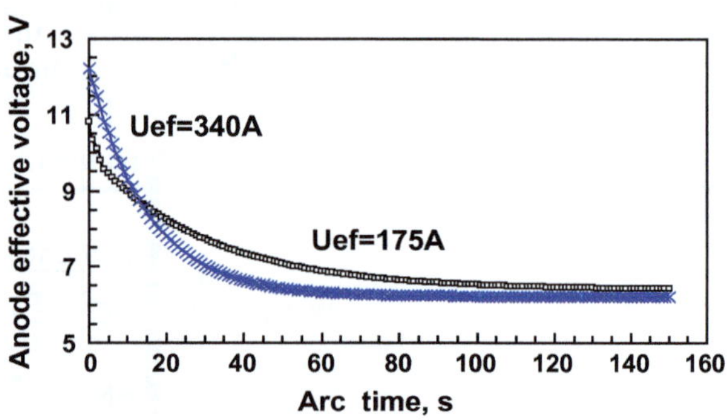

Fig. 9 Effective anode voltages

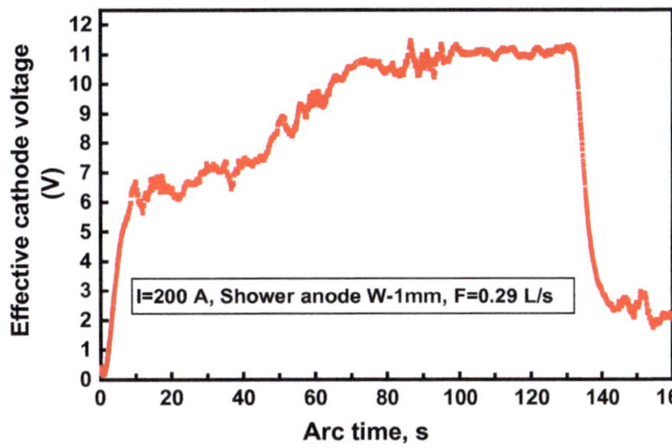

Fig. 10 Increasing of the cathode voltage

4.1.2 VABBA Effective Cathode Voltage [18]

Figure 10 presented measurements of the cathode effective voltage dependence on arc time. It can be seen that in the arc beginning this voltage is relatively low, 6 V, while with arc time it was increased to a steady state of 11–12 V.

In VABBA the total effective voltage is 11–12 V consists of:

(1) Ucef = 6.5 V (conventional from the spot) and
(2) Part of returned plasma jet energy Uj (12 V) that has not expanded into the surroundings and remains within in the VABBA.

4.1.3 Plasma Density Inside of VABBA

The time dependent ion current density j_i and plasma density n were measured in a VABBA with one-hole Ta anode. The probe was mounted inside the 4 mm diam hole in the anode and positioned near the flat anode surface, facing the plasma inside of the closed cavity, to estimate the plasma density therein. The probe was connected to ground through a R_{ion}=1 Ω shunt resistor and a DC power supply (GW INSTEK SPS-606) producing a probe bias of V_{bias}=−10 V with respect to the grounded anode.

After ignition of an I=200 A arc, and with the end of the probe inserted to a depth of 4 mm beyond the inside surface of the anode, the ion current was about 50 mA for a period of ~40 s. Then it started to increase rapidly and reached a maximal value of ~500 mA before the arc was turned off at 60 s as shown in Fig. 11. This corresponds to an ion current density j_i which increased from 2.8 to about 28.3 A/cm^2. The estimated plasma density can be estimated from n~j_i/0.4eν_i (e-electron charge, $\nu_i = (2T_e/m_i)^{1/2}$, T_e is the electron temperature of ~1 eV) increased from 2.5 × 10^{14} to 2.5 × 10^{15} cm^{-3}. The probe was heated by the incident plasma, so that thermionic electron emission could have in principle influenced the measurements. The probe surface temperature was estimated using the one-dimensional heat conduction equation to

Fig. 11 Ion current time vs time inside one-hole Ta anode of VABBA

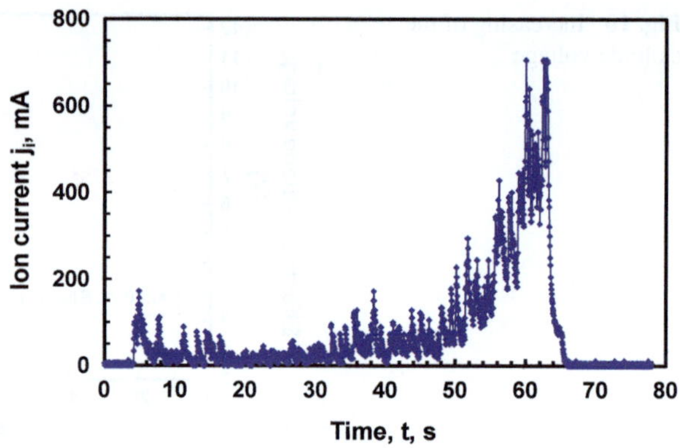

Fig. 12 Anode surface temperature at I = 175A, gap = 10 mm

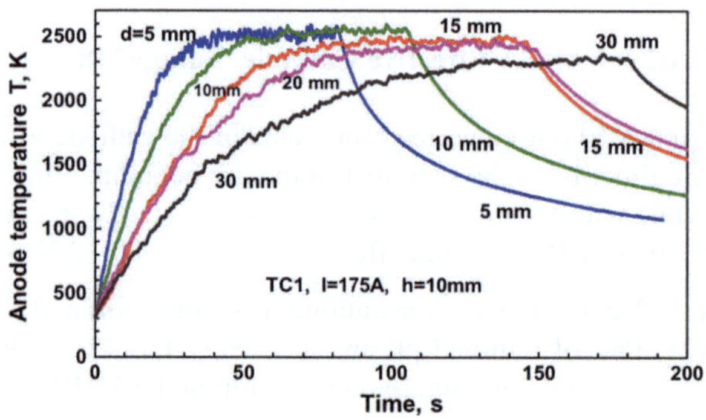

reach about 1600 K. In this case, the thermionic electron emission was estimated as ~10^{-6}A/cm^2 and therefore its influence can be neglected.

4.2 HRAVA Parameters. Film Deposition

1. Time-dependent W anode surface temperature T versus time with anode thickness d as a parameter is shown in Fig. 12. The larger anode temperature reaches at d = 5 mm, and it is weakly decreased with d. The arc time for steady state T is lower at shorter anode thickness.
2. Ion current fraction (HRAVA) [19].

Figure 13 shows that the ion current fraction at steady state and the initial stage (8%, conventional arc). This fraction increases with gap distance and reaches a high value of 11% at HRAVA stage.

3. Plasma parameters, HRAVA.

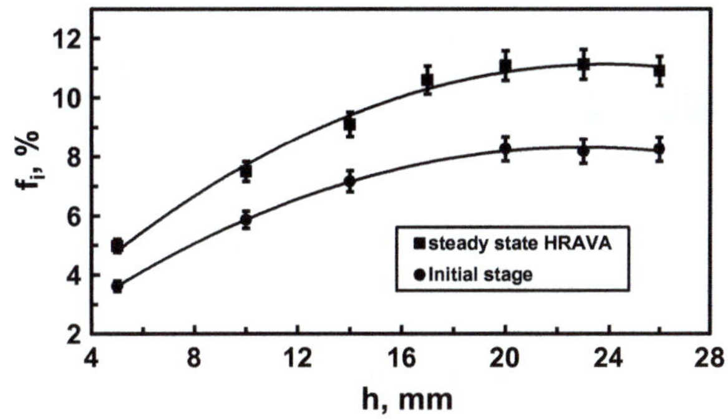

Fig. 13 Ion current fraction as function on gap distance

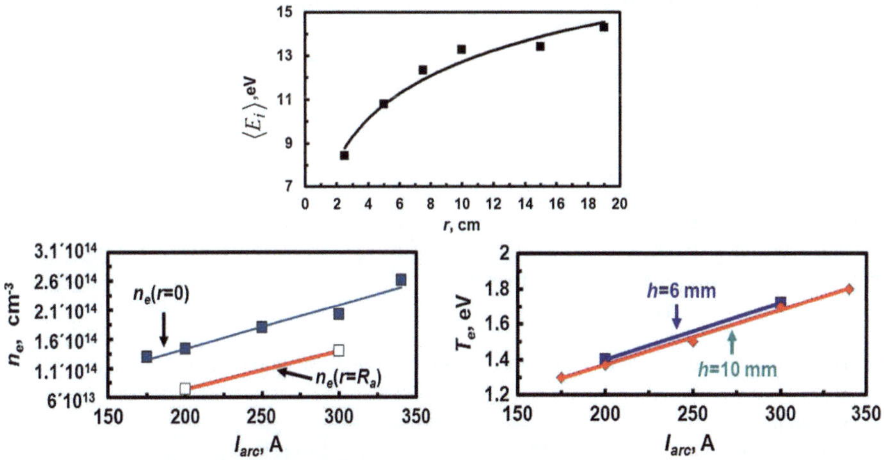

Fig. 14 Ion energy, plasma density and electron temperature

Average ion energy increase with distance from the arc axis, plasma density and electron temperature as a function on arc current are shown in Fig. 14. The electron density and temperature linearly increases with arc current.

4. Metallic film deposition.

Figure 15 demonstrate the quality and Fig. 16 the rate of thin film deposition as dependence on arc time and arc current. The steady-state deposition rate reaches about 80 s of arcing. Figure 17 demonstrates Cu film is filling in the trenches of interconnections in microelectronic wafers [20].

5. VABBA deposition.

The Nb (cathode) deposition rate as a function of arc current is presented in Fig. 18 for different anodes: a single-hole Ta anode, a shower-head W anode, and a shower-head Nb anode in the VABBA configuration. The Nb deposition rate with a cylinder W anode by HRAVA is also indicated. It can be seen that the deposition rate significantly depends on the anode configuration, material and arc mode.

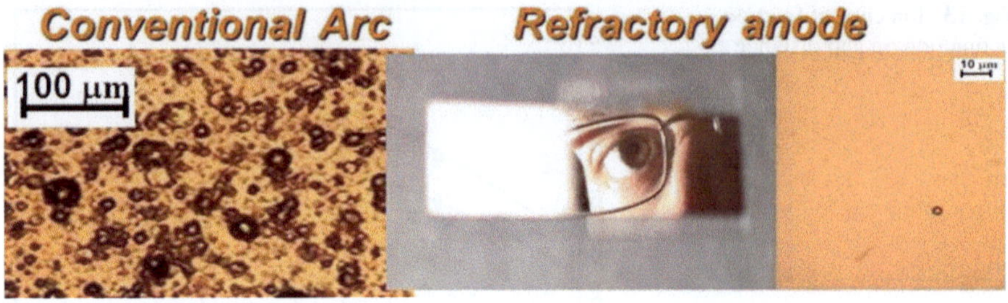

Fig. 15 Comparison of Cu film with conventional and by HRAVA deposition

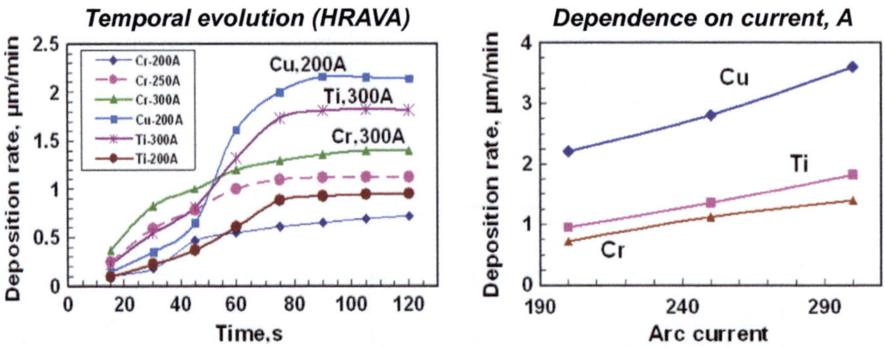

Fig. 16 Rate of film deposition as function on time and steady state on arc current

Fig. 17 Wafer interconnection filling

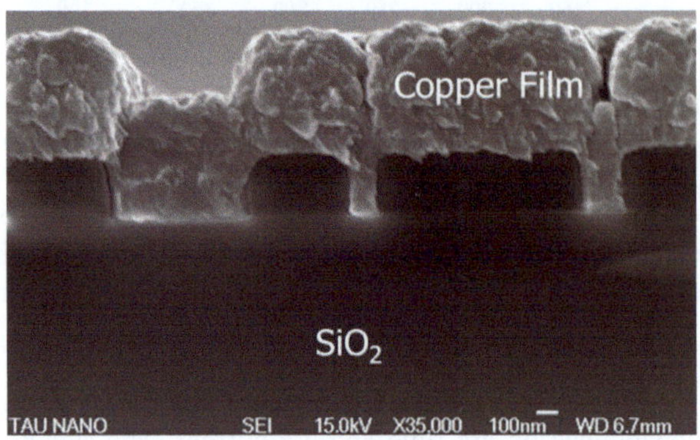

WABBA with W film deposition rate is shown in Fig. 19 using tungsten hot shower (0.85 mm holes) anode for L = 60 mm, 100 s arcing (closed shutter) and 60 s exposition.

6. VABBA Image of Cu deposition on glass [21].

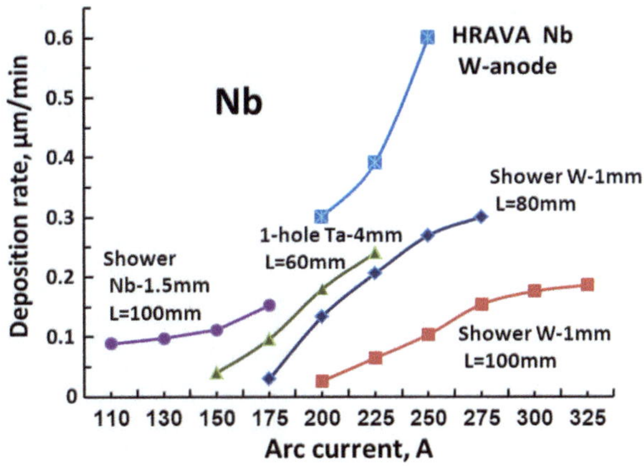

Fig. 18 Thin film deposition of Nb

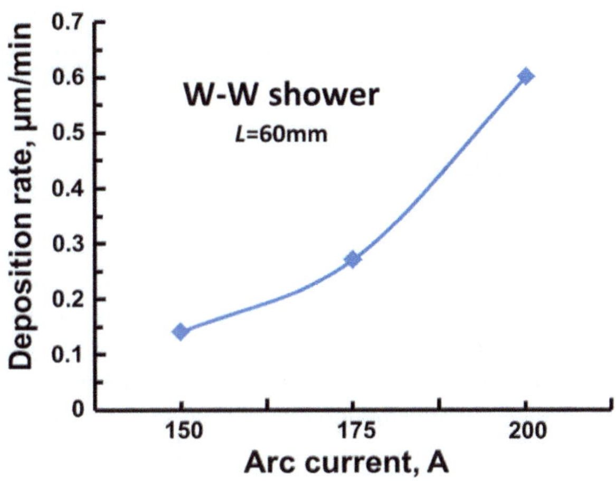

Fig. 19 Tungsten film deposition by VABBA

XRD pattern was analyzed for a Cu thin film deposited with shower Ta anode, I = 200 A, Ls = 100 mm, and with an exposure time of 20 s beginning 60 s after arc initiation. Polycrystalline Cu was detected, and the grain size was 30 nm.

SEM was used to analyze the Cu thin film, Ta (0.6 mm holes) anode, I = 200 A. The shutter was closed for 60 s and deposition time was 20 s (shutter was opened) at distance of Ls = 100 mm from the anode front surface. An SEM image nanosize grain was presented in Fig. 20. It was also observed that for Cu film a low number of nanoparticles were deposited for times about 30 s, while at more than 60 s after arc ignition, no MPs were detected on the substrate area (25 × 75 mm^2).

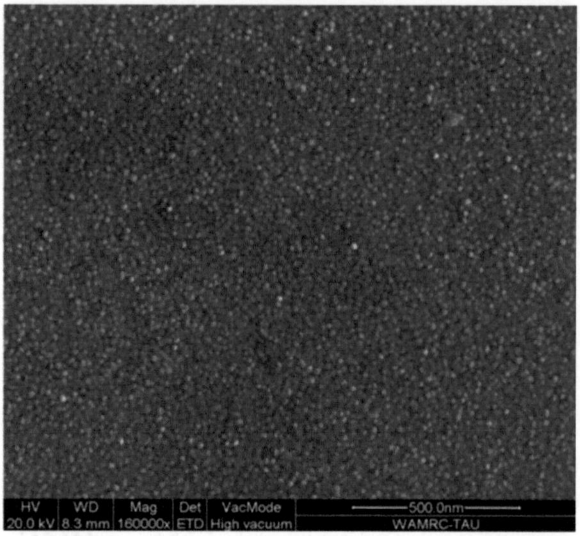

Fig. 20 SEM image, Cu deposit, nanosize grain

5 Conclusions

The refractory anode arc produced a plasma plume that expanded radially from arcs with an open electrode configuration (HRAVA) or direct flux in closed electrode configuration (VABBA). Plasma in both cases consisted of the cathode material.

In the HRAVA, the effective anode voltage was about 11–12 V at the arc beginning and about 6 V during steady state arc operation.

The effective cathode voltage Ucef in the VABBA increased to ~6.5 V for cold anodes (like in the conventional cathodic arc) and then increased to a steady state of ~11–12 V during the hot anode arc stage.

The ion flux fraction (11%) in arcs with refractory anodes (HRAVA) was larger than in conventional arc (8%).

The plasma temperature was Te ~ 1.3–1.8 eV, the density was >1014 cm^{-3}, and the plasma was accelerated up to ~15 eV during its expansion (HRAVA).

The arc is a simple metallic plasma source for coatings. The high efficiency is attributed to converting MPs into plasma. The steady-state HRAVA deposition rate reached 3.6 (Cu), 1.8 (Ti) and 1.4 μm/min (Cr) for I = 300 A, while PVD, CVD, FVAD reach about 0.1 μm/min.

Acknowledgements The authors gratefully acknowledge S. Goldsmith, H. Rosenthal, M. Keidar, A. Shashurin, Y. Koulik, D. Arbilly, A. Nemirovsky, A. Shnaiderman and D. Grach for their contributions at different stages of the investigations.

References

1. Hopwood JA (2000) Ionized physical vapor deposition, vol 27. Academic Press, San Diego, New York
2. Makhlouf ASH (2011) Current and advanced coating technologies for industrial applications. In: Nanocoatings and ultra-thin films. Woodhead publishing series in metals and surface engineering, pp 3–23
3. Clavero C, Slack JL, Anders A (2013) Size and composition-controlled fabrication of thermochromic metal oxide nanocrystals. J Phys D Appl Phys 46(36):362001
4. Pogrebnjak A, Ivashchenko V, Bondar O (2017) Multilayered vacuum-arc nanocomposite TiN/ZrN coatings before and after annealing: structure, properties, first-principles calculations. Mater Charact 134:55–63
5. Sanders DM, Anders A (2000) Review of cathodic arc deposition technology at the start of the new millennium. Surf Coat Technol 133–134:78–90
6. Anders A (2012) The evolution of ion charge states in cathodic vacuum arc plasmas: a review. Plasma Sources Sci Technol 21(3):035014
7. Witke T, Siemroth P (2009) Deposition of droplet-free films by vacuum arc evaporation—results and applications. IEEE Trans Plasma Sci 27(4):1039–1044
8. Zöhrer S, Anders A, Franz R (2018) Time-resolved ion energy and charge state distributions in pulsed cathodic arc plasmas of Nb–Al cathodes in high vacuum. Plasma Sources Sci Technol 27(5):055007
9. Pogrebnjak AD, Beresnev VM (2012) Hard nanocomposite coatings, their structure and properties. In: Nanocomposites—new trends and developments, InTech. http://dx.doi.org/10.5772/50567
10. Boxman RL, Martin P, Sanders D (eds) (1995) Handbook of vacuum arc science and technology. Noyes Publishing, Ridge Park, NJ
11. Beilis II, Boxman RL (2009) Metallic film deposition using a vacuum arc plasma source with a refractory anode. Surf Coat Technol 204(6–7):865–871
12. Beilis II, Koulik Y, Boxman RL (2011) Evolution of a plasma plume from a shower anode in a vacuum arc with a black-body electrode configuration. IEEE Trans Plasma Sci 39(11):2838–2839
13. Reece MP (1963) The vacuum switch. Proc IEEE Inst Electr Electron Eng 110:793–811
14. Beilis II (1977) Cathode spots on metallic electrode of a vacuum arc. High Temp 15:818–824
15. Daalder JE (1078) A cathode spot model and its energy balance for metal vapor arcs. J Phys D Appl Phys 11(12):1667–1682
16. Beilis II (2001) State of the theory of vacuum arcs. IEEE Trans Plasma Sci 29(5):657–670 (Part I)
17. Rosenthall H, Beilis II, Goldsmith S et al (1995) Heat fluxes during the development of hot anode vacuum arc. J Phys D Appl Phys 28(2):353–363
18. Beilis II, Koulik Y, Boxman RL (2013) Effective cathode voltage in a vacuum arc with a black body electrode configuration. IEEE Trans Plasma Sci 41(8):1992–1995 (Part II)
19. Beilis II, Shashurin A, Boxman RL et al (2006) Total ion current fraction in a hot refractory anode vacuum arc. Appl Phys Lett 88:071501
20. Beilis D II, Grach Shashurin A et al (2008) Filling trenches on a SiO_2 substrate with Cu using a hot refractory anode vacuum arc. Microelectron Eng 85:1713–1716
21. Beilis II, Koulik Y, Boxman RL (2014) Cu film deposition using a vacuum arc with a black-body electrode assembly. Surf Coat Technol 258:908–912

Multilayer Design of CrN/MoN Superhard Protective Coatings and Their Characterisation

B. O. Postolnyi, O. V. Bondar, K. Zaleski, E. Coy, S. Jurga, L. Rebouta and J. P. Araujo

Abstract Multilayer CrN/MoN transition metal nitride coatings were studied in this research. Films were deposited by vacuum arc deposition (Arc-PVD) from Cr and Mo cathodes in nitrogen atmosphere $p_N = 0.4$ Pa. Three series of samples with different values of negative bias voltage (−20, −150, and −300 V) applied to the surface were fabricated. Each series has samples with 11, 22, 44, 88, 180 and 354 layers while total thickness was maintained with the same value. Samples were studied by scanning electron microscopy (SEM) on cross-sections and coatings surface, energy-dispersive X-ray spectroscopy (EDS), electron backscatter diffraction (EBSD), high-resolution transmission electron microscopy (HRTEM), X-ray diffraction (XRD), micro-indentation. Two main cubic phases of γ-Mo_2N and cubic CrN were detected. It was observed that the crystal growth orientation changes while the negative bias voltage of the substrate decreases. The maximum values of hardness (38–42 GPa) among the studied samples were obtained for coatings with a minimal individual layer thickness of 20 nm deposited at $U_b = -20$ V.

Keywords Multilayers · Microstructure · Diffraction · Hardness · Coatings

B. O. Postolnyi (✉) · O. V. Bondar
Sumy State University, Sumy, Ukraine
e-mail: b.postolnyi@gmail.com

B. O. Postolnyi · J. P. Araujo
IFIMUP-IN, Department of Physics and Astronomy, Faculty of Sciences, University of Porto, Porto, Portugal

K. Zaleski · E. Coy · S. Jurga
NanoBioMedical Centre, Adam Mickiewicz University, Poznań, Poland

L. Rebouta
Centre of Physics, University of Minho, Guimarães, Portugal

© Springer Nature Singapore Pte Ltd. 2019

A. D. Pogrebnjak and V. Novosad (eds.), *Advances in Thin Films, Nanostructured Materials, and Coatings*, Lecture Notes in Mechanical Engineering,
https://doi.org/10.1007/978-981-13-6133-3_2

1 Introduction

Improvement of properties of materials involved in engineering and industry is one of the main tasks of material science. In the production of machines, working parts of mechanisms, drilling tools, etc. it is crucial to increase the wear resistance of used materials and, thus, extend their lifetime. Among others approaches, the methods of surface modification are the most promising and economically effective [1–7]. The authors are convinced that for mechanical properties enhancement the best fits the deposition of superhard protective coatings on the surface of working parts [8–24]. This method is quite used and well known, but the diversity of deposition techniques, structures and materials give plenty of opportunities to design new coatings with superior properties [25–42].

Considering new European and world policy regarding critical raw materials and recycling strategies it is expected to observe a new wave of interest and focus on protective coatings such as proposed in this paper [43–46]. When the coating is already destroyed on the used tool, the surface can be easily cleaned, and new films are deposited to put the machine in service for a new working period.

As a solution in the presented work, it is proposed to employ mainly two approaches and get the benefit of this combination: Hall-Petch strengthening, which is based on the effect of gaining size control, multilayer design of coatings [47]. The first assumes the increase of hardness values when the grain size reduces up to around 10 nm. The second approach, multilayer architecture of coatings, improve protective properties of the multilayer films by reduction of cracks propagation towards the substrate by the deflection of cracks on the interlayer interfaces. Additionally, by control of individual layers thickness in coatings it's possible to grain size in coatings: the thinner layers, the smaller grains may be obtained.

Multilayer design of the coatings also allows benefiting from the combination of outstanding properties of different materials used deposited in individual layers of the films [48–64].

Chromium nitride has better thermal stability (up to 600 °C) [65–68] in comparison to Titanium nitride, the most widely used transition metal nitride protective coating since the 1960s [69–71]. While the hardness of CrN is on the same level as TiN, it demonstrates higher corrosion and wear resistance, and extremely strong adhesion to metal surface.

Molybdenum nitride is the hardest superconductive metal nitride (28–34 GPa) but still has been less studied. The diversity of possible phases with a wide range of stoichiometry and lattice structures (cubic γ-$Mo_2N_{1\pm x}$, tetragonal β-$Mo_2N_{1\pm x}$, hexagonal δ-MoN and metastable MoN_x phase of NaCl-B1-type cubic structure) makes it interesting for the current research [72].

2 Deposition of Coatings

Multilayer CrN/MoN coatings were deposited by vacuum arc deposition (Arc-PVD) using special vacuum unit Bulat-6 in a reactive atmosphere of nitrogen from two cathodes of Cr ($I_{arc} = 100$ A) and Mo ($I_{arc} = 120$ A). Polished discs of stainless steel 12X18H9T were used as the substrates. The cathodes were placed on the opposite sides of the chamber, and the rotatable sample holder in the middle exposed the substrates alternately to the chromium and molybdenum. While the substrate holder was in motion, the cathodes were turned off. By decreasing twice the deposition time of individual layer for each next sample, but maintaining the same total deposition time, it was possible to get samples with similar total thickness but different individual layer thickness. The samples with 11, 22, 44, 88, 180 and 354 layers have been deposited. The used negative bias voltage applied to the substrates during the deposition was −20, −150 and −300 V. Working pressure of reactive nitrogen gas was 0.4 Pa.

3 Experimental Details

Morphology of surface and cross-sections of coatings were studied by scanning electron microscopy (SEM) using Quanta 200 3D microscope and FEI Quanta 400 FEG environmental scanning electron microscope. Elemental composition was estimated by energy-dispersive X-ray spectroscopy (EDS) using mentioned microscopes. Electron beam energy used in elemental composition analysis by EDS was 10 keV. Cross-section samples for SEM were prepared by cutting with further grinding, polishing and finishing with colloidal silica polishing suspension.

Microstructure, grains size, and growth were studied by electron backscatter diffraction (EBSD) analysis using the unit of EDAX EBSD forward scatter detector system and high-resolution DigiView III camera attached to the FEI Quanta 400 FEG ESEM. The processing of the obtained data, reconstruction of grains and the evaluation of their size were performed using specialized software "OIM (Orientation Imaging Microscopy) Analysis" of EDAX (AMETEK, Inc.). The grain tolerance angle of 5° was used for grains determination. Grains at edges of scans were not included in statistics.

For the more precise study of the multilayer structure, the analysis by high-resolution transmission electron microscopy (HRTEM) was performed using JEOL-ARM 200F equipped with an EDS detector. Accelerating voltage was 200 kV. Thin sections to be studied by HRTEM were prepared using a focused ion beam (FIB) using JEOL JIB-4000 with Ga^+ ions and were coated by thin carbon film before etching to prevent damages of the samples.

The structural studies and phase evaluation were performed by X-ray diffraction (XRD) analysis using high-resolution diffractometer Rigaku SmartLab with the 9 kW rotating anode (voltage of 45 kV and current of 200 mA) in Cu-K_{α} radiation. For most of the scans, the continuous mode of the acquisition was used with the diffraction angular range from 20° to 120° with a step of 0.01° and scan speed of 0.3°/min. For low-angle XRD methods, the incident angle ω was 0.4°, 0.6°, 0.8°, and 1.2° while the scan speed was 0.5°/min, the angular range was 32–84° and step was 0.024°. Phase identification was done using ICDD Powder Diffraction Files No. 00-006-0694 for Cr (bcc), No. 00-042-1120 for Mo (bcc), No. 00-011-0065 for CrN (fcc), No. 00-035-0803 for β-Cr_2N (hexagonal), No. 00-025-1366 for γ–Mo_2N (fcc) and No. 00-025-1368 for β-Mo_2N (tetragonal). The phase identification and other data processing were performed using Crystal Impact's software "Match!". The average crystallite size was calculated from XRD data using the Scherrer equation. Instrumental effects, microstrain, solid solution inhomogeneity or any other possible contributions to the width of a diffraction peak were ignored during crystallite size calculations.

Micro-indentation of samples was performed using NanoTest instrument from Micro Materials company to evaluate hardness and Young's modulus (elastic modulus) of coatings. The maximum load has reached the values of 600 mN. Up to 10 indentations oriented in line with intervals of 50 μm were performed by Berkovich indenter with controlled maximum penetration depth in the range 0.6–1.3 μm, but no more than 10% of total thickness of the coating.

4 Results and Discussion

The morphology of the sample with an individual layer thickness of 1.1 μm and the top layer of chromium nitride is shown in Fig. 1. Presented rocky, rough and highly structured surface is typical for CrN thick films. Despite the special filtration system installed in the deposition unit Bulat-6, non-significant droplet incorporations are observed on the surface of the coating, which is a feature of the Arc-PVD method.

Figure 2 shows the cross-section image of the same sample presented in Fig. 1 and deposited at a negative bias voltage of −20 V with an individual layer thickness of 1.1 μm. It demonstrates good quality of the deposited films and a sharp interface between layers. Dark layers correspond to the CrN films and brighter zones—MoN layers. One may notice that MoN layers are slightly thinner then CrN, which is caused by a small difference in Cr and MoN cathodes evaporation speed and nitrides deposition rates.

Additional information about microstructure and crystal growth was obtained by EBSD analysis. Registered Kikuchi patterns (see Fig. 3), which are diffraction patterns produced by Bragg reflections of inelastically scattered (thermal diffuse scattering) electrons in a specimen, were analyzed by specialized software to study crystal orientation in each control point of the sample. The resulting unique color

Fig. 1 Scanning electron microscopy image of CrN/MoN surface, top layer is CrN with individual layer thickness of 1.1 μm

grain map is for CrN layer of sample deposited at a negative bias voltage of −20 V with an individual layer thickness of 1.1 μm is shown in Fig. 4.

Results of XRD analysis for multilayer CrN/MoN coatings deposited at a negative bias voltage of −20, −150 and −300 V for Series 1, 2 and 3 respectively are shown in Fig. 5. It was observed mainly two phases in the studied CrN/MoN coatings: cubic (NaCl type) high-temperature phase of γ-Mo_2N and cubic CrN phase. With increasing of the absolute value of negative bias voltage from −20 to −150 V and to −300 V the change of preferential crystal growth from (311) to (111) and to (200) respectively was observed.

HRTEM analysis demonstrated good planarity of multi-layered coatings with a sharp interface between layers. The results of XRD study about the presence of cubic γ-Mo_2N and CrN phases were confirmed. Smaller values of calculated lattice constants a, in comparison to stress-free values for CrN (4.149 Å) and γ–Mo_2N (4.163

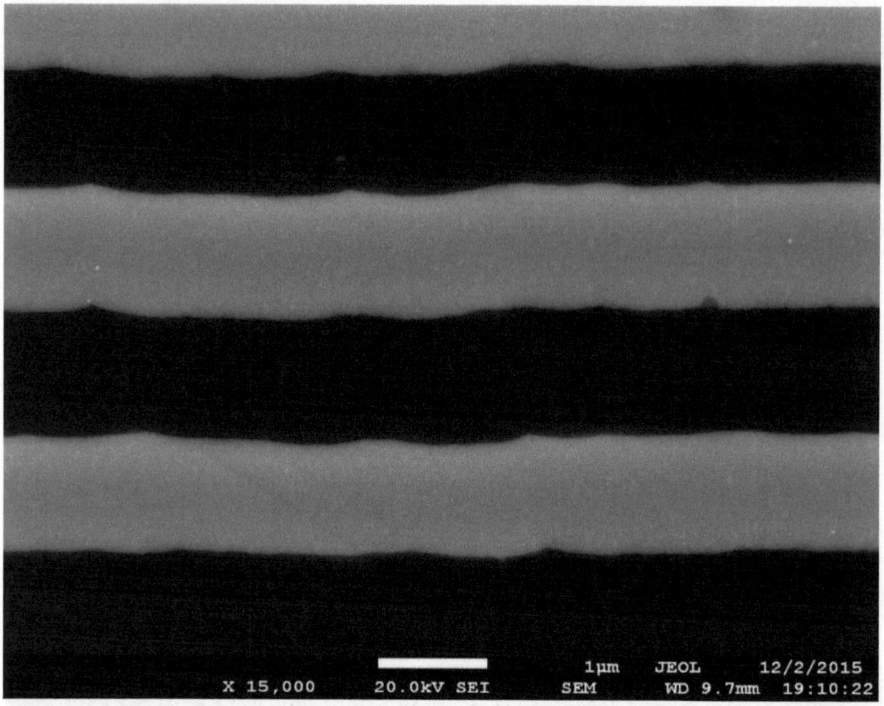

Fig. 2 SEM-image of CrN/MoN cross-section, $U_b = -20$ V

Å), have shown the presence of compressive stress in the deposited coatings. Elemental composition study by HRTEM-EDS gives an almost equal atomic concentration of Cr and Mo up to 41–42 at.%.

Study of mechanical properties by hardness measurements using micro-indentation has shown the highest values for samples deposited at $U_b = -20$ V (Series 1). The growth of the hardness values from 22 to 39 GPa with decreasing of individual layer thickness from 1.1 μm to 20 nm was observed.

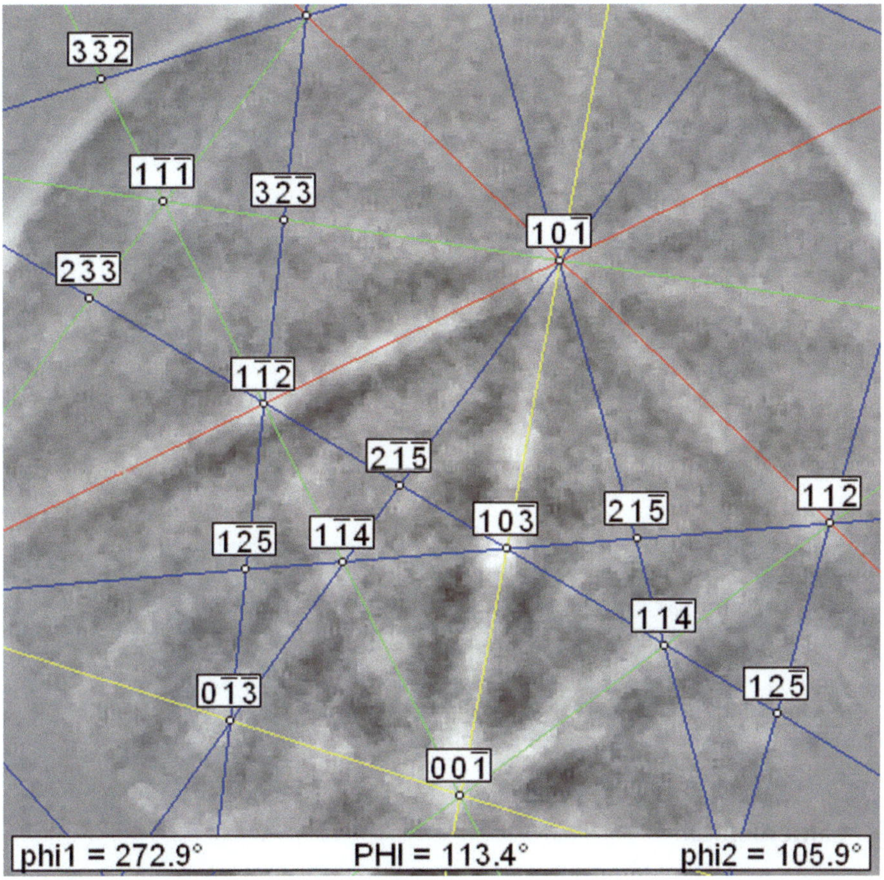

Fig. 3 Kikuchi pattern obtained by EBSD analysis from CrN layer on cross-section sample of CrN/MoN coating

Fig. 4 Unique colour grain map for CrN layer obtained by EBSD analysis on cross-section sample of CrN/MoN coating

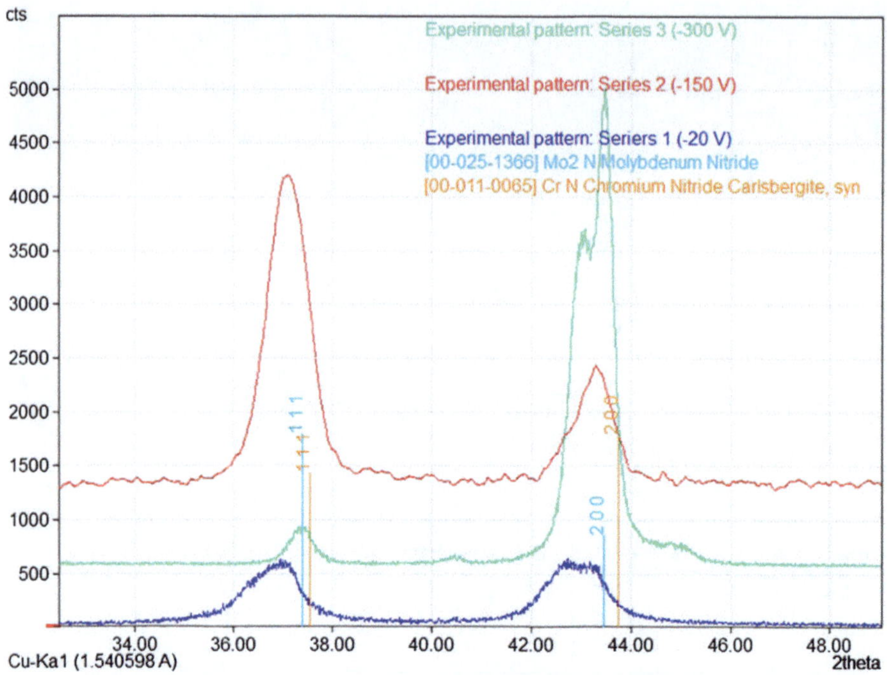

Fig. 5 X-ray diffraction patterns in range $\theta/2\theta = 33$–$49°$ for multilayer CrN/MoN coatings deposited at Ub = −20 V (Series 1), −150 V (Series 2) and −300 V (Series 3)

5 Conclusions

Multilayer CrN/MoN transition metal nitride coatings have been deposited by Arc-PVD, and their microstructure, elemental, phase composition and mechanical properties have been studied. Coatings have typical morphology for Arc-PVD metal nitride coatings; they are characterized by good quality, the high planarity of layers and sharp interlayer interface. Two main cubic phases of γ-Mo_2N and CrN were detected. When the negative bias voltage decreases from −20 to –150 V and –300 V the change of preferential crystal growth from (311) to (111) and to (200) appears. Grains size decrease was observed with individual layer thickness reduction. With the reduction of grains size the hardness values increases. The best mechanical properties were demonstrated by samples deposited at $U_b = -20$ V, where the highest hardness of 38–42 GPa was registered for the sample with an individual layer thickness of 20 nm. Achieved properties demonstrate that CrN/MoN coatings are very promising in various protective applications and belong to the superhard class of materials.

Acknowledgements This work was partly financed by the Foundation of Science and Technology (FCT) of Portugal [references NORTE-01-0145-FEDER-022096, SFRH/BD/129614/2017], Network of Extreme Conditions Laboratories (NECL) and by Ukrainian state budget programs [No. 0116U006816, 0118U003579 and 0116U002621]. Partial support by COST Action CA15102 is also greatly appreciated. The authors are very thankful to Prof. Alexander Pogrebnjak from Sumy State University and Prof. Vyacheslav Beresnev from V. N. Karazin National University in Ukraine for their support and help.

References

1. Pogrebnjak AD, Ponomarev AG, Shpak AP et al (2012) Application of micro- and nanoprobes to the analysis of small-sized 3D materials, nanosystems, and nanoobjects. Physics-Uspekhi 55(3):270–300. https://doi.org/10.3367/UFNe.0182.201203d.0287
2. Pogrebnjak AD, Bondar O V., Abadias G et al (2016) Structural and mechanical properties of NbN and Nb-Si-N films: experiment and molecular dynamics simulations. Ceram Int 42(10):11743–11756. https://doi.org/10.1016/j.ceramint.2016.04.095
3. Pogrebnjak AD, Lebed AG, Ivanov YF (2001) Modification of single crystal stainless steel structure (Fe-Cr-Ni-Mn) by high-power ion beam. Vacuum 63(4): 483–486. https://doi.org/10.1016/S0042-207X(01)00225-1
4. Pogrebnjak AD, Bazyl EA (2001) Modification of wear and fatigue characteristics of Ti-V-Al alloy by Cu and Ni ion implantation and high-current electron beam treatment. Vacuum 64(1):1–7. https://doi.org/10.1016/S0042-207X(01)00160-9
5. Boiko O, Koltunowicz TN, Zukowski P et al. (2017) The effect of sputtering atmosphere parameters on dielectric properties of the ferromagnetic alloy—ferroelectric ceramics nanocomposite $(FeCoZr)_x(PbZrTiO3)_{(100-x)}$. Ceram Int 43(2):2511–2516. https://doi.org/10.1016/j.ceramint.2016.11.052
6. Li Z, Ludwig A, Savan A et al (2018) Combinatorial metallurgical synthesis and processing of high-entropy alloys. J Mater Res, 1–14. https://doi.org/10.1557/jmr.2018.214
7. Matizamhuka W (2016) Structure-properties relationships. Microstructure-property correlations for hard, superhard, and ultrahard materials. Springer International Publishing, Cham, pp 75–103. https://doi.org/10.1007/978-3-319-29291-5_3
8. Musil J (2000) Hard and superhard nanocomposite coatings. Surf Coat Technol 125(1–3):322–330. https://doi.org/10.1016/S0257-8972(99)00586-1
9. Postolnyi BO, Beresnev VM, Abadias G et al (2017) Multilayer design of CrN/MoN protective coatings for enhanced hardness and toughness. J Alloys Compd 725:1188–1198. https://doi.org/10.1016/j.jallcom.2017.07.010
10. Kumar CS, Patel SK (2018) Application of surface modification techniques during hard turning: Present work and future prospects. Int J Refract Met Hard Mater 76:112–127. https://doi.org/10.1016/j.ijrmhm.2018.06.003
11. Pogrebnjak AD, Bagdasaryan AA, Yakushchenko I V et al (2014) The structure and properties of high-entropy alloys and nitride coatings based on them. Russ Chem Rev 83(11):1027–1061. https://doi.org/10.1070/RCR4407
12. Pogrebnjak AD, Shpak AP, Azarenkov NA et al (2009) Structures and properties of hard and superhard nanocomposite coatings. Physics-Uspekhi 52(1): 29–54. https://doi.org/10.3367/UFNe.0179.200901b.0035
13. Ferreira F, Aijaz A, Kubart T et al (2018) Hard and dense diamond like carbon coatings deposited by deep oscillations magnetron sputtering. Surf Coat Technol 336:92–98. https://doi.org/10.1016/j.surfcoat.2017.10.055
14. Schalk N, Simonet Fotso JFT, Holec D et al (2016) Influence of varying nitrogen partial pressures on microstructure, mechanical and optical properties of sputtered TiAlON coatings. Acta Mater 119:26–34. https://doi.org/10.1016/j.actamat.2016.08.007
15. Klimashin FF, Mayrhofer PH (2017) Ab initio-guided development of super-hard Mo–Al–Cr–N coatings. Scr Mater 140:27–30. https://doi.org/10.1016/j.scriptamat.2017.06.052

16. Mourlas A, Psyllaki P, Chaliampalias D et al (2016) Tribological behaviour of gradient TiAlSiN superhard coatings. Key Eng Mater 674:207–212. https://doi.org/10.4028/www.scientific.net/KEM.674.207
17. Pogrebnjak AD, Bondar OV, Abadias G et al. (2015) Investigation of nanoscale TiN/MoN multilayered systems, fabricated using arc evaporation. Acta Phys Pol A 128(5):836–841. https://doi.org/10.12693/APhysPolA.128.836
18. Musil J, Jirout M (2007) Toughness of hard nanostructured ceramic thin films. Surf Coat Technol 201(9–11 SPEC. ISS.):5148–5152. https://doi.org/10.1016/j.surfcoat.2006.07.020
19. Veprek S, Veprek-Heijman MGJ, Karvankova P et al (2005) Different approaches to superhard coatings and nanocomposites. Thin Solid Films 476(1):1–29. https://doi.org/10.1016/j.tsf.2004.10.053
20. Mironov V, Kolbe M, Lapkovskis V et al (2014) Application of pulse electromagnetic field for metal coatings manufacturing. Key Eng Mater 604:269–272. https://doi.org/10.4028/www.scientific.net/KEM.604.269
21. Chayeuski VV, Zhylinski VV, Rudak PV et al (2018) Characteristics of ZrC/Ni-UDD coatings for a tungsten carbide cutting tool. Appl Surf Sci 446:18–26. https://doi.org/10.1016/j.apsusc.2018.02.239
22. Abadias G, Daniliuk AY, Solodukhin IA et al (2018) Thermal stability of TiZrAlN and TiZrSiN films formed by reactive magnetron sputtering. Inorg Mater Appl Res 9(3):418–426. https://doi.org/10.1134/S2075113318030024
23. Abadias G, Chason E, Keckes J et al (2018) Review article: stress in thin films and coatings: current status, challenges, and prospects. J Vac Sci Technol A Vac Surf Film 36(2):020801. https://doi.org/10.1116/1.5011790
24. Fernandes F, Danek M, Polcar T et al (2018) Tribological and cutting performance of TiAlCrN films with different Cr contents deposited with multilayered structure. Tribol Int 119:345–353. https://doi.org/10.1016/j.triboint.2017.11.008
25. Pogrebnjak AD, Rogoz VM, Bondar OV et al (2016) Structure and physicomechanical properties of NbN-based protective nanocomposite coatings: a review. Prot Met Phys Chem Surf 52(5):802–813. https://doi.org/10.1134/S2070205116050191
26. Bondar O V., Postol'nyi BA, Beresnev VM et al (2015) Composition, structure and tribotechnical properties of TiN, MoN single-layer and TiN/MoN multilayer coatings. J Superhard Mater 37(1):27–38. https://doi.org/10.3103/S1063457615010050
27. Pogrebnjak AD, Isakov IF, Opekunov MS et al (1987) Increased wear resistance and positron annihilation in Cu exposed to high power ion beam. Phys Lett A 123(8)
28. Koltunowicz TN, Zukowski P, Bondariev V et al (2015) The effect of annealing on induction like properties of $(FeCoZr)_x(CaF\ 2)_{(100-x)}$ nanocomposite films produced by ion-beam sputtering in the vacuum environment. Vacuum 120:44–50. https://doi.org/10.1016/j.vacuum.2015.01.030
29. Ivashchenko VI, Veprek S, Turchi PEA et al (2012) First-principles study of TiN/SiC/TiN interfaces in superhard nanocomposites. Phys Rev B 86(1):014110. https://doi.org/10.1103/PhysRevB.86.014110
30. Kadyrzhanov DB, Zdorovets M, Kozlovskiy AL et al (2018) Influence of ionizing irradiation on the parameters of Zn nanotubes arrays for design of flexible electronics elements. Devices Methods Meas 9(1):66–73. https://doi.org/10.21122/2220-9506-2018-9-1-66-73
31. Zhang YJ, Qin YG, Qing YA et al. (2018) TiCuN solid solution coating: excellent wear-resistant biocompatible material to protect artificial joint. Mater Lett 227:145–148. https://doi.org/10.1016/j.matlet.2018.05.061
32. Johansson K, Riekehr L, Fritze S et al. (2018) Multicomponent Hf-Nb-Ti-V-Zr nitride coatings by reactive magnetron sputter deposition. Surf Coat Technol 349:529–539. https://doi.org/10.1016/j.surfcoat.2018.06.030
33. van Hove RP, Sierevelt IN, van Royen BJ et al. (2015) Titanium-nitride coating of orthopaedic implants: a review of the literature. Biomed Res Int 2015:1–9. https://doi.org/10.1155/2015/485975

34. Dobrozhan O, Kurbatov D, Danilchenko P et al (2018) Nanostructured ZnO, Cu_2ZnSnS_4, $Cd_{1-x}Zn_xTe$ thin films obtained by spray pyrolysis method. In: Semiconductors—growth and characterization (InTech). https://doi.org/10.5772/intechopen.72988
35. Pogrebnjak AD, Beresnev VM, Kolesnikov DA et al (2013) Multicomponent (Ti-Zr-Hf-V-Nb)N nanostructure coatings fabrication, high hardness and wear resistance. Acta Phys Pol A 123(5):816–818. https://doi.org/10.12693/APhysPolA.123.816
36. Pogrebnjak AD, Postol'nyi BA, Kravchenko YA et al (2015) Structure and properties of (Zr-Ti-Cr-Nb)N multielement superhard coatings. J Superhard Mater 37(2):101–111. https://doi.org/10.3103/S1063457615020045
37. Postolnyi BO, Konarski P, Komarov FF et al (2014) Study of elemental and structural phase composition of multilayer nanostructured TiN/MoN coatings, their physical and mechanical properties. J Nano-Electron Phys 6(4):04016
38. Bondar OV, Postolnyi BO, Kravchenko YA et al (2015) Fabrication and research of superhard (Zr-Ti-Cr-Nb)N coatings. Acta Phys Pol A 128(5):867–871. https://doi.org/10.12693/APhysPolA.128.867
39. Pogrebnyak AD, Kylyshkanov MK, Tyurin YN et al (2012) Properties and structure of oxidized coatings deposited onto Al-Cu and Al-Mg alloys. Tech Phys 57(6):840–848. https://doi.org/10.1134/S1063784212060217
40. Postolnyi B, Bondar O, Opielak M et al (2016) Structural analysis of multilayer metal nitride films CrN/MoN using electron backscatter diffraction (EBSD). In: Vladescu M, Tamas R, Cristea I (eds) Proceedings of SPIE—the international society for optical engineering, p 100100E. https://doi.org/10.1117/12.2243279
41. Shishkin A, Hussainova I, Kozlov V et al. (2018) Metal-coated cenospheres obtained via magnetron sputter coating: a new precursor for syntactic foams. JOM 70(7):1319–1325. https://doi.org/10.1007/s11837-018-2886-0
42. Pogrebnjak AD, Bagdasaryan AA, Pshyk A et al (2017) Adaptive multicomponent nanocomposite coatings in surface engineering. Physics-Uspekhi 60(6):586–607. https://doi.org/10.3367/UFNe.2016.12.038018
43. European Commission (2018) Report on critical raw materials and the circular economy
44. Grilli M, Bellezze T, Gamsjäger E et al (2017) Solutions for critical raw materials under extreme conditions: a review. Materials (Basel) 10(3):285. https://doi.org/10.3390/ma10030285
45. Blengini GA, Nuss P, Dewulf J et al (2017) EU methodology for critical raw materials assessment: policy needs and proposed solutions for incremental improvements. Resour Policy 53:12–19. https://doi.org/10.1016/j.resourpol.2017.05.008
46. Gaustad G, Krystofik M, Bustamante M et al (2018) Circular economy strategies for mitigating critical material supply issues. Resour Conserv Recycl 135:24–33. https://doi.org/10.1016/j.resconrec.2017.08.002
47. Veprek S (2013) Recent search for new superhard materials: go nano! J Vac Sci Technol A Vac Surf Film 31(5):050822. https://doi.org/10.1116/1.4818590
48. Silva HG, Pereira AM, Teixeira JM et al (2010) Magnetic field strength and orientation effects on Co-Fe discontinuous multilayers close to percolation. Phys Rev B 82(14):144432. https://doi.org/10.1103/PhysRevB.82.144432
49. Romano Brandt L, Salvati E, Papadaki C et al (2017) Probing the deformation and fracture properties of Cu/W nano-multilayers by in situ SEM and synchrotron XRD strain microscopy. Surf Coat Technol 320:158–167. https://doi.org/10.1016/j.surfcoat.2017.01.065
50. Chen W, Lin Y, Zheng J et al. (2015) Preparation and characterization of CrAlN/TiAlSiN nano-multilayers by cathodic vacuum arc. Surf Coatings Technol 265:205–211. https://doi.org/10.1016/j.surfcoat.2015.01.023
51. Wang H, Zeng H, Li Q et al (2016) Superlattice super toughness of TiN/*M*N (M = V, Nb, Ta, Mo, and W): first-principles study. Thin Solid Films 607:59–66. https://doi.org/10.1016/j.tsf.2016.03.061
52. Li W, Zheng K, Liu P et al (2016) Microstructure and super hardness effect of CrAlN/SiO 2 nanomultilayered film synthesized by reactive magnetron sputtering. Mater Charact 118:79–84. https://doi.org/10.1016/j.matchar.2016.05.016

53. Badiger PV, Desai V, Ramesh MR (2017) Development and characterization of Ti/TiC/TiN coatings by cathodic arc evaporation technique. Trans Indian Inst Met 70(9):2459–2464. https://doi.org/10.1007/s12666-017-1107-9
54. Lei Z, Zhang Q, Zhu X et al (2018) Corrosion performance of ZrN/ZrO_2 multilayer coatings deposited on 304 stainless steel using multi-arc ion plating. Appl Surf Sci 431:170–176. https://doi.org/10.1016/j.apsusc.2017.06.273
55. Seidl WM, Bartosik M, Kolozsvári S et al (2018) Mechanical properties and oxidation resistance of Al-Cr-N/Ti-Al-Ta-N multilayer coatings. Surf Coat Technol 347:427–433. https://doi.org/10.1016/j.surfcoat.2018.05.025
56. Caicedo JC, Amaya C, Yate L et al (2010) Hard coating performance enhancement by using $[Ti/TiN]_n$, $[Zr/ZrN]_n$ and $[TiN/ZrN]_n$ multilayer system. Mater Sci Eng B 171(1–3):56–61. https://doi.org/10.1016/j.mseb.2010.03.069
57. Uglov VV, Abadias G, Zlotski SV et al (2018) Blister formation in ZrN/SiN multilayers after He irradiation. Surf Coat Technol 344:170–176. https://doi.org/10.1016/j.surfcoat.2018.02.095
58. Vilarinho PM, Mahajan A, Sterianou I et al (2012) Layered composite thick films for dielectric applications. J Eur Ceram Soc 32(16):4319–4326. https://doi.org/10.1016/j.jeurceramsoc.2012.05.026
59. Ivashchenko VI, Veprek S, Argon AS et al (2015) First-principles quantum molecular calculations of structural and mechanical properties of TiN/SiN_x heterostructures, and the achievable hardness of the nc-TiN/SiN_x nanocomposites. Thin Solid Films 578:83–92. https://doi.org/10.1016/j.tsf.2015.02.013
60. Ivashchenko VI, Veprek S, Turchi PEA et al (2014) First-principles molecular dynamics investigation of thermal and mechanical stability of the TiN(001)/AlN and ZrN(001)/AlN heterostructures. Thin Solid Films 564:284–293. https://doi.org/10.1016/j.tsf.2014.05.036
61. Fernandes F, Morgiel J, Polcar T et al (2015) Oxidation and diffusion processes during annealing of TiSi(V)N films. Surf Coat Technol 275:120–126. https://doi.org/10.1016/j.surfcoat.2015.05.031
62. Matlak J, Rismaniyazdi E, Komvopoulos K (2018) Nanostructure, structural stability, and diffusion characteristics of layered coatings for heat-assisted magnetic recording head media. Sci Rep 8(1):9807. https://doi.org/10.1038/s41598-018-27688-4
63. Kelly PJ, Li H, Benson PS et al (2010) Comparison of the tribological and antimicrobial properties of CrN/Ag, ZrN/Ag, TiN/Ag, and TiN/Cu nanocomposite coatings. Surf Coat Technol 205(5):1606–1610. https://doi.org/10.1016/j.surfcoat.2010.07.029
64. Daniel R, Meindlhumer M, Baumegger W et al (2017) Grain boundary design of thin films: using tilted brittle interfaces for multiple crack deflection toughening. Acta Mater 122:130–137. https://doi.org/10.1016/j.actamat.2016.09.027
65. Polcar T, Martinez R, Vítů T et al (2009) High temperature tribology of CrN and multilayered Cr/CrN coatings. Surf Coat Technol 203(20–21):3254–3259. https://doi.org/10.1016/j.surfcoat.2009.04.005
66. Nouveau C, Djouadi M., Decès-Petit C et al (2001) Influence of Cr_xN^y coatings deposited by magnetron sputtering on tool service life in wood processing. Surf Coat Technol 142–144:94–101. https://doi.org/10.1016/S0257-8972(01)01092-1
67. Milošev I, Strehblow H-H, Navinšek B (1997) Comparison of TiN, ZrN and CrN hard nitride coatings: electrochemical and thermal oxidation. Thin Solid Films 303(1–2):246–254. https://doi.org/10.1016/S0040-6090(97)00069-2
68. Wang Y, Chen Y, Zhao D et al (2018) Deformation mechanism of CrN/nitriding coated steel in wear and nano-scratch experiments under heavy loading conditions. Appl Surf Sci 447:100–106. https://doi.org/10.1016/j.apsusc.2018.03.213
69. Subramanian B, Muraleedharan CV, Ananthakumar R et al (2011) A comparative study of titanium nitride (TiN), titanium oxy nitride (TiON) and titanium aluminum nitride (TiAlN), as surface coatings for bio implants. Surf Coat Technol 205(21–22):5014–5020. https://doi.org/10.1016/j.surfcoat.2011.05.004
70. Chim YC, Ding XZ, Zeng XT et al (2009) Oxidation resistance of TiN, CrN, TiAlN and CrAlN coatings deposited by lateral rotating cathode arc. Thin Solid Films 517(17):4845–4849. https://doi.org/10.1016/j.tsf.2009.03.038

71. Musil J, Jaroš M, Kos Š et al (2018) Hard TiN_2 dinitride films prepared by magnetron sputtering. J Vac Sci Technol A 36(4):040602. https://doi.org/10.1116/1.5038555
72. Jauberteau I, Bessaudou A, Mayet R et al (2015) Molybdenum nitride films: crystal structures, synthesis, mechanical, electrical, and some other properties. Coatings 5(4):656–687. https://doi.org/10.3390/coatings5040656

Structure and Properties of Combined Multilayer Coatings Based on Alternative Triple Nitride and Binary Metallic Layers

O. V. Bondar, Alexander D. Pogrebnjak, Y. Takeda, B. Postolnyi, P. Zukowski, R. Sakenova, V. Beresnev and V. Stolbovoy

Abstract Combined multilayered coatings based on alternative triple nitride and binary metallic layers were deposited using vacuum-arc evaporation of a cathode. (TiMo)N/TiMo, (CrMo)N/CrMo, (CrZr)N/CrZr, (TiCr)N/TiCr and (MoZr)N/MoZr multilayer coatings were fabricated under the same deposition conditions, while bias potential was −200 V. Total thickness of the coatings was around 54 μm, while bilayer thickness was around 900 nm and we had 60 bilayers in each coating. Thicknesses of triple nitride and binary metallic layers were 750 and 150 nm respectively. Various methods of analysis were used for coatings characterization, including, but not limited to, XRD, SEM, EDS, TEM, HR-TEM, SIMS, as well as indentation tests. Forming of two-phase state with (111) and (200) preferable orientation was found in the coatings. Vickers hardness HV0.1, HV0.5 and HV1 of the coatings varied from 2347 to 2912, 2077 to 2584 and from 1369 to 2327 respectively, which makes them perspective for application as hard protective coatings.

Keywords Multilayered coatings · Nitrides · Hardness · Mechanical properties

O. V. Bondar (✉) · A. D. Pogrebnjak
Sumy State University, Sumy, Ukraine
e-mail: oleksandr.v.bondar@gmail.com

O. V. Bondar · Y. Takeda
National Institute for Materials Science (NIMS), Tsukuba, Ibaraki Prefecture, Japan

B. Postolnyi
IFIMUP and IN-Institute of Nanoscience and Nanotechnology,
University of Porto, Porto, Portugal

P. Zukowski
Lublin University of Technology, Lublin, Poland

R. Sakenova
East-Kazakhstan State Technical University, Ust'-Kamenogorsk, Kazakhstan

V. Beresnev
V.N. Karazin National University, Kharkiv, Ukraine

V. Stolbovoy
National Science Center "Kharkiv Institute of Physics and Technology", Kharkiv, Ukraine

© Springer Nature Singapore Pte Ltd. 2019
A. D. Pogrebnjak and V. Novosad (eds.), *Advances in Thin Films, Nanostructured Materials, and Coatings*, Lecture Notes in Mechanical Engineering,
https://doi.org/10.1007/978-981-13-6133-3_3

1 Introduction

The fabrication of protective coatings by different deposition methods, such as chemical processes, physical vapour deposition, ion-assisted deposition, allows to solve a large variety of problems in machine-building industry, as well as in nuclear and rocket-building industries and medicine to improve wear and corrosion resistance, reduce friction, increase fatigue strength of cutting tools and mechanical products [1–8]. Among this variety of strengthening technologies, a specific attention is paid to vacuum-arc deposition usually named also as cathodic arc physical vapour deposition (Arc-PVD) [9], which makes it possible to fabricate coatings with improved physical-mechanical and tribological properties, resistance to the influence of irradiation and high temperatures [10–14], aggressive environment etc. [15–17].

The nitride coatings based on transition metals are widely used in modern materials science because of their exceptional physical and tribological properties, such as high hardness, high melting temperatures, good chemical and physical stability, etc. The most typical areas of application of such materials are wear protection, as well as corrosion and abrasive protection. The most extensively studied single-layer coatings are TiN, CrN, MoN and ZrN, but their properties are not good enough for modern demands in materials science. It has already been proved that the multilayer coatings demonstrate better properties, including magnetic and electrical [18, 19], in comparison with the single-layer ones [14, 20–26]. The combination of alternative layers made of binary or even triple nitrides of different refractory metals can provide superior performance and productivity of various tool components [27–30], their enhanced resistance to wear and oxidation under the influence of high temperatures [31–39], as well as good biocompatibility [40], which makes them perspective to use in biomedical applications.

The substitutional defects may occur along the interfaces between adjacent layers in multilayer films, when some of the elements of one layer enter the crystal lattice of the adjacent one, thus replacing its atoms. This process usually leads to the generation of strain energy proportional to the shear modulus of the material. The layers with different shear modulus prevent the movement of dislocations, thus preventing the destruction of the coating material. Firstly such type of model to describe hardness enhancement was proposed by Koehler [41] and then approved and followed by many experimental and theoretical works, as well as by review articles [42–54]. Additionally, deviations or redistribution of dislocations and cracks at the grain boundaries help to increase the coatings resistance to stress, wear, and destruction. The multilayer structure significantly reduces the influence of interlayer cracking and allows it's employing under large dynamic loads. The alternation of nanoscale layers with dissimilar physical-mechanical characteristics allows to change significantly the properties of multilayer coatings, such as the concentration of internal stresses, crack propagation [55–57] and, hence, to increase the fracture toughness of such materials [58, 59]. However, some works have been published recently where the enhancement of hardness and toughness in multilayer thin films is considered mainly due to the grain rotation for the nanocrystals and grain boundary sliding for larger grains [58, 60].

This present paper deals with fabrication and investigation of a novel nanocomposite combined multilayered coatings based on alternative triple nitride and binary metallic layers were deposited using vacuum-arc evaporation of a cathode, such as (TiMo)N/TiMo, (CrMo)N/CrMo, (CrZr)N/CrZr, (TiCr)N/TiCr and (MoZr)N/MoZr, which were fabricated by Arc-PVD. First studies of structure, elemental and phase composition, as well as physical-mechanical properties of such coatings, are described and discussed.

2 Experimental Details

Polished substrates of A 570 Grade 36 stainless steel with a surface roughness Ra up to 0.09 μm were used for the deposition. The multilayer coatings were deposited by vacuum-arc evaporation of cathodes in a vacuum-arc device Bulat-6 with two evaporators, which allows deposition of nanostructured multilayer coatings. Figure 1 shows a principal scheme of the deposition system. The vacuum chamber (1) (base pressure in the chamber was 0.001 Pa) was equipped with a system of automatic nitrogen pressure control (2) and two evaporators consisting of appropriate metals for each coating (purity of metallic target was 99.8%). The substrate holder (5) was mounted on a rotating stainless steel plate (300 × 300 mm) on which the substrates (6) were placed. BULAT-6 was also equipped with DC voltage source (7), the value of which can be varied between 5 and 1000 V, and high-voltage impulse generator (8) with adjustable voltage pulse amplitude of 0.5–2 kV and repetition frequency of 5–7 kHz.

The substrate cleaning process was carried out before coatings deposition while applying a 1 kV substrate potential. Further, nitrogen was injected into the chamber to fabricate nitrides of appropriate refractory or transition metals. Rotation of the substrate holder allowed deposition of alternating layers, while injection and stopping the injection of nitrogen into the deposition chamber allowed deposition of nitride and pure metallic layers.

Main deposition parameters for multilayer coatings are presented in a Table 1. All coatings were deposited for 6 h in a continuous rotation of substrates mode. Distance from cathodes to substrates was 200 mm for each sample.

The structure-phase state of the deposited coatings was analyzed using X-Ray diffraction (XRD) in terms of θ-2θ scans in Bregg-Bertrano geometry. Scanning electron microscopy with Energy-dispersive spectra (SEM with EDX) was used for studies of coatings surface and elemental composition, as well as coatings cross-sections. Also, a laser digital scanning was used to study surface roughness of the coatings. Time of flight secondary ion mass spectrometry (ToF SIMS) was used for studies of the distribution of elements along the depth. The hardness of the deposited coatings was studied by the micro-Vickers method. At least ten indentations were made for each sample and each loading.

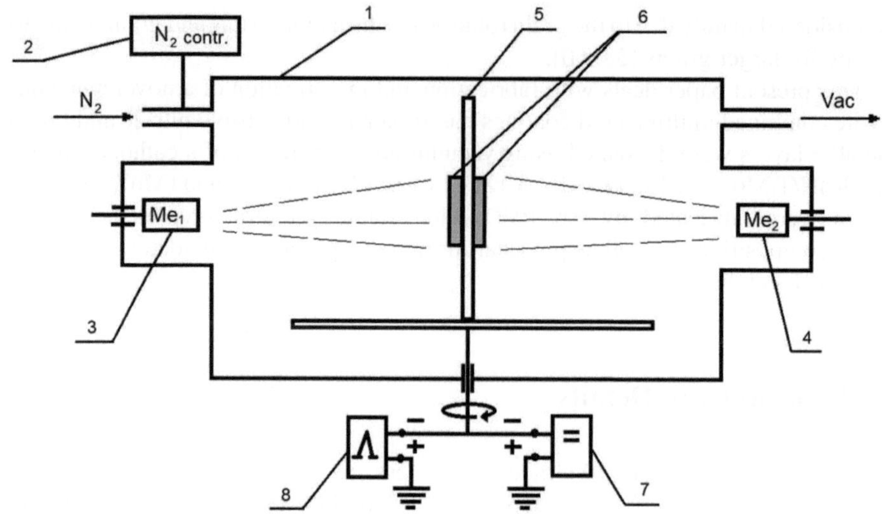

Fig. 1 A principal scheme of the Bulat-6 deposition system

Table 1 Composition and main deposition parameters of studied coatings

No	Composition	I_{arc_Me1}/I_{arc_Me2}, A	U_b, V	Dep. time for each layer and P_N (Pa)
877	(TiMo)N/(TiMo)	100/130	−200	5 min, 4×10^{-3}, 1 min, 2×10^{-5}
878	(CrMo)N/(CrMo)	100/130	−200	5 min, 4×10^{-3}, 1 min, 2×10^{-5}
891	(CrZr)N/(CrZr)	100/100	−200	5 min, 4×10^{-3}, 1 min, 2×10^{-5}
893	(TiCr)N/(TiCr)	100/100	−200	5 min, 4×10^{-3}, 1 min, 2×10^{-5}
894	(MoZr)N/(MoZr)	125/100	−200	5 min, 4×10^{-3}, 1 min, 2×10^{-5}

3 Results and Discussion

Studies of the surface morphology of the as-deposited samples were carried out using Digital Microscopy unit Keyence VK-X100 Series. Images of the surface of the samples 877 and 893 are presented in Fig. 2. One can see quite a smooth surface with some droplet fractions, which is typical for coatings deposited using the vacuum-arc method. Average roughness (mean height) of the samples did not exceed the value of 0.3 μm, thus pointing at the fact of quite good quality of the surfaces of the studied coatings.

(a) **(b)**

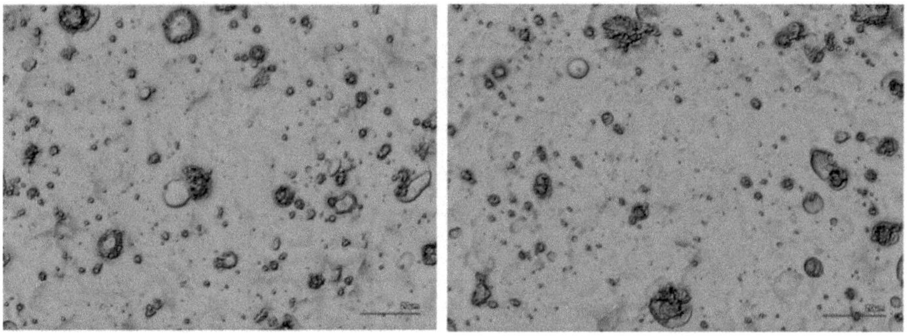

Fig. 2 Surface morphology of the samples 877 (**a**) and 893 (**b**)

Typical XRD spectra of the studied coatings are presented in Fig. 3. Corresponding planes are indicated below the curves. A spectrum for each sample has its corresponding color. As we can see from the Fig. 3, the most intensive peaks correspond to (111) and (200) planes, but we can also observe weaker reflections from (311) and (220) planes.

Two-phase state (nitride and metallic phases) was formed in the coatings, and each phase corresponds to a certain layer. For sample No 877 and 893 we can observe overlapping of peaks from Ti and Mo (sample 877) and Ti and Cr (sample 893). Broadening of the peaks from (222) plane is also observed indicating at the high disorder of polycrystallites, probably from molybdenum phase. Shape and intensity of diffraction peaks from nitride layers can point on the pretty good crystalline structure of the nitride layers. In according to the results of the EDS studies, top layers of the coatings are metallic ones, the ratio between metallic components is in the range 0.9–0.92.

Cross-sectional view under magnification at ×800 (total view) and ×5500 (magnified nitride and pure metallic layers) of the sample No 891 are presented in Fig. 4a, b accordingly. The presented cross-sectional view is typical for all samples. We can observe good planarity of layers; they do not intersect and have clear visible borders. The total thickness of the coating is around 54 μm, while alternative layers have the thickness of around 750 nm (nitride layers) and 150 nm (for pure metallic layers). The obtained data is in a good agreement with the deposition parameters, where one can see 5 and 1 min deposition times for nitride and pure metallic layers.

The hardness of the coatings in the as-deposited state was studied by Micro- Vickers method using Shimadzu HMV-G Micro Vickers hardness tester. Mean HV0.1 hardness of the coatings was 2700, while HV0.5 and HV1 equaled 2250 and 1700 accordingly, which makes them perspective for application as hard protective coatings.

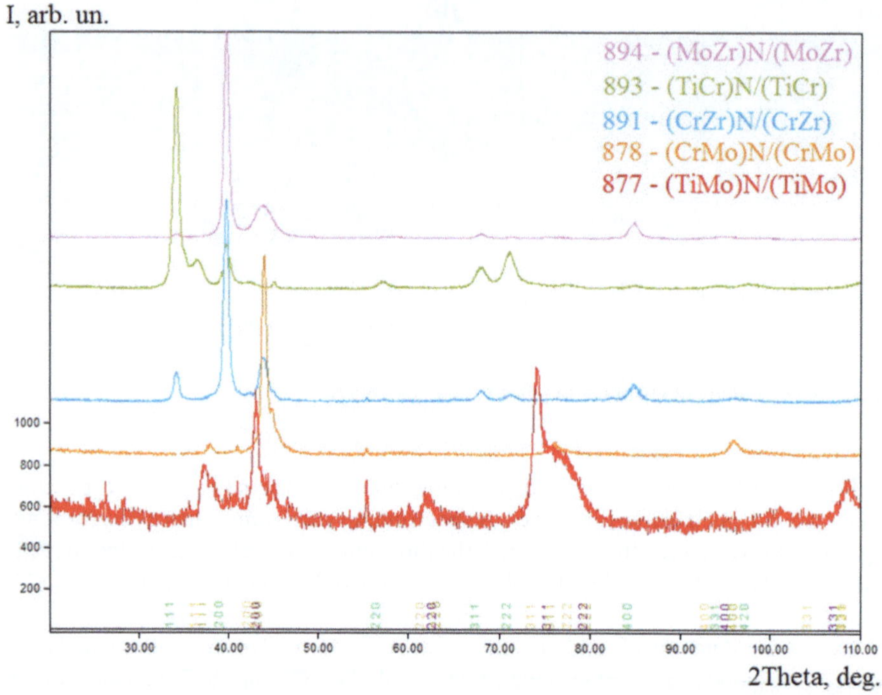

Fig. 3 XRD spectra of the samples No 877, 878, 891, 893 and 894. Corresponding planes are indicated in the figure

4 Conclusions

Multilayered coatings, based on a combination of nitrides of refractory metals and pure metallic layers were deposited using vacuum-arc evaporation of a cathode method. Two-phase state (nitride and metallic phases) was formed in the coatings, and each phase corresponds to a certain layer. All deposited coatings demonstrated good adhesion to the substrate. Mean Vickers hardness of the coatings was above 2000, which makes them perspective for application as hard protective coatings.

Acknowledgements This work was done under the aegis of Ukrainian state budget programs No. 0116U006816 "Development of perspective nanostructured multilayered coatings with enhanced physical-mechanical and tribological properties", 0118U003579 "Multilayer and multicomponent coatings with adaptive behavior in wear and friction conditions" and 0116U002621 "Physical basics of forming the composition and properties of transition metals boride, nitride and boride-nitride films for application in machine-building", as well as by Science and Technology Center in Ukraine (STCU) program entitled "A first-principle approach for the design of new superhard nanocomposite coatings" (Project No 6372-C).

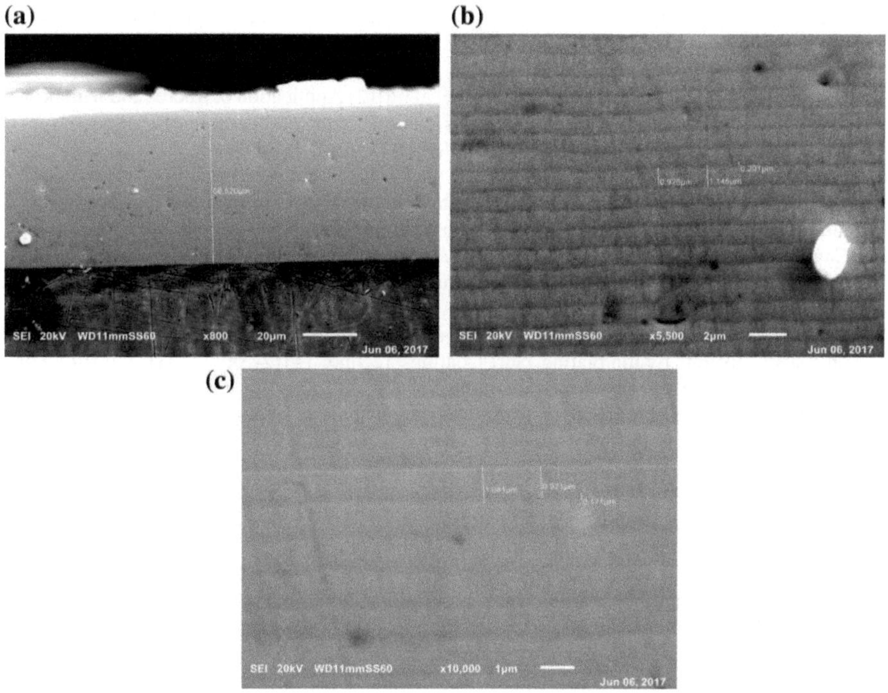

Fig. 4 SEM images of the cross-section of the sample 891: total view (**a**) and layers under higher magnifications ×5500 (**b**) and ×10000 (**c**)

References

1. Musil J (2000) Hard and superhard nanocomposite coatings. Surf Coatings Technol 125(1–3):322–330. https://doi.org/10.1016/S0257-8972(99)00586-1
2. Musil J (2007) Properties of hard nanocomposite thin films. In: Nanocomposite thin films and coatings (Published by Imperial College Press and Distributed by World Scientific Publishing Co.), pp 281–328. https://doi.org/10.1142/9781860949975_0005
3. Hao S, Delley B, Veprek S et al (2006) Superhard nitride-based nanocomposites: role of interfaces and effect of impurities. Phys Rev Lett 97(8):086102. https://doi.org/10.1103/PhysRevLett.97.086102
4. Veprek S, Veprek-Heijman MGJ, Karvankova P et al (2005) Different approaches to superhard coatings and nanocomposites. Thin Solid Films 476(1):1–29. https://doi.org/10.1016/j.tsf.2004.10.053
5. Pogrebnjak AD (2013) Structure and properties of nanostructured (Ti–Hf–Zr–V–Nb)N Coatings. J Nanomater 1–12. https://doi.org/10.1155/2013/780125
6. Pogrebnjak AD, Kravchenko YO, Bondar OV et al (2018) Structural features and tribological properties of multilayer coatings based on refractory metals. Prot Met Phys Chem Surfaces 54(2):240–258. https://doi.org/10.1134/S2070205118020107
7. Pogrebnjak AD, Shpak AP, Azarenkov NA et al (2009) Structures and properties of hard and superhard nanocomposite coatings. Phys.-Uspekhi 52(1):29–54. https://doi.org/10.3367/UFNe.0179.200901b.0035

8. Pogrebnyak AD, Tyurin YN (2005) Modification of material properties and coating deposition using plasma jets. Phys.-Uspekhi 48(5):487–514. https://doi.org/10.1070/PU2005v048n05ABEH002055
9. Pogrebnjak AD, Ponomarev AG, Shpak AP et al (2012) Application of micro- and nanoprobes to the analysis of small-sized 3D materials, nanosystems, and nanoobjects. Phys.-Uspekhi 55(3):270–300. https://doi.org/10.3367/UFNe.0182.201203d.0287
10. Sangiovanni DG, Hultman L, Chirita V et al (2016) Effects of phase stability, lattice ordering, and electron density on plastic deformation in cubic TiWN pseudobinary transition-metal nitride alloys. Acta Mater 103:823–835. https://doi.org/10.1016/j.actamat.2015.10.039
11. Cavaleiro AJ, Ramos AS, Martins RMS et al (2017) The effect of heating rate on the phase transformation of Ni/Ti multilayer thin films. Vacuum 139:23–25. https://doi.org/10.1016/j.vacuum.2017.02.004
12. Lee DB, Kim MH, Lee YC et al (2001) High temperature oxidation of a CrN coating deposited on a steel substrate by ion plating. Surf Coatings Technol 141(2–3):227–231. https://doi.org/10.1016/S0257-8972(01)01238-5
13. Luridiana S, Miotello A (1996) Spectrophotometric study of oxide growth on arc evaporated TiN and ZrN coatings during hot air oxidation tests. Thin Solid Films 290–291:289–293. https://doi.org/10.1016/S0040-6090(99)80009-1
14. Mei AB, Howe BM, Zhang C et al (2013) Physical properties of epitaxial ZrN/MgO(001) layers grown by reactive magnetron sputtering. J Vac Sci Technol A Vacuum, Surfaces, Film 31(6):061516. https://doi.org/10.1116/1.4825349
15. Pogrebnjak AD, Bor'ba SO, Kravchenko YO et al (2016) Effect of the high doze of N+(1018 cm–2) ions implantation into the (TiHfZrVNbTa)N nanostructured coating on its microstructure, elemental and phase compositions, and physico-mechanical properties. J Superhard Mater 38(6):393–401. https://doi.org/10.3103/S1063457616060034
16. Berladir KV, Budnik OA, Dyadyura KA et al (2016) Physicochemical principles of the technology of formation of polymer composite materials based on polytetrafluoroethylene—a Review. High Temp Mater Process 20(2):157–184. https://doi.org/10.1615/HighTempMatProc.2016017875
17. Pogrebnjak AD, Lebed AG, Ivanov YF (2001) Modification of single crystal stainless steel structure (Fe-Cr-Ni-Mn) by high-power ion beam. Vacuum 483–486. https://doi.org/10.1016/S0042-207X(01)00225-1
18. Kasiuk JV, Fedotova JA, Koltunowicz TN et al (2014) Correlation between local Fe states and magnetoresistivity in granular films containing FeCoZr nanoparticles embedded into oxygen-free dielectric matrix. J Alloys Compd 586(Suppl 1):S432–S435. https://doi.org/10.1016/j.jallcom.2012.09.058
19. Boiko O, Koltunowicz TN, Zukowski P et al (2017) The effect of sputtering atmosphere parameters on dielectric properties of the ferromagnetic alloy – ferroelectric ceramics nanocomposite (FeCoZr)x(PbZrTiO3)(100−x). Ceram Int 43(2):2511–2516. https://doi.org/10.1016/j.ceramint.2016.11.052
20. Ivashchenko VI, Veprek S, Argon AS et al (2015) First-principles quantum molecular calculations of structural and mechanical properties of TiN/SiNxheterostructures, and the achievable hardness of the nc-TiN/SiNxnanocomposites. Thin Solid Films 578:83–92. https://doi.org/10.1016/j.tsf.2015.02.013
21. Ivashchenko VI, Veprek S, Turchi PEA et al (2012) First-principles study of TiN/SiC/TiN interfaces in superhard nanocomposites. Phys Rev B—Condens Matter Mater Phys 86(1):014110. https://doi.org/10.1103/PhysRevB.86.014110
22. Maksakova O, Simoẽs S, Pogrebnjak A et al (2018) The influence of deposition conditions and bilayer thickness on physical-mechanical properties of CA-PVD multilayer ZrN/CrN coatings. Mater Charact 140:189–196. https://doi.org/10.1016/j.matchar.2018.03.048
23. Pogrebnjak AD, Beresnev VM, Bondar OV et al (2018) Superhard CrN/MoN coatings with multilayer architecture. Mater Des 153:47–59. https://doi.org/10.1016/j.matdes.2018.05.001
24. Pogrebnjak AD, Ivashchenko VI, Skrynskyy PL et al (2018) Experimental and theoretical studies of the physicochemical and mechanical properties of multi-layered TiN/SiC films:

Temperature effects on the nanocomposite structure. Compos Part B Eng 142:85–94. https://doi.org/10.1016/j.compositesb.2018.01.004
25. Pogrebnjak A, Ivashchenko V, Bondar O et al (2017) Multilayered vacuum-arc nanocomposite TiN/ZrN coatings before and after annealing: Structure, properties, first-principles calculations. Mater Charact 134:55–63. https://doi.org/10.1016/j.matchar.2017.10.016
26. Pogrebnjak AD, Bondar OV, Erdybaeva NK et al (2015) Influence of thermal annealing and deposition conditions on structure and physical-mechanical properties of multilayered nano-sized TiN/ZrN coatings. Prz Elektrotechniczny 1(12):228–232. https://doi.org/10.15199/48.2015.12.59
27. Bondar OV, Postol'nyi BA, Beresnev VM et al (2015) Composition, structure and tribotechnical properties of TiN, MoN single-layer and TiN/MoN multilayer coatings. J Superhard Mater 37(1):27–38. https://doi.org/10.3103/S1063457615010050
28. Yang S, Yan X, Yang K et al (2016) Effect of the addition of nano-Al2O3 on the microstructure and mechanical properties of twinned Al0.4FeCrCoNi1.2Ti0.3 alloys. Vacuum 131:69–72. https://doi.org/10.1016/j.vacuum.2016.05.019
29. Pogrebnjak AD, Rogoz VM, Bondar OV et al (2016) Structure and physicomechanical properties of NbN-based protective nanocomposite coatings: a review. Prot Met Phys Chem Surfaces 52(5):802–813. https://doi.org/10.1134/S2070205116050191
30. Pogrebnjak AD, Beresnev VM, Kolesnikov DA et al (2013) Multicomponent (Ti-Zr-Hf-V-Nb)N nanostructure coatings fabrication, high hardness and wear resistance. Acta Phys Pol A 123(5):816–818. https://doi.org/10.12693/APhysPolA.123.816
31. Ming J, Li M, Kumar P et al (2016) Multilayer approach for advanced hybrid lithium battery. ACS Nano 10(6):6037–6044. https://doi.org/10.1021/acsnano.6b01626
32. Wedig A, Luebben M, Cho D-Y et al (2015) Nanoscale cation motion in TaOx, HfOx and TiOx memristive systems. Nat Nanotechnol 11(1):67–74. https://doi.org/10.1038/nnano.2015.221
33. Ishida H, Campbell S, Blackwell J (2000) General approach to nanocomposite preparation. Chem Mater 12(5):1260–1267. https://doi.org/10.1021/cm990479y
34. Söderberg H, Odén M, Larsson T et al (2006) Epitaxial stabilization of cubic-SiN[sub x] in TiN/SiN[sub x] multilayers. Appl Phys Lett 88(19):191902. https://doi.org/10.1063/1.2202145
35. Hultman L, Bareño J, Flink A et al (2007) Interface structure in superhard TiN-SiN nanolaminates and nanocomposites: film growth experiments and ab initio calculations. Phys Rev B 75(15):155437. https://doi.org/10.1103/PhysRevB.75.155437
36. Fallqvist A, Ghafoor N, Fager H et al (2013) Self-organization during growth of ZrN/SiNx multilayers by epitaxial lateral overgrowth. J Appl Phys 114(22):224302. https://doi.org/10.1063/1.4838495
37. Setoyama M, Nakayama A, Tanaka M et al (1996) Formation of cubic-AlN in TiN/AlN superlattice. Surf Coatings Technol 86–87(PART 1):225–230. https://doi.org/10.1016/S0257-8972(96)03033-2
38. Ghafoor N, Lind H, Tasnádi F et al (2014) Anomalous epitaxial stability of (001) interfaces in ZrN/SiNx multilayers. APL Mater 2(4):046106. https://doi.org/10.1063/1.4870876
39. Ma S, Xu B, Wu G et al (2008) Microstructure and mechanical properties of SiCN hard films deposited by an arc enhanced magnetic sputtering hybrid system. Surf Coatings Technol 202(22):5379–5382. https://doi.org/10.1016/j.surfcoat.2008.06.057
40. van Raay JJAM, Rozing PM, van Blitterswijk CA et al (1995) Biocompatibility of wear-resistant coatings in orthopedic surgery in vitro testing with human fibroblast cell cultures. J Mater Sci Mater Med 6(2):80–84. https://doi.org/10.1007/BF00120412
41. Koehler JS (1970) Attempt to design a strong solid. Phys Rev B 2(2):547–551. https://doi.org/10.1103/PhysRevB.2.547
42. Ivashchenko VI, Veprek S, Turchi PEA et al (2014) First-principles molecular dynamics investigation of thermal and mechanical stability of the TiN(001)/AlN and ZrN(001)/AlN heterostructures. Thin Solid Films 564:284–293. https://doi.org/10.1016/j.tsf.2014.05.036
43. Zhitomirsky VN (2007) Structure and properties of cathodic vacuum arc deposited NbN and NbN-based multi-component and multi-layer coatings. Surf Coatings Technol 201(13):6122–6130. https://doi.org/10.1016/j.surfcoat.2006.08.125

44. Söderberg H, Odén M, Molina-Aldareguia JM et al (2005) Nanostructure formation during deposition of TiN/SiN[sub x] nanomultilayer films by reactive dual magnetron sputtering. J Appl Phys 97(11):114327. https://doi.org/10.1063/1.1935135
45. Lin S, Zhou K, Dai M et al (2015) Influence of modulation period on mechanical behavior of Ti-TiN-Zr-ZrN multi-layered coatings. Zhenkong Kexue yu Jishu Xuebao/J Vac Sci Technol 35(1):114–118. https://doi.org/10.13922/j.cnki.cjovst.2015.01.21
46. Kong M, Dai J, Lao J et al (2007) Crystallization of amorphous SiC and superhardness effect in TiN/SiC nanomultilayers. Appl Surf Sci 253(10):4734–4739. https://doi.org/10.1016/j.apsusc.2006.10.050
47. López-Vidrier J, Löper P, Schnabel M et al (2016) Silicon nanocrystals embedded in silicon carbide as a wide-band gap photovoltaic material. Sol Energy Mater Sol Cells 144:551–558. https://doi.org/10.1016/j.solmat.2015.10.006
48. Huang S-H, Chen S-F, Kuo Y-C et al (2011) Mechanical and tribological properties evaluation of cathodic arc deposited CrN/ZrN multilayer coatings. Surf Coatings Technol 206(7):1744–1752. https://doi.org/10.1016/j.surfcoat.2011.10.029
49. Wu MK, Lee JW, Chan YC et al (2011) Influence of bilayer period and thickness ratio on the mechanical and tribological properties of CrSiN/TiAlN multilayer coatings. Surf Coatings Technol 206(7):1886–1892. https://doi.org/10.1016/j.surfcoat.2011.07.045
50. Silva HG, Pereira AM, Teixeira JM et al (2010) Magnetic field strength and orientation effects on Co-Fe discontinuous multilayers close to percolation. Phys Rev B—Condens Matter Mater Phys 82(14):144432. https://doi.org/10.1103/PhysRevB.82.144432
51. Tavares CJ (1998) Deposition and characterization of multilayered TiNrZrN coatings. Thin Solid Films 317(1–2):8–12. https://doi.org/10.1016/S0040-6090(97)00607-X
52. Krishna H, Shirato N, Yadavali S et al (2011) Self-organization of nanoscale multilayer liquid metal films: experiment and theory. ACS Nano 5(1):470–476. https://doi.org/10.1021/nn1022632
53. Pogrebnjak A, Maksakova O, Kozak C et al (2016) Physical and mechanical properties of nanostructured (Ti-Zr-Nb)N coatings obtained by vacuum-arc deposition method. Prz Elektrotechniczny 92(8):180–183. https://doi.org/10.15199/48.2016.08.49
54. Maksakova OV, Grankin SS, Bondar OV et al (2015) Nanostructured (Ti-Zr-Nb)N coatings obtained by vacuum-arc deposition method: structure and properties. J Nano- Electron Phys 7(4):04098-1–04098-7
55. Martínez-Martínez D, López-Cartes C, Fernández A et al (2009) Influence of the microstructure on the mechanical and tribological behavior of TiC/a-C nanocomposite coatings. Thin Solid Films 517(5):1662–1671. https://doi.org/10.1016/j.tsf.2008.09.091
56. Chen S-F, Kuo Y-C, Wang C-J et al (2013) The effect of Cr/Zr chemical composition ratios on the mechanical properties of CrN/ZrN multilayered coatings deposited by cathodic arc deposition system. Surf Coatings Technol 231:247–252. https://doi.org/10.1016/j.surfcoat.2012.03.002
57. Braic M, Balaceanu M, Parau AC et al (2015) Investigation of multilayered TiSiC/NiC protective coatings. Vacuum 120(PA):60–66. https://doi.org/10.1016/j.vacuum.2015.06.019
58. Bobzin K, Brögelmann T, Kruppe NC et al (2017) Plastic deformation behavior of nanostructured CrN/AlN multilayer coatings deposited by hybrid dcMS/HPPMS. Surf Coatings Technol 332:253–261. https://doi.org/10.1016/j.surfcoat.2017.06.092
59. Musil J, Jirout M (2007) Toughness of hard nanostructured ceramic thin films. Surf Coatings Technol 201(9–11):5148–5152. https://doi.org/10.1016/j.surfcoat.2006.07.020
60. Jian J, Lee JH, Liu Y et al (2016) Plastic deformation in nanocrystalline TiN at ultra-low stress: An in situ nanoindentation study. Mater Sci Eng A 650:445–453. https://doi.org/10.1016/j.msea.2015.10.002

DSC Investigations of the Effect of Annealing Temperature on the Phase Transformation Behaviour in (Zr–Ti–Nb)N Coatings Deposited by CA-PVD

O. V. Maksakova, M. K. Kylyshkanov and S. Simoẽs

Abstract Changes in thermal transformation properties due to annealing and consequent cooling within the temperature ranged from 30 to 1400 °C were studied for (Zr–Ti–Nb)N coatings by differential scanning calorimetry (DSC) measurements in an argon atmosphere. Temperature and phase transformations in investigated coatings occurred in two stages: at intermediate temperature region (>670 °C) and high-temperature region (>1100 °C). There were also noticeable changes in values of heat capacity depending on nitrogen pressure applied during a deposition process.

Keywords Nitrides · PVD · Heat capacity · Enthalpy

1 Introduction

Transition metal nitride (TMN) coatings belong to the group of perspective functional materials, frequently also termed as "refractory metal nitrides" [1–12]. Depending on the application, the purpose of the coating is to provide increased surface hardness, abrasion/wear resistance, and corrosion resistance, lower friction, as well as resistant to high temperatures and oxidizing ambients. Among simple TMN combinations, TiN, ZrN and NbN are by far the most studied and characterized materials [13–16]. Industrial interest and significant research in modern materials science caused the development of complex ternary nitride coatings. Ternary nitride coatings are achieved by addition of a third material such as Vanadium (V), Aluminium (Al), Molybdenum (Mo), Silicon (Si), Zirconium (Zr) or other to a binary nitride com-

O. V. Maksakova (✉)
Sumy State University, Sumy, Ukraine
e-mail: maksakova.tereshenko@gmail.com

M. K. Kylyshkanov
Sarsen Amanzholov East Kazakhstan State University, Ust-Kamenogorsk, Republic of Kazakhstan

S. Simoẽs
University of Porto, Porto, Portugal

© Springer Nature Singapore Pte Ltd. 2019
A. D. Pogrebnjak and V. Novosad (eds.), *Advances in Thin Films, Nanostructured Materials, and Coatings*, Lecture Notes in Mechanical Engineering,
https://doi.org/10.1007/978-981-13-6133-3_4

pound (e.g. TiN, MoN, VN, ZrN, CrN, SiN, etc.). The presence of a small quantity of the third material has a better impact on the morphology, structure, and bonding of the film and substrate material [17–23]. Thus, very probably, one can suppose that properties and performance of thin coatings consisting of three compounds of transition metals from IV and V groups and nitride can be superior to commonly used ternary systems. It is already reported about quaternary thin nitride films as: TiZrSiN [24], TiAlYN [25], TiAlTaN [26], TiZrAlN [27], TiAlSiN [28] etc. Such coatings comply with the most challenging requirements (high hardness and elastic recovery, thermal stability, wear–corrosion–oxidation resistance, low friction coefficient), demanded in a wide variety industry-specific application [29–31].

The choice of components for quaternary (Zr–Ti–Nb)N coatings was based on exceptionally advantageous properties that demonstrate each metal element or its combination with nitride. In our previous investigations, we had already explored the mechanical and tribological properties of the coatings above [32, 33].

This paper reports on the study of heat capacity, thermal effects (enthalpy of phase transition), temperatures of phase transitions of (Zr–Ti–Nb)N coatings obtained by the cathodic arc physical vapor deposition (CA-PVD).

2 Experimental Details

Experimental samples were deposited onto AISI321 stainless steel substrates in a vacuum-arc device Bulat-6. The atomic percentages of constituent elements of a cathode were 30 at. % of Ti and 35 at.% each of Zr and Nb. Prior to coatings' production substrates were polished and cleaned to provide excellent adhesion between thin films and steel substrates.

Parameters that were specifying in deposition process were: (1) the current of the arc on the cathode (I_A); (2) the negative electric bias potential applied to the substrate (U_S); (3) the substrate temperature (T_S); (4) target to substrate distance (d_{T-S}); (5) the nitrogen pressure in a chamber (P_N). Values of deposition parameters are presented in Table 1.

Microstructure and chemical composition of the coatings were characterized using scanning electron microscopy (SEM) with energy dispersive spectroscopy (EDS) in a FEI Quanta 400 FEG ESEM/EDAX Genesis X4M. The coatings were observed in surface and cross-section views.

The phase identification, crystal structure, and substructural parameters were also studied by X-ray diffraction (XRD) in a Panalytical X'Pert Pro MPD. CuKα radiation

Table 1 Deposition conditions of the production of (Zr–Ti–Nb)N coatings

Sample series	T_S, °C	$d_{T\text{-}S}$, mm	I_A, A	U_S, V	P_N, Pa
a	450	250	95	−100	0.04
b					0.53

was used to collect patterns from 10 to 80° (2θ) in θ-2θ Bragg-Brentano mode. The identification of phases was determined using the Joint Committee on Powder Diffraction and Standards (JCPDS) database.

The thermal characteristics extraction in thin coatings was conducted by differential scanning calorimeter (DSC) using TG-DTA/DSC equipment. The applied temperature regime for each of specimens was the following: heating from +30 to +1400 °C and then cooling to approximately +30 °C at constant heating/cooling rates of 10 °C/min. The average weight of samples was 55 mg. The thermal flow of the samples consisted of (Zr–Ti–Nb)N coating, and steel substrate was investigated. One heating/cooling cycle was applied to all samples. To avoid disturbance from oxidation on the phase transformations, the DSC measurements are performed in a protective atmosphere. Here an argon flow (99.9% purity) was typically used. Calorimetric and phase diagrams were used for the study of phase transformation courses, the determination of the critical transformation temperatures and the measurement of the transformation energies. The start and finish temperatures of each phase transformation were determined by the intersections of a baseline and the tangents to a thermal peak in Infinity Pro Thermal Analysis software.

3 Results and Discussions

The surfaces of experimental coatings had different structure due to increasing concentration of reactive nitrogen ion in the chamber. The surface morphology of sample *a* deposited at $P_N = 0.04$ Pa showed an irregular structure and a rather rough appearance due to a large number of macro- and microdroplets. While the structure of sample *b* deposited at $P_N = 0.53$ Pa was compact but with a number of the particles of larger dimensions. The thickness of the coatings ranged from 7 to 10 μm. The investigation of the elemental composition of the deposited coatings showed following atomic percent of components: 14–16 at.% Ti, 13–14 at.% Zr, 13–14 at.% Nb and 55–59 at.% N.

X-ray spectra analysis showed that defining phase was a phase with face-centered cubic crystal lattice that means NaCl-structure of the coatings. The stronger preferred orientation along (111) plane of NbN phase was observed in all multilayers (see Fig. 1). The intensity of the indexed (111) peak for the NbN phase was several times higher than that indexed (222) peak. The low-intensity peak observed at 2θ of 37.84° told about the slight inclusion of bcc lattice, which was typical for the droplet phase at vacuum-arc deposition method. Increasing the pressure of reaction gas did not increase the peaks from the family of planes {111} but decreases the intensity of other peaks (200), (220), (311) which was defined by a significant decrease of droplet phase contents in the coatings and correlates with the results of surface characterization.

The results of the calorimetric tests after annealing at a temperature exceeding 1400 °C are submitted in Fig. 2. Extreme growth of DSC curves from 30 to 100 °C was evidenced by the exothermic effect on DSC curves. It was attributed to nitridation

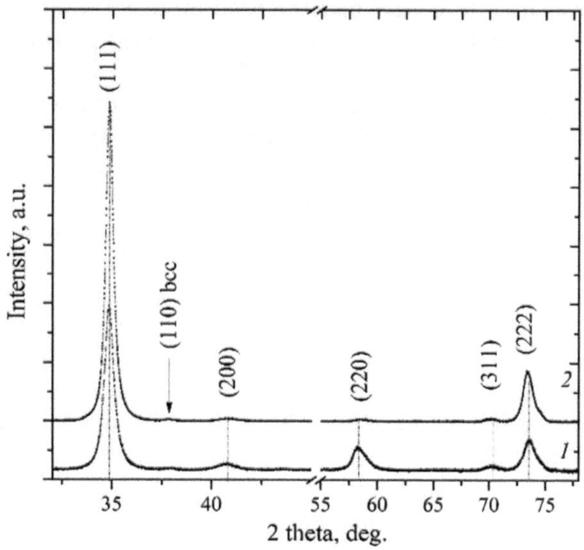

Fig. 1 XRD spectra of (Zr-Ti-Nb)N coatings

of coatings and a sharp increase in the mass of samples. The evaporation of water and sorbed gases at approximately 100 °C was evidenced by the small endothermic effect, according to DSC curves, after that a slight decrease in the mass of samples was registered.

It follows from the curves that the heating phase and structural transformations in the coatings were in two-stages: (1) at intermediate temperature region (>670 °C); (2) at high-temperature region (>1100 °C). Phase diagram of NbN showed that transformation of cubic δ-NbN or hexagonal ε-NbN phases into tetragonal γ-Nb_4N_3 occurred approximately at 1500 °C [34]. Thus, none of DSC peaks could be connected with niobium nitride. According to the phase diagrams of Cr–N [35] and Zr–N [36] systems, the first peak on curves was related with CrN and Cr_2N transformations. These exothermic peaks (low intense and broad) also corresponded to a cold crystallization effect, when the transition from amorphous to the crystalline state occurred in the coating upon heating. The enthalpy of this effect varied from −4.5 to −5.0 J/g. With the increase in the heat treatment temperature, the second stage of transformation caused distinct endothermic peaks that related to the process of formation of crystal lattice defects in investigated films. Some satellites of the endothermic peaks could be noticeable pointing at a trace amount of β-Cr_2N+gas transformation with the enthalpy value ranged from 0.03 to 0.064 J/g. Intensive oxidation and nitridation of coatings started at this stage as intensive and broad peaks observed on DSC curves [37, 38].

By using Infinity Pro Thermal Analysis software, the transformation temperatures and heats were assessed and summarized in Table 2. Moreover, heat capacity C_p of investigated samples was calculated according to formula reported in [39].

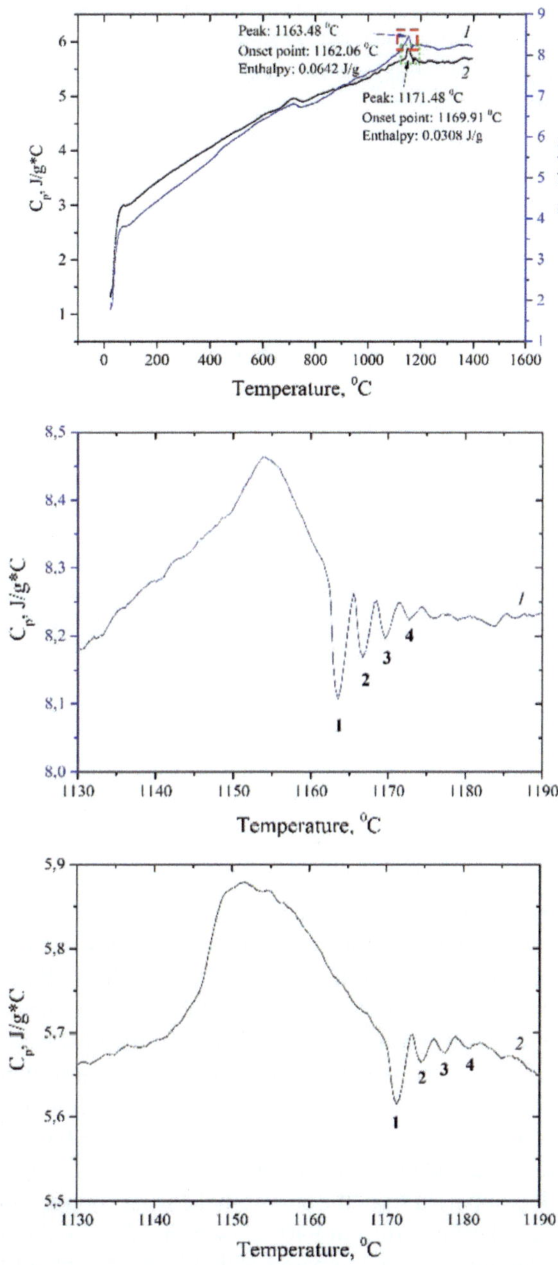

Fig. 2 DSC heating curves of (Zr–Ti–Nb)N coatings annealed from +30 to + 1400 °C

Table 2 Heat capacity, thermal effects (enthalpy of phase transition), temperature phase transition of (Zr–Ti–Nb)N coatings after heating

Sample series	Peak, °C	Onset point, °C	Enthalpy, J/g	Effect	Heat capacity, J/g °C
a (line 1)	P1: 700	695	−5	↑ Exothermic	6.74
	P2: 1163	1162	0.064	↓ Endothermic	8.46
b (line 2)	P1: 705	690	−4.5	↑ Exothermic	4.9
	P2: 1171	1169	0.03	↓ Endothermic	5.87

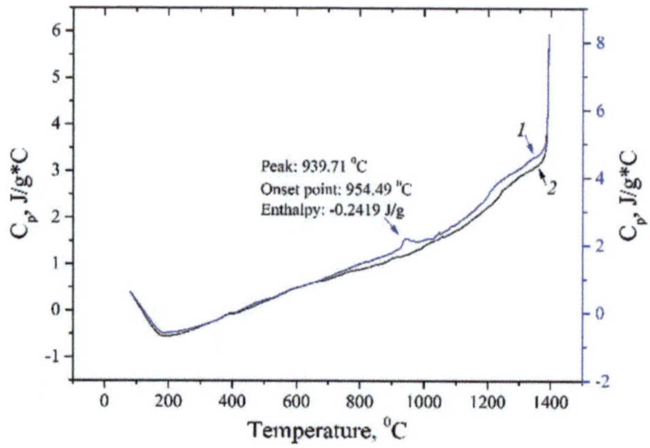

Fig. 3 DSC cooling curves of (Zr–Ti–Nb)N coatings annealed from +1400 to + 30° C

Heat capacity is the fundamental thermodynamic characteristic of the material, which can be used for evaluating the changes in the structure. For example, if the molecules in a material can make only small oscillations near the equilibrium position, as in a crystal, then the value of C_p is low. If the molecules can rotate and move, then C_p rises. By a jump-like change in the heat capacity, one can judge the phase transition in the material. Depending on the nitrogen pressure applied during the deposition process, there was some difference in the heat capacity values of investigated coatings. Sample *a* deposited at $P_N = 0.04$ Pa demonstrated higher $C_p =$ 8.46 J/g°C and consequently had better thermal properties compared to sample *b*.

It is apparent from the curves that during cooling only one sample *a* demonstrated the exothermic effect at 940 °C (see Fig. 3). Cooling peak was shifted to the lower temperature region compared with a heating cycle of this sample. The shape of the peak entailed minor changes in phase transformation and structure.

4 Conclusions

The results described in this work determined the heat and temperature of the phase transition by measuring the heat flow and heat capacity of (Zr–Ti–Nb)N coatings. It was established that no abnormal changes were observed on DSC curves. There were small changes in the mass of studied samples during heating and cooling cycles. Consequently, the change in the heat capacities and different thermal events on DSC curves can be attributed to changes in the structure and phase state of the samples.

Acknowledgements The authors gratefully to Ministry of Education and Science of Ukraine for financial support (Project No. 0118U003579 and 0117U003923). Authors are very thankful to Prof. A. D. Pogrebnjak from Sumy State University for project supervision and analysis of results and Prof. V. M. Beresnev from V. N. Karazin Kharkiv National University for deposition of the samples.

References

1. Navinsek B, Seal S (2001) Transition metal nitride functional coatings. JOM 53:51–54. https://doi.org/10.1007/s11837-001-0072-1
2. Chen JG (1996) Carbide and nitride overlayers on early transition metal surfaces: preparation, characterization, and reactivities. Chem Rev 96:1477–1498. https://doi.org/10.1021/cr950232u
3. Pogrebnyak AD, Shpak AP, Azarenkov NA et al (2009) Structures and properties of hard and superhard nanocomposite coatings. Phys Usp 52:29–54. https://doi.org/10.3367/UFNr.0179.200901b.0035
4. Pogrebnjak AD, Beresnev VM, Kolesnikov DA et al (2013) Multicomponent (Ti-Zr-Hf-V-Nb)N nanostructure coatings fabrication, high hardness and wear resistance. Acta Phys. Polonica A 123:816–818. https://doi.org/10.12693/APhysPolA.123.816
5. Pogrebnjak AD, Bor'ba SO, Kravchenko YaO et al (2016) Effect of the high doze of N+ (10^{18} cm^{-2}) ions implantation into the (TiHfZrVNbTa)N nanostructured coating on its microstructure, elemental and phase compositions, and physico-mechanical properties. J Superhard Mater 38:393–401. https://doi.org/10.3103/S1063457616060034
6. Maksakova O, Simoẽs S, Pogrebnjak A et al (2018) The influence of deposition conditions and bilayer thickness on physical-mechanical properties of CA-PVD multilayer ZrN/CrN coatings. Mater Charact 140:189–196. https://doi.org/10.1016/j.matchar.2018.03.048
7. Pogrebnjak AD, Bazyl EA (2001) Modification of wear and fatigue characteristics of Ti-V-Al alloy by Cu and Ni ion implantation and high-current electron beam treatment. Vacuum 64(1):1–7. https://doi.org/10.1016/S0042-207X(01)00160-9
8. Pogrebnjak AD, Beresnev VM, Smyrnova KV et al (2018) The influence of nitrogen pressure on the fabrication of the two-phase superhard nanocomposite (TiZrNbAlYCr)N coatings. Mater Lett 211:316–318. https://doi.org/10.1016/j.matlet.2017.09.121
9. Berladir KV, Budnik AO, Dyadyura KA et al (2016) Physicochemical principles of the technology of formation of polymer composite materials based on polytetraflouroethylene – a review. High Temp Mater Process 20(2):157–184. https://doi.org/10.1615/HighTempMatProc2016017875
10. Cramer CJ, Truhlar DG (2009) Density functional theory for transition metals and transition metal chemistry. Phys Chem 11:10757–10816. https://doi.org/10.1039/b907148b
11. Ivashchenko VI, Veprek S, Argon AS et al (2015) First-principles quantum molecular calculations of structural and mechanical properties of TiN/SiN_x heterostructures, and the achievable hardness of the nc-TiN/SiN_x nanocomposites. Thin Solid Films 578:83–92. https://doi.org/10.1016/j.tsf.2015.02.013

12. Ivashchenko VI, Veprek S, Turchi PEA et al (2014) First-principles molecular dynamics investigation of thermal and mechanical stability of the TiN(001)/AlN and ZrN(001)/AlN heterostructures. Thin Solid Films 564:284–293. https://doi.org/10.1016/j.tsf.2014.05.036
13. Milošev I, Strehblow HH, Navinšek B (1997) Comparison of TiN, ZrN and CrN hard nitride coatings: Electrochemical and thermal oxidation. Thin Solid Films 303(1–2):246–254. https://doi.org/10.1016/S0040-6090(97)00069-2
14. Bondar OV, Postol'nyi BA, Beresnev VM et al (2015) Composition, structure and tribotechnical properties of TiN, MoN single-layer and TiN/MoN multilayer coatings. J. Superhard Mater 37(1):27–38. https://doi.org/10.3103/S1063457615010050
15. Mercier F, Coindeau S, Lay S et al (2014) Niobium nitride thin films deposited by high temperature chemical vapor deposition. Surf Coatings Technol 260:126–132. https://doi.org/10.1016/j.surfcoat.2014.08.084
16. Benkahoul M, Martinez E, Karimi A et al (2004) Structural and mechanical properties of sputtered cubic and hexagonal NbN_x thin films. Surf Coatings Technol 180–181:178–183. https://doi.org/10.1016/j.surfcoat.2003.10.040
17. McIntyre D, Greene JE, Håkansson G et al (1990) Oxidation of metastable single-phase polycrystalline Ti0.5Al0.5 N films: kinetics and mechanisms. J Appl Phys 67:1542–1553. https://doi.org/10.1063/1.345664
18. Deeleard T, Buranawong A, Choeysuppaket A et al (2012) Structure and composition of TiVN thin films deposited by reactive DC magnetron co-sputtering. Proc Eng 32:1000–1005. https://doi.org/10.1016/j.proeng.2012.02.045
19. Cheng YH, Browne T, Heckerman B et al (2010) Mechanical and tribological properties of nanocomposite TiSiN coatings. Surf Coatings Technol 204(14):2123–2129. https://doi.org/10.1016/j.surfcoat.2009.11.034
20. Rogström L, Ghafoor N, Schroeder J et al (2015) Thermal stability of wurtzite $Zr_{1-x}Al_xN$ coatings studied by in situ high-energy x-ray diffraction during annealing. J Appl Phys 118:035309. https://doi.org/10.1063/1.4927156
21. Jiang X, Yang FC, Chen WC et al (2017) Effect of nitrogen-argon flow ratio on the microstructural and mechanical properties of AlSiN thin films prepared by high power impulse magnetron sputtering. Surf Coatings Technol 320:138–145. https://doi.org/10.1016/j.surfcoat.2017.01.085
22. Kasiuk JV, Fedotova JA, Koltunowicz TN et al (2014) Correlation between local Fe states and magnetoresistivity in granular films containing FeCoZr nanoparticles embedded into oxygen-free dielectric matrix. J Alloys Compd 586:S432–S435. https://doi.org/10.1016/j.jallcom.2012.09.058
23. Boiko O, Koltunowicz TN, Zukowski P et al (2017) The effect of sputtering atmosphere parameters on dielectric properties of the ferromagnetic alloy—ferroelectric ceramics nanocomposite $(FeCoZr)_x(PbZrTiO_3)_{(100-x)}$. Ceram Int 43(2):2511–2516. https://doi.org/10.1016/j.ceramint.2016.11.052
24. Saladukhin IA, Abadias G, Michel A et al (2015) Structure and hardness of quaternary TiZrSiN thin films deposited by reactive magnetron co-sputtering. Thin Solid Films 581:25–31. https://doi.org/10.1016/j.tsf.2014.11.020
25. Riedl H, Holec D, Rachbauer R et al (2013) Phase stability, mechanical properties and thermal stability of Y alloyed Ti-Al-N coatings. Surf Coatings Technol 235:174–180. https://doi.org/10.1016/j.surfcoat.2013.07.030
26. Koller CM, Hollerweger R, Sabitzer C et al (2014) Thermal stability and oxidation resistance of arc evaporated TiAlN, TaAlN, TiAlTaN, and TiAlN/TaAlN coatings. Surf Coatings Technol 259:599–607. https://doi.org/10.1016/j.surfcoat.2014.10.024
27. Abadias G, Saladukhin IA, Uglov VV et al (2013) Thermal stability and oxidation behavior of quaternary TiZrAlN magnetron sputtered thin films: Influence of the pristine microstructure. Surf Coatings Technol 237:187–195. https://doi.org/10.1016/j.surfcoat.2013.07.055
28. Miletić A, Panjan P, Škorić B et al (2014) Microstructure and mechanical properties of nanostructured Ti-Al-Si-N coatings deposited by magnetron sputtering. Surf Coatings Technol 241:105–111. https://doi.org/10.1016/j.surfcoat.2013.10.050

29. Mitterer C (2014) PVD and CVD Hard Coatings. In: Sarin VK, Llanes L, Mari D (eds) Comprehensive Hard Materials, 1st edn, vol 2. Elsevier, pp 449–467. https://doi.org/10.1016/B978-0-08-096527-7.00035-0
30. Pogrebnjak AD, Lebed AG, Ivanov YF (2001) Modification of single crystal stainless steel structure (Fe-Cr-Ni-Mn) by high-power ion beam. Vacuum 63(4):483–486. https://doi.org/10.1016/S0042-207X(01)00225-1
31. Kadyrzhanov DB, Zdorovets MV, Kozlovskiy AL et al (2018) Influence of ionizing irradiation on the parameters of Zn nanotubes arrays for design of flexible electronics elements. Devices Methods Meas 9(1):66–73. https://doi.org/10.21122/2220-9506-2018-9-1-66-73
32. Beresnev VM, Sobol OV, Grankin SS et al (2016) Physical and mechanical properties of (Ti–Zr–Nb)N coatings fabricated by vacuum-arc deposition. Inorg Mater Appl Res 7(3):388–394
33. Pogrebnjak A, Maksakova O, Kozak C et al (2016) Physical and mechanical properties of nanostructured (Ti-Zr-Nb)N coatings obtained by vacuum-arc deposition method. Prz Elektrotechniczny (8):180–183. https://doi.org/10.15199/48.2016.08.49
34. Pogrebnjak AD, Rogoz VM, Bondar OV et al (2016) Structure and physicomechanical properties of NbN-based protective nanocomposite coatings: a review. Prot Met Phys Chem Surfaces 52:802–813. https://doi.org/10.1134/S2070205116050191
35. Pogrebnjak AD, Bondar OV, Abadias G et al (2016) Structural and mechanical properties of NbN and Nb-Si-N films: experiment and molecular dynamics simulations. Ceram Int 42(10):11743–11756. https://doi.org/10.1016/j.ceramint.2016.04.095
36. Keogh DW (2011) Encyclopedia of inorganic and bioinorganic chemistry. https://doi.org/10.1002/9781119951438
37. Gribaudo L, Arias D, Abriata J (1994) The N-Zr (Nitrogen-Zirconium) System. J Phase Equilibria 15(4):441–449. https://doi.org/10.1007/BF02647575
38. Pogrebnjak AD, Bagdasaryan AA, Yakushchenko IV et al (2014) The structure and properties of high-entropy alloys and nitride coatings based on them. Russ Chem Rev 83(11):1027–1061
39. Hultman L (2000) Thermal stability of nitride thin films. Vacuum 57(1):1–30. https://doi.org/10.1016/S0042-207X(00)00143-3

Microstructure and Mechanical Properties of Multilayered α-AlN/α-BCN Coatings Depending on Flux Density During Target B4C Sputtering

V. I. Ivashchenko, V. M. Rogoz, T. N. Koltunowicz and A. I. Kupchishin

Abstract Multilayered AlN/BCN coatings with nanoscale layers were fabricated by magnetron sputtering of Al and B_4C targets on Si substrate. Deposited amorphous AlN/BCN coatings have demonstrated increased nano- and Knoop hardnesses, Young's modulus in compare with AlN and BCN coatings, which explained by strain modulation in amorphous layers of AlN and BCN. The application of flux density I_{B4C} (100 mA) has led to significant increasing of hardness from 18 to 27 GPa due to the formation of α-BCN phase according to Fourier spectra. Nanolayered coatings have been thermally stable up to 600 °C due to the slow diffusion processes in amorphous sublayer, which indicates higher oxidation resistance then nanocrystalline.

Keywords Multilayers · Coatings · Mechanical properties · XRD investigation

1 Introduction

It is well known that multilayer nanoscaled coatings based on nitrides of transition metals have good physic-mechanical and tribological properties due to the formation of at least two phases and a large number of interphase boundaries [1–7].

Aluminum nitride (AlN) is a material with remarkable properties with high hardness and good thermal conductivity [8–19]. The deposition of AlN films with con-

---

V. I. Ivashchenko
Frantsevich Institute for Materials Science Problems, National Academy of Sciences of Ukraine, Kiev, Ukraine

V. M. Rogoz (✉)
Sumy State University, Sumy, Ukraine
e-mail: vrogoz2009@gmail.com

T. N. Koltunowicz
Lublin University of Technology, Lublin, Poland

A. I. Kupchishin
Abai Kazakh National Pedagogical University, Almaty, Kazakhstan

© Springer Nature Singapore Pte Ltd. 2019
A. D. Pogrebnjak and V. Novosad (eds.), *Advances in Thin Films, Nanostructured Materials, and Coatings*, Lecture Notes in Mechanical Engineering,
https://doi.org/10.1007/978-981-13-6133-3_5

trolled morphology and high crystalline quality [20–34] is one of the key issues to be addressed to fabricate high-performance AlN based devices.

Boron carbon nitride (BCN) compounds have been expected to combine the excellent properties of boron carbide (B4C), boron nitride (BN) and carbon nitride (C_3N_4), with their properties adjustable, depending on composition and structure [35–43]. Excellent mechanical properties such as adhesion, high hardness, and good wear resistance have been reported in the case of sputtered BCN thin films [44–51].

Such internal structure impedes the movement of dislocations and leads to an increase of the hardness of the material [52–54]. At the same time, according to other works [54, 55], nanocomposite coatings should have nanocrystalline grains (nanometer-sized) of one phase, which are embedded into another phase—amorphous interlayer (α), which favorably affected by the high-temperature stability [53, 56–62]. In addition, the amorphous interlayer also prevents the movement of dislocations, which makes this coatings more ductile and reduces the propensity to fracture [63–67]. In this paper, two types of layers of AlN and BCN are used which tend amorphization at low deposition temperatures, which makes them as potential candidates for obtaining protective coatings, which are capable of operating under severe conditions.

Therefore, this work aimed to obtain multilayer nanoscale coatings in which at least one layer (of two) consists of an amorphous phase, and also to investigate the microstructure and properties of obtained coatings.

2 Experimental Methods

Multilayer AlN/BCN nanoscale coatings (116 layers) were obtained by alternating magnetron sputtering of Al and B_4C targets in argon-nitrogen atmosphere onto a polished Si (100) substrate. The pressure in the chamber was maintained within 1.4 mTorr, the bias potential U was -50 V, the flow rate of argon and nitrogen was 47 and 12 cm^3/min, respectively. During the deposition of the coatings, current (I_{B4C}) was changed from 30 to 100 mA. The deposition parameters of AlN/BCN coatings (samples of the ABC series) are presented in Table 1. Obtained coatings were annealed in vacuum at temperatures $T_a = 600$, 900, and 1000 °C.

The coating structure was studied using X-ray diffraction (XRD, Dron 3M diffractometer, CuK_α radiation). PowderCell 2.4 software was used for profiles separation in the case of overlaying of the complex profiles. The substructure characteristics (crystallite size and microstrain) were determined by an approximation method using the Cauchi function as the approximating function.

Nanoindentation of the films was carried out in the continuous stiffness measurement (CSM) mode using the NanoIndenter-G200 system (Agilent Technologies). All nanoindentation measurements were performed using a Berkovich diamond tip with a nominal radius of 340 nm. The load and displacement were continuously recorded up to a maximum displacement of 200 nm at a constant indentation rate of 0.05 s^{-1}, the amplitude of the oscillations was 2 nm. Knoop hardness (HK) was determined

Table 1 Deposition parameters: T_S is the temperature of the substrate, U is the potential bias, I is the current at the target, and t is the deposition time of one layer

Series ABC	T_S (°C)	U_{B4C} (V)	I_{B4C} (mA)	t_{B4C} (min)	U_{Al} (V)	I_{Al} (mA)	t_{Al} (min)
1	350	460	100	1	300	200	2
2	350	460	80	1	300	200	2
3	350	400	50	1	300	200	2
4	350	420	30	1	300	200	2
5	350	–	–	–	250	200	60
6	350	400	100	60	–	–	–

using the MICROMET2103 microhardness instrument (BUEHLER, USA) at a load of 100 mN. The hardness value for each sample was averaged over 10 measurements.

3 Results and Discussion

As a result of the X-ray diffraction analysis (Fig. 1) it was found that the BCN coating has a completely amorphous structure. In the coatings based on nitride aluminum, microcrystallites w-AlN (type ZnS, #186, P63mc) are formed with parameters of a crystal lattice a = 0.331 нм та c = 0.511 нм. The size of crystallites with preferred orientation (110) less than L = 3 nm. The average value of the microstress is $<\varepsilon> = 0.98 \times 10^{-2}$.

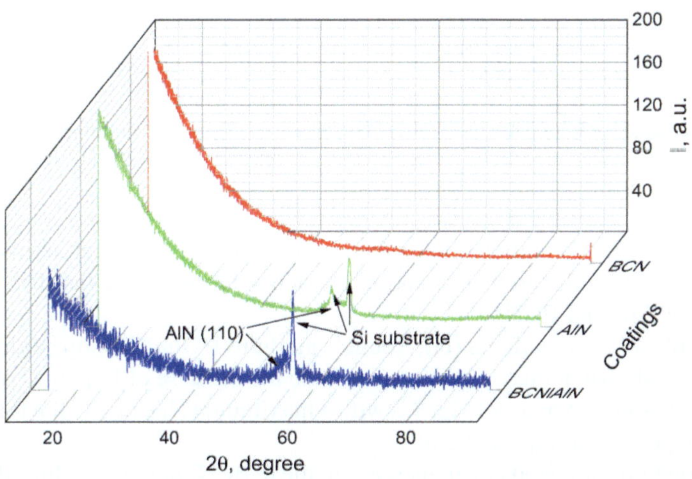

Fig. 1 X-ray spectra multilayered coatings BCN/AlN (series 1) and monolayered coatings Al-N (series 5) and BCN (series 6)

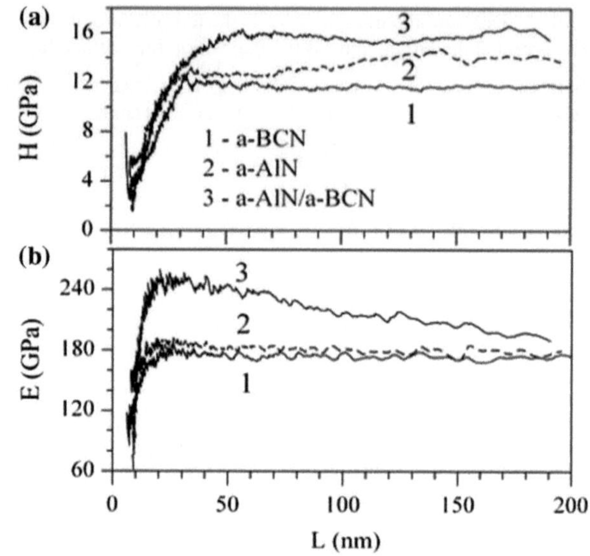

Fig. 2 **a** Nanohardness (H); **b** Young's modulus (E) from the penetration depth of the nanoindenter for monolayer α-BCN (1), α-AlN (2) and multilayer α-AlN/α-BCN (3) films, which are deposited at $I_{B4C} = 100$ mA

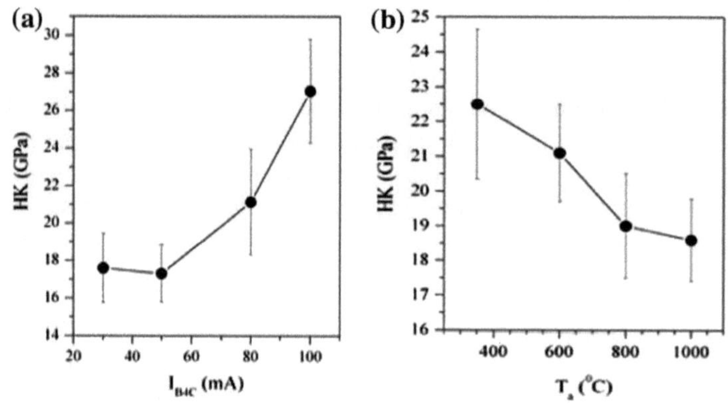

Fig. 3 Dependence of Knoop hardness (HK): **a** on the current I_{B4C}; **b** on the annealing temperature T_a

As a result of the formation multilayers coating, the size of crystallites increase up to 10%, most likely due to the formation of solid solutions at the interface between the layers of AlN and BCN.

The results of nanohardness (H), Young's modulus (E) and Knoop hardness measurement are shown in Figs. 2 and 3 (a, b). With the increase of I_{B4C}, the hardness of multilayer coatings increases (Fig. 3a) with a maximum value of 27 GPa for coating, which is deposited at $I_{B4C} = 100$ mA.

The growth of I_{B4C} leads to an increase in the α-BCN interlayer. Thus, the difference in the α-AlN and α-BCN amorphous layers enhances the stress modulation in the coating. It should be noted that the compressive stresses develop in the α-AlN layer, whereas in α-BCN layer there are tension stresses. The hardness of single-layer coatings of α-AlN and α-BCN is 12.2 and 12.8 GPa, respectively. Therefore, we are still inclined to the idea that, despite the amorphous structure of multilayer coatings, there is an alternation of stresses, which is responsible for increasing the hardness. This issue requires further study.

Figure 3b shows that annealing of coatings after 600 °C leads to a significant decrease in hardness. It can be assumed that the AlN/BCN coatings are thermally stable up to 600 °C. With a further increase of annealing temperature, it is obvious that the layers are mixed, which leads to a reduction in stress and hardness. Another explanation may be that the increase of the fraction of the more softer aluminum and boron oxides with increasing annealing temperature Ta leads to the decrease in hardness. Based on our results and previously published data [54, 63, 68–71], it can be argued that the oxidation of amorphous nanolayer coatings is prevented by the low rate of oxygen diffusion in the amorphous matrix as compared with nanocrystalline. Thus, amorphous coatings more effectively prevent oxidation than nanocrystalline ones [72–76].

4 Conclusions

The deposited of multilayer amorphous AlN/BCN coatings demonstrated an increase in nanohardness, Young's modulus and Knoop hardness as compared with monolayer AlN and BCN coatings, which is explained by the presence of modulation of stresses associated with amorphous AlN and BCN layers.

During the deposition of multi-layer AlN/ BCN coatings a nanocomposite structure in the form of microcrystallites w-AlN, less than 3 nm was formed inside an amorphous matrix.

With the increasing of I_{B4C} current density to 100 mA, a significant increase in the hardness of nanolayer coatings from 18 to 27 GPa is observed. Such an effect can be explained due to the increase on the amorphous BCN interlayer, which is confirmed by the obtained Fourier spectra.

Multilayer coatings are thermally stable up to 600 °C due to the slowing of diffusion in an amorphous interlayer. Thus, the amorphous multilayer coatings are more stable to oxidation than nanocrystalline coatings.

Acknowledgements This work was done under the aegis of Ukrainian state budget program «Physical basis of forming of composition and properties of nanostructured boride, nitride and boron-nitride films of refractory metals for their usage in machine building» (No 0116U002621). Authors are grateful to Prof. O. D. Pogrebnyak and A.A. Bagdasaryan for discussing the results of the research.

References

1. Pogrebnjak AD, Beresnev VM, Smyrnova KV et al (2018) The influence of nitrogen pressure on the fabrication of the two-phase superhard nanocomposite (TiZrNbAlYCr)N coatings. Mater Lett 211:316–318. https://doi.org/10.1016/j.matlet.2017.09.121
2. Pogrebnjak AD, Beresnev VM, Bondar OV et al (2018) Superhard CrN/MoN coatings with multilayer architecture. Mater Des 153:47–59. https://doi.org/10.1016/j.matdes.2018.05.001
3. Maksakova O, Simoẽs S, Pogrebnjak A et al (2018) The influence of deposition conditions and bilayer thickness on physical-mechanical properties of CA-PVD multilayer ZrN/CrN coatings. Mater Charact 140:189–196. https://doi.org/10.1016/j.matchar.2018.03.048
4. Kadyrzhanov EY, Zdorovets DB, Kozlovskiy M et al (2018) Influence of ionizing irradiation on the parameters of Zn nanotubes arrays for design of flexible electronics elements. Devices Methods Meas 9(1):66–73. https://doi.org/10.21122/2220-9506-2018-9-1-66-73
5. Ivashchenko VI, Veprek S, Turchi PEA et al (2012) First-principles study of TiN/SiC/TiN interfaces in superhard nanocomposites. Phys Rev B 86(1):14110. https://doi.org/10.1103/PhysRevB.86.014110
6. Pogrebnjak AD, Ponomarev AG, Shpak AP et al (2012) Application of micro- and nanoprobes to the analysis of small-sized 3D materials, nanosystems, and nanoobjects. Phys Usp 55(3):270. Available at: http://stacks.iop.org/1063-7869/55/i=3/a=A04
7. Boiko O, Koltunowicz TN, Zukowski P et al (2017) The effect of sputtering atmosphere parameters on dielectric properties of the ferromagnetic alloy—ferroelectric ceramics nanocomposite (FeCoZr)x(PbZrTiO3)(100−x). Ceram Int 43(2):2511–2516. https://doi.org/10.1016/j.ceramint.2016.11.052
8. Van Bui H, Wiggers FB, Gupta A et al (2015) Initial growth, refractive index, and crystallinity of thermal and plasma-enhanced atomic layer deposition AlN films. J Vac Sci Technol A Vac, Surfaces, Film 33(1):01A111. https://doi.org/10.1116/1.4898434
9. Banal RG, Imura M, Liu J et al (2016) Structural properties and transfer characteristics of sputter deposition AlN and atomic layer deposition Al2O3 bilayer gate materials for H-terminated diamond field effect transistors. J Appl Phys 120(11):115307. https://doi.org/10.1063/1.4962854
10. Aissa KA, Achour A, Camus J et al (2014) Comparison of the structural properties and residual stress of AlN films deposited by dc magnetron sputtering and high power impulse magnetron sputtering at different working pressures. Thin Solid Films 550:264–267. https://doi.org/10.1016/j.tsf.2013.11.073
11. Shin I, Kim J, Lee D et al (2018) Epitaxial growth of single-crystalline AlN layer on Si(111) by DC magnetron sputtering at room temperature. Jpn J Appl Phys 57(6):060306. https://doi.org/10.7567/JJAP.57.060303
12. Su X, Zhang J, Huang J et al (2017) Defect structure of high temperature hydride vapor phase epitaxy–grown epitaxial (0 0 0 1) AlN/sapphire using growth mode modification process. J Cryst Growth 467(1):82–87. https://doi.org/10.1016/j.jcrysgro.2017.03.031
13. Trant M, Fischer M, Thorwarth K et al (2018) Tunable ion flux density and its impact on AlN thin films deposited in a confocal DC magnetron sputtering system. Surf Coatings Technol 348:159–167. https://doi.org/10.1016/j.surfcoat.2018.04.091
14. Wang H, Sodabanlu H, Daigo Y et al (2016) Initial growth control of GaN on Si with physical-vapor-deposition-AlN seed layer for high-quality GaN templates. Appl Phys Express 9(5):055503. https://doi.org/10.7567/APEX.9.055503
15. Murillo AE, Melo-Máximo L, García-Farrera B et al (2018) Development of AlN thin films for breast cancer acoustic biosensors. J Mater Res Technol. https://doi.org/10.1016/j.jmrt.2018.02.007
16. Welz R, Howell K, Luis P (2018) Optimization of sputter deposition Process for piezoelectric AlN ultra-thin Films. Semester Project Advanced NEMS group, Autumn Semester 2017, Losanna, Jan 2018
17. Riah B, Ayad A, Camus J et al (2018) Textured hexagonal and cubic phases of AlN films deposited on Si (100) by DC magnetron sputtering and high power impulse magnetron sputtering. Thin Solid Films 655(100):34–40. https://doi.org/10.1016/j.tsf.2018.03.076

18. Pu K, Dai X, Miao D et al (2017) A kinetics model for MOCVD deposition of AlN film based on Grove theory. J Cryst Growth 478:42–46. https://doi.org/10.1016/j.jcrysgro.2017.08.012
19. Wistrela E, Schneider M, Bittner A et al (2016) Impact of the substrate dependent polarity distribution in c-axis oriented AlN thin films on the etching behaviour and the piezoelectric properties. Microsyst Technol 22(7):1691–1700. https://doi.org/10.1007/s00542-015-2799-6
20. Balaji M, Ramesh R, Arivazhagan P et al (2015) Influence of initial growth stages on AlN epilayers grown by metal organic chemical vapor deposition. J Cryst Growth 414:69–75. https://doi.org/10.1016/j.jcrysgro.2014.10.055
21. Aissa KA, Achour A, Elmazria O et al (2015) AlN films deposited by dc magnetron sputtering and high power impulse magnetron sputtering for SAW applications. J Phys D Appl Phys 48(14):145307. https://doi.org/10.1088/0022-3727/48/14/145307
22. Duta L, Stan GE, Stroescu H et al (2016) Multi-stage pulsed laser deposition of aluminum nitride at different temperatures. Appl Surf Sci 374:143–150. https://doi.org/10.1016/j.apsusc.2015.10.093
23. D'Acapito F, Torrengo S, Xenogiannopoulou E et al (2016) Evidence for Germanene growth on epitaxial hexagonal (h)-AlN on Ag(1 1 1). J Phys Condens Matter 28(4):45002. https://doi.org/10.1088/0953-8984/28/4/045002
24. Wang H, Liu Y, Li M et al (2010) Multifunctional TiO2nanowires-modified nanoparticles bilayer film for 3D dye-sensitized solar cells. Optoelectron Adv Mater Rapid Commun 4(8):1166–1169. https://doi.org/10.1039/b000000x
25. Kohout J, Qian J, Schmitt T et al (2017) Hard AlN films prepared by low duty cycle magnetron sputtering and by other deposition techniques. J Vac Sci Technol A Vacuum, Surfaces, Film 35(6):61505. https://doi.org/10.1116/1.4999460
26. Caban P, Rudzinski M, Wojcik M et al (2015) Growth of aluminium nitride with linear change of ammonia flow. J Cryst Growth 414:81–86. https://doi.org/10.1016/j.jcrysgro.2014.11.015
27. Iqbal A, Walker G, Iacopi A et al (2016) Controlled sputtering of AlN (002) and (101) crystal orientations on epitaxial 3C-SiC-on-Si (100) substrate. J Cryst Growth 440(2):76–80. https://doi.org/10.1016/j.jcrysgro.2016.01.037
28. Ji Z, Wang L, Zhao G et al (2017) Growth and characterization of AlN epilayers using pulsed metal organic chemical vapor deposition. Chin Phys B 26(7):078102. https://doi.org/10.1088/1674-1056/26/7/078102
29. Huang WC, Chu CM, Wong YY et al (2016) Investigations of GaN growth on the sapphire substrate by MOCVD method with different AlN buffer deposition temperatures. Mater Sci Semicond Process 45:1–8. https://doi.org/10.1016/j.mssp.2016.01.008
30. Motamedi P, Cadien K (2015) Structural and optical characterization of low-temperature ALD crystalline AlN. J Cryst Growth 421:45–52. https://doi.org/10.1016/j.jcrysgro.2015.04.009
31. Mazur MM, Pianaro SA, Portella KF et al (2015) Deposition and characterization of AlN thin films on ceramic electric insulators using pulsed DC magnetron sputtering. Surf Coatings Technol 284:247–251. https://doi.org/10.1016/j.surfcoat.2015.06.082
32. Li XH, Wei YO, Wang S et al (2015) Temperature dependence of the crystalline quality of AlN layer grown on sapphire substrates by metalorganic chemical vapor deposition. J Cryst Growth 414:76–80. https://doi.org/10.1016/j.jcrysgro.2014.10.007
33. Moreira MA, Törndahl T, Katardjiev I et al (2015) Deposition of highly textured AlN thin films by reactive high power impulse magnetron sputtering. J Vac Sci Technol A Vacuum, Surfaces, Film 33(2):21518. https://doi.org/10.1116/1.4907874
34. Schmerler S, Kortus J (2014) Ab initio study of AlN: anisotropic thermal expansion, phase diagram, and high-temperature rocksalt to wurtzite phase transition. Phys Rev B—Condens Matter Mater Phys 89(6):064109. https://doi.org/10.1103/PhysRevB.89.064109
35. Sun X, Li JS, Cheng HF (2015) Influence of temperatures on the formation of SiBCN powders prepared by CVD Using Borazine and Liquid Polycarbosilane. Mater Sci Forum 816:182–185. https://doi.org/10.4028/www.scientific.net/MSF.816.182
36. Prakash A, Sundaram KB (2016) Deposition and XPS studies of dual sputtered BCN thin films. Diam Relat Mater 64:80–88. https://doi.org/10.1016/j.diamond.2016.01.014

37. Xiao JL, Wang CB, Shen Q et al (2015) Influence of nitrogen pressure on bonding structure and mechanical properties of pulsed laser deposited BCN thin films. Surf Coatings Technol 276:141–144. https://doi.org/10.1016/j.surfcoat.2015.06.070
38. Wang CB, Xiao JL, Shen Q et al (2016) Bonding structure and mechanical properties of B-C-N thin films synthesized by pulsed laser deposition at different laser fluences. Thin Solid Films 603:323–327. https://doi.org/10.1016/j.tsf.2016.02.053
39. Tavsanoglu T, Jeandin M, Addemir O (2016) Synthesis and characterisation of thin films in the B–C–N triangle. Surf Eng 32(10):755–760. https://doi.org/10.1080/02670844.2016.1143197
40. Zhang T, Zhang J, Wen G et al (2018) Ultra-light h-BCN architectures derived from new organic monomer with tunable electromagnetic wave absorption. Carbon 136:345–358. https://doi.org/10.1016/j.carbon.2018.05.001
41. Xu S, Ma X, Tang G et al (2015) Air annealing effect on scratch behaviour of BCN films. Surf Eng 31(7):1–7. https://doi.org/10.1179/1743294414y.0000000447
42. Xing M, Li B, Yu Z et al (2016) Elastic anisotropic and thermodynamic properties of I-4m2-BCN. Acta Phys Pol A 129(6):1124–1130. https://doi.org/10.12693/aphyspola.129.1124
43. Chen D, Huang Y, Hu X et al (2018) Synthesis and characterization of "Ravine-Like" BCN compounds with high capacitance. Materials (Basel) 11(2):209. https://doi.org/10.3390/ma11020209
44. Hirte T, Feuerfeil R, Perez-Solorzano V et al (2015) Influence of composition on the wear properties of boron carbonitride (BCN) coatings deposited by high power impulse magnetron sputtering. Surf Coatings Technol 284:94–100. https://doi.org/10.1016/j.surfcoat.2015.07.077
45. Deng X, Kousaka H, Tokoroyama T et al (2014) Deposition and tribological behaviors of ternary BCN coatings at elevated temperatures. Surf Coatings Technol 259:2–6. https://doi.org/10.1016/j.surfcoat.2014.08.087
46. Mannan A, Baba Y, Kida T et al (2015) Synthesis of Hexagonal boron carbonitride without nitrogen void defects. MSA 6(05):353–359. https://doi.org/10.4236/msa.2015.65041
47. Li Y, Jia X, Shi W et al (2014) The preparation of new "BCN" diamond under higher pressure and higher temperature. Int J Refract Met Hard Mater 43:147–149. https://doi.org/10.1016/j.ijrmhm.2013.11.010
48. Houska J, Steidl P, Vlcek J et al (2016) Thermal, mechanical and electrical properties of hard B4C, BCN, ZrBC and ZrBCN ceramics. Ceram Int 42(3):4361–4369. https://doi.org/10.1016/j.ceramint.2015.11.115
49. Jin Y, Yasuhara S, Shimizu T et al (2015) Deposition of Boron Nitride Films by Filament-Assisted CVD Using Tris(Bimethylamino)Borane Precursor. Key Eng Mater 661:142–148. 10.4028/www.scientific.net/KEM.661.142
50. Prakash A, Sundaram KB, Campiglia AD (2016) Photoluminescence studies on BCN thin films synthesized by RF magnetron sputtering. Mater Lett 183:355–358. https://doi.org/10.1016/j.matlet.2016.07.140
51. Petrović M, Horn-von Hoegen M, Meyer zu Heringdorf FJ (2018) Lateral heterostructures of hexagonal boron nitride and graphene: BCN alloy formation and microstructuring mechanism. Appl Surf Sci 455(111):1086–1094. https://doi.org/10.1016/j.apsusc.2018.06.057
52. Pogrebnjak AD, Bagdasaryan AA, Pshyk A et al (2017) Adaptive multicomponent nanocomposite coatings in surface engineering. Uspekhi Fiz Nauk 187(6):586–607. https://doi.org/10.3367/ufnr.2016.12.038018
53. Pogrebnjak AD, Ivashchenko VI, Skrynskyy PL et al (2018) Experimental and theoretical studies of the physicochemical and mechanical properties of multi-layered TiN/SiC films: temperature effects on the nanocomposite structure. Compos Part B Eng 142:85–94. https://doi.org/10.1016/j.compositesb.2018.01.004
54. Musil J, Sklenka J, Cerstvy R (2012) Transparent Zr-Al-O oxide coatings with enhanced resistance to cracking. Surf Coatings Technol 206(8–9):2105–2109. https://doi.org/10.1016/j.surfcoat.2011.09.035
55. Veprek S, Veprek-Heijman MGJ, Karvankova P et al (2005) Different approaches to superhard coatings and nanocomposites. Thin Solid Films 476(1):1–29. https://doi.org/10.1016/j.tsf.2004.10.053

56. Ivashchenko VI, Veprek S, Argon AS et al (2015) First-principles quantum molecular calculations of structural and mechanical properties of TiN/SiNx heterostructures, and the achievable hardness of the nc-TiN/SiNx nanocomposites. Thin Solid Films 578:83–92. https://doi.org/10.1016/j.tsf.2015.02.013
57. Kasiuk J V, Fedotova JA, Koltunowicz TN et al (2014) Correlation between local Fe states and magnetoresistivity in granular films containing FeCoZr nanoparticles embedded into oxygen-free dielectric matrix. J Alloys Compd 586:S432–S435. https://doi.org/10.1016/j.jallcom.2012.09.058
58. Pogrebnjak AD, Rogoz VM, Bondar OV et al (2016) Structure and physicomechanical properties of NbN-based protective nanocomposite coatings: a review. Prot Met Phys Chem Surfaces 52(5):802–813. https://doi.org/10.1134/s2070205116050191
59. Pogrebnyak AD, Shpak A, Azarenkov NA et al (2009) Structures and properties of hard and superhard nanocomposite coatings. Phys Usp 52(1):29–54. https://doi.org/10.3367/ufne.0179.200901b.0035
60. Berladir KV, Budnik OA, Dyadyura KA et al (2016) Physicochemical principles of the technology of formation of polymer composite materials based on polytetrafluoroethylene—a review. High Temp Mater Process Int Q High-Technol Plasma Process 20(2):157–184. https://doi.org/10.1615/hightempmatproc.2016017875
61. Pogrebnjak AD, Bondar OV, Abadias G et al (2016) Structural and mechanical properties of NbN and Nb-Si-N films: experiment and molecular dynamics simulations. Ceram Int 42(10):11743–11756. https://doi.org/10.1016/j.ceramint.2016.04.095
62. Pogrebnjak AD (2013) Structure and properties of nanostructured (Ti-Hf-Zr-V-Nb)N coatings. J Nanomater 2013:1–12. https://doi.org/10.1155/2013/780125
63. Pogrebnjak A, Ivashchenko V, Bondar O et al (2017) Multilayered vacuum-arc nanocomposite TiN/ ZrN coatings before and after annealing: structure, properties, first-principles calculations. Mater Charact 134:55–63. https://doi.org/10.1016/j.matchar.2017.10.016
64. Mayrhofer PH, Hovsepian PE, Mitterer C et al (2004) Calorimetric evidence for frictional self-adaptation of TiAlN/VN superlattice coatings. Surf Coatings Technol 177:341–347. https://doi.org/10.1016/j.surfcoat.2003.09.024
65. Ou YX, Lin J, Tong S et al (2016) Structure, adhesion and corrosion behavior of CrN/TiN superlattice coatings deposited by the combined deep oscillation magnetron sputtering and pulsed dc magnetron sputtering. Surf Coatings Technol 293:21–27. https://doi.org/10.1016/j.surfcoat.2015.10.009
66. Pogrebnyak AD, Bondar OV, Zhollybekov B et al (2017) Influence of the bilayer thickness of nanostructured multilayer MoN/CrN coating on its microstructure, hardness, and elemental composition. Phys Solid State 59(9):1798–1802. https://doi.org/10.1134/s1063783417090232
67. Kumar DD, Kumar N, Kalaiselvam S et al (2017) Wear resistant super-hard multilayer transition metal-nitride coatings. Surfaces Interfaces 7:74–82. https://doi.org/10.1016/j.surfin.2017.03.001
68. Chang Y-Y, Chiu W-T, Hung J-P (2016) Mechanical properties and high temperature oxidation of CrAlSiN/TiVN hard coatings synthesized by cathodic arc evaporation. Surf Coatings Technol 303:18–24. https://doi.org/10.1016/j.surfcoat.2016.02.047
69. Yousaf MI, Pelenovich VO, Yang B et al (2015) Effect of bilayer period on structural and mechanical properties of nanocomposite TiAlN/MoN multilayer films synthesized by cathodic arc ion-plating. Surf Coatings Technol 282:94–102. https://doi.org/10.1016/j.surfcoat.2015.10.018
70. Yu R-S, Huang R-H, Lee C-M et al (2012) Synthesis and characterization of multi-element oxynitride semiconductor film prepared by reactive sputtering deposition. Appl Surf Sci 263:58–61. https://doi.org/10.1016/j.apsusc.2012.08.109
71. Bagdasaryan AA, Pshyk AV, Coy LE et al (2018) A new type of (TiZrNbTaHf) N/MoN nanocomposite coating: microstructure and properties depending on energy of incident ions. Compos Part B Eng 146:132–144. https://doi.org/10.1016/j.compositesb.2018.04.015
72. Pogrebnjak AD, Isakov IF, Opekunov MS et al (1987) Increased wear resistance and positron annihilation in Cu exposed to high power ion beam. Phys Lett A 123(8):410–412. https://doi.org/10.1016/0375-9601(87)90043-0

73. Pogrebnjak AD, Bazyl EA (2001) Modification of wear and fatigue characteristics of Ti-V-Al alloy by Cu and Ni ion implantation and high-current electron beam treatment. Vacuum 64(1):1–7. https://doi.org/10.1016/s0042-207x(01)00160-9
74. Pogrebnjak AD, Lebed AG, Ivanov YF (2001) Modification of single crystal stainless steel structure (Fe-Cr-Ni-Mn) by high-power ion beam. Vacuum 63(4):483–486. https://doi.org/10.1016/s0042-207x(01)00225-1
75. Demianenko A, Smyrnova K, Zhollybekov B et al (2015) Process of formation of spheroidal gold particles and of nanophases in AlN-TiB2-TiSi2 coatings after annealing with subsequent implantation. High Temp Mater Process 19(2):189–200. https://doi.org/10.1615/hightempmatproc.2016016017
76. Bagdasaryan AA, Pshyk AV, Coy LE et al (2018) Structural and mechanical characterization of (TiZrNbHfTa) N/WN multilayered nitride coatings. Mater Lett 229:364–367. https://doi.org/10.1016/j.matlet.2018.07.048

Mass Transfer Model of Sputtering from Rod-Like Targets for Synthesis of Multielement Nanocoatings

Yu. O. Kosminska and V. I. Perekrestov

Abstract This work develops the mathematical model that allows calculating element concentration depending on substrate location at low working gas pressures for coatings deposited by new magnetron sputtering device on the basis of hollow cathode and rod-like target. In this work, a target composed of two semicylindrical constituents is considered. As the rod-like target can be made of multiple materials in any geometry, the model can be adjusted for any particular case. The calculations explain the general trend of experimental data behavior.

Keywords Sputtering · Mathematical model · Mass transfer · Composite target · Rod target · Element concentration · Multielement coatings

1 Introduction

Modern material science and nanotechnology highly depend on the development of technological approaches to the synthesis of new functional coatings, thin films, and nanostructures. Therefore, researchers pay much attention to both creation of new synthesis methods and improvement of existing techniques, including plasma-based sputtering approach. Thus, magnetron sputtering has proved to be efficient for multiple tasks, and a number of its new variants have recently appeared. Along with classical variants of dc and rf magnetron sputtering, reactive and non-reactive sputtering, such new trends as unbalanced magnetron sputtering, high power impulse magnetron sputtering, etc., can be mentioned which are widely investigated [1–6].

Authors of the present work developed an alternative approach to dc magnetron sputtering and created a patented series of sputtering devices featuring hollow cathodes [7–14]. The sputterers work in highly pure inert ambient under working gas pressure of a few Pa. The operational principle of these devices is based on a combination of the magnetron and hollow cathode effects. As typical for these sputterers, they

Yu. O. Kosminska (✉) · V. I. Perekrestov
Department of Nanoelectronics, Sumy State University, Sumy, Ukraine
e-mail: y.kosminska@phe.sumdu.edu.ua

© Springer Nature Singapore Pte Ltd. 2019

A. D. Pogrebnjak and V. Novosad (eds.), *Advances in Thin Films, Nanostructured Materials, and Coatings*, Lecture Notes in Mechanical Engineering,
https://doi.org/10.1007/978-981-13-6133-3_6

have axial symmetry, and their magnet system is separated from a cathode assembly. The cathode can be made in the form of a half-closed cylinder or a truncated cone of various dimensions. Hence, the area of intersected electric and magnetic fields changes its form depending on specific device geometry. Correspondingly, the sputtering target can be placed in different locations and made in different forms. Among all possible configurations, the present article deals with the device that features half-closed cylindrical cathode and sputtered target in the form of a composite cylindrical rod placed at the central axis [10–12]. Plasma of the magnetron discharge is concentrated around the rod and inside the hollow cathode which provides sputtering of the upper and central part of the rod. Sputtered atoms move towards external substrate holders located concentrically to the hollow cathode. Depending on the geometry of the rod components one can expect the formation of coatings of various composition, for example, multicomponent high-entropy alloy coatings. Therefore, it is necessary to adjust all the technological parameters of deposition and geometrical characteristics of the sputterers to fabricate coatings with predictable properties.

The present work is aimed to develop a mathematical model to estimate the quantitative elemental composition of growing layers for the above sputtering device in an approximation of low working gas pressures.

2 Mass Transfer Model

In this section, the mathematical model is developed and tested on existing experimental results on the concentration distribution of two-component coatings depending on the location of substrates. Typical design of the sputterer with the rod-like target is shown in Fig. 1a. Main parts of the sputterer are cylindrical hollow cathode with the composite rod-like target along the central axis of the device. In this particular case, the rod consists of two semi-cylinders of Cu and Ni. The magnet system is separated from the cathode and located inside the anode assembly. Figure 1b shows the principle of substrate locations on the inner surface of the cylinder substrate holder. The mathematical model considers locations marked with capital letters. Locations filled with color correspond to those experimentally studied.

Uniform sputtering of each part of the rod is assumed. To simplify, rectilinear motion of sputtered atoms in the target-substrate region is considered. This can take place for low working gas pressures with high enough mean free path of atoms. This is due to the small diameters of the device (4–10 cm). The formulation of the model starts with an angular distribution of sputtered atoms which is chosen below.

It is well-known that in case of evaporation the angular distribution of evaporated atom flux can be satisfactorily described as a cosine distribution as for the ideal Knudsen cell [15]. In the case of sputtering the cosine distribution is predicted by the collision cascade theory for normally incident bombarding particles and at the same time, it is distorted in experiments and simulations. First of all, the real picture of the angular distribution depends on the incident energy of bombarding particles [16]. As the incident energy increases, the distribution changes from heart-shaped

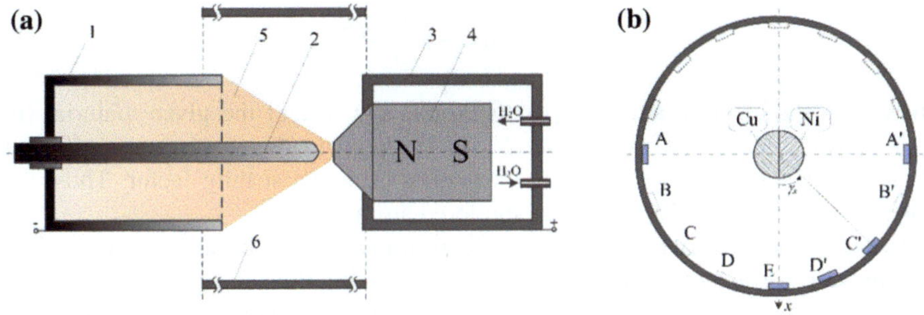

Fig. 1 Schematic representation of the sputtering device: **a** axial cross-section of the main components: 1—the cathode, 2—the rod-like target, 3—the anode, 4—the water-cooled magnet system with the magnetic conductor, 5—plasma area, 6—cylindrical substrate holder; **b** diametrical cross-section that shows possible location of substrates relatively to the target (here the rod-like target is composed of two identical semi-cylinders each of which is made of Cu or Ni, γ_x is the polar angle that determines location of each substrate relatively to the fixed x-axis, each location is marked with a letter, filled locations correspond to experimentally used, a stroke at a letter means that the location is repeated symmetrically from geometrical point of view but is asymmetrical relatively to the sputtered target components)

and under-cosine to over-cosine type. Also, the angular distribution depends on the surface structure. For different surfaces and incident energies, a number of fitting expressions is proposed [16]. In this work, we assume cosine distribution since it is quite sufficient for the case of a finely dispersed polycrystalline target and a surface without preferable orientation [17].

Thus, the number of atoms d^2N_{sp} sputtered to form a target surface element dS_t per a time unit within the solid angle $d\omega$ (see Fig. 2) can be written as

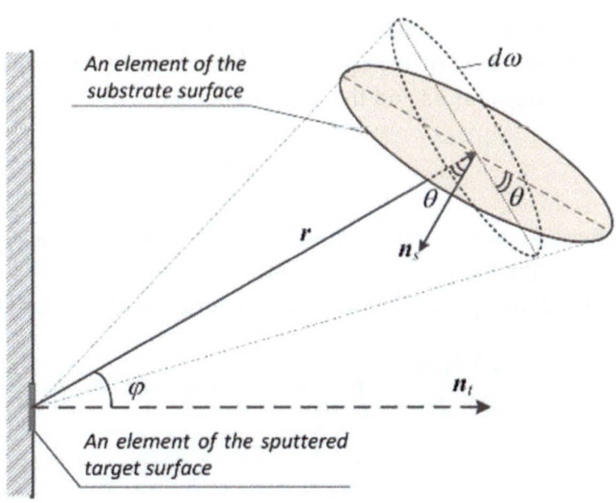

Fig. 2 To determination of relation between the target and the substrate surface elements through the solid angle $d\omega$ in (1). Here $\boldsymbol{n}_t$ is the normal to the target surface element dS_t and $\boldsymbol{n}_s$ is the normal to the element of the substrate surface dS_s. $\boldsymbol{r}$ denotes the direction of the flux that is sputtered from dS_t and incident to dS_s. The angle φ is the ejection angle and the angle θ is the incident angle of sputtered flux

$$d^2N_{sp} = N_0 \cos\varphi \frac{d\omega}{\pi} dS_t, \tag{1}$$

where φ is the ejection angle, N_0 is the sputtering rate of the given material (in $m^{-2}s^{-1}$). In calculations we assume N_0 to be equal to the sputtering coefficient of the target material with accuracy to a certain relative scaling factor. Thus, (1) represents the angular distribution of sputtered atoms.

To determine the arrival rate of atoms, let us find a relation between the target and the substrate surface elements dS_t and dS_s considering their position and the solid angle $d\omega$ as shown in Fig. 2. Their mutual position is determined by the distance r between the source dS_t and the receiver dS_s, as well as by the incident angle θ.

From Fig. 2 it follows that

$$d\omega = dS_s \frac{\cos\theta}{r^2}. \tag{2}$$

Substituting (2) into (1), we get

$$d^2N_{sp} = \frac{N_0}{\pi r^2} \cos\varphi \, \cos\theta dS_s dS_t. \tag{3}$$

Let us consider an arbitrary point S on a cylindrical substrate holder (Fig. 3) and define its angular position by the angle γ_x counted from the immovable x-axis. Let the x-axis be attached to the line (surface) separating two components of the sputtering target (see Figs. 1 and 3). Our final task is to calculate the concentration of atoms of the two target materials at S-points depending on their position γ_x. So, the sputtered substance can arrive at the point S from any ejecting point T within the viewing field. The mutual position of the points S and T can be described by the angle γ_T counted from movable x'-axis (see Fig. 3).

Let us take into account the axial symmetry of the system. Then we can use a cylindrical coordinate system with the polar angle γ_x and z-axis placed along the symmetry axis of the sputterer. The element dS_t can be found as a product of the height element dz and the arc length that rests upon γ_T:

$$dS_t = dz \cdot R_t d\gamma_T, \tag{4}$$

where R_t is the target radius (Fig. 3). Substituting (4) into (3), we get a general expression for the number of atoms arriving at a substrate surface unit (the point S) from all the points T within the viewing field and length dz per a time unit:

$$\frac{dN(\gamma_x)}{dS_s dz} = \int\limits_{S_t} \frac{N_0}{\pi r^2} \cos\varphi \, \cos\theta dS_t = \int\limits_{\gamma_T} \frac{N_0 R_t}{\pi r^2} \cos\varphi \, \cos\theta d\gamma_T. \tag{5}$$

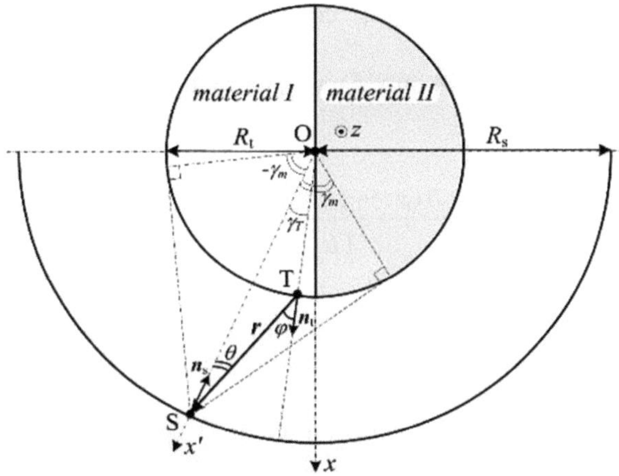

Fig. 3 Simplified cross-section plane of the system shown in Fig. 1b and main geometrical parameters used for determination of the number of atoms arriving at a substrate point S. The arrival point S corresponds to any location of the substrates in Fig. 1b and is determined by the polar angle γ_x. $\boldsymbol{r}$ is the direction of mass transfer from the floating ejection point T to the point S. x' is the movable coordinate axis that is connected with the position of the point S

One can see that integration over the target surface is replaced by integration over γ_T, i.e., the viewing field. To take the integral, r, cosφ and cosθ should be expressed through γ_T.

First, r can be easily found from the triangle STO (Fig. 3) by cosine theorem as

$$r^2 = R_s^2 + R_t^2 - 2R_s R_t \cos \gamma_T, \tag{6}$$

where R_s is the substrate holder radius (Fig. 3). To make further calculations dependent only on relative sizes, let us introduce the ratio of the radii R_s and R_t as $\rho = R_t/R_s \in (0, 1)$.

Then the distance r can be represented as

$$r = R_s\sqrt{1 + \rho^2 - 2\rho \cos \gamma_T}. \tag{7}$$

From the same triangle STO it follows by sine theorem that

$$\frac{\sin(\pi - \varphi)}{R_s} = \frac{\sin \gamma_T}{r} \quad \Rightarrow \quad \sin \varphi = \frac{\sin \gamma_T}{\sqrt{1 + \rho^2 - 2\rho \cos \gamma_T}}, \tag{8}$$

$$\cos \varphi = \sqrt{1 - \frac{\sin^2 \gamma_T}{1 + \rho^2 - 2\rho \cos \gamma_T}}. \tag{9}$$

Sum of the angles in STO is $\pi = \theta + \gamma_T + \pi - \varphi$. Hence

$$\cos\theta = \cos(\varphi - \gamma_T) = \cos\varphi\,\cos\gamma_T + \sin\varphi\,\sin\gamma_T. \tag{10}$$

Finally, substituting (7)–(10) into (5), we get the final dependence of the number of arriving atoms in this form:

$$\frac{dN(\gamma_x)}{dS_s dz'} = \int_{-\gamma_m}^{\gamma_m} \frac{N_0\rho}{\pi} \frac{B(\rho,\gamma_T)\cos\gamma_T + \sqrt{B(\rho,\gamma_T)}\sin^2\gamma_T}{\left(B(\rho,\gamma_T) + \sin^2\gamma_T\right)^2} d\gamma_T, \tag{11}$$

where $B(\rho, \gamma_\mathrm{T}) = 1 + \rho^2 - 2\rho\cos\gamma_\mathrm{T} - \sin^2\gamma_\mathrm{T}$; z' is z normalized to R_s; $\pm\gamma_\mathrm{m}$ are the integration limits defined by the viewing field (Fig. 4). N_0 characterizes the sputtered material.

Possible locations of arrival point S on the arc K′M′ are not equivalent even if symmetrical relatively to x-axis because of the structure of the rod-like target. KK′ and MM′ are shaded from material at the opposite side. Thus, for KK′ and MM′ the integral in (11) should be taken over $\gamma_\mathrm{m} \in [-\arccos\rho; \arccos\rho]$ for materials 1 and 2 correspondingly. For the arc KM, the integral in (11) turns into a sum of two integrals over both materials (see Fig. 4).

Figures 5 and 6 show calculated relative concentration of Cu and Ni depending on substrate location for different ratios ρ. Additionally, Fig. 5 features comparison with experimental measurements from [18]. One can see that the distribution of deposited substance is not symmetrical, and the amount of Cu exceeds that of Ni. The central E-position is characterized with a constant fraction of elements independent from

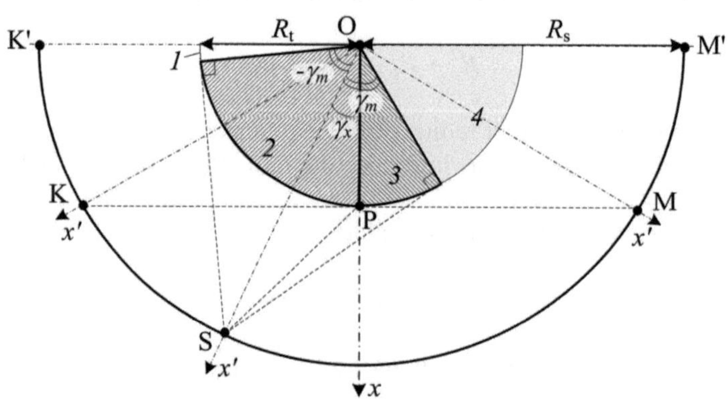

Fig. 4 Simplified cross-section plane of the system shown in Fig. 3 explaining determination of the limits of integration γ_m in (5), (11). The regions 1, 2 belong to the material I and the regions 3, 4 belong to the material II (see Fig. 3). For the given location of the arrival point S integration in (5), (11) should be represented as a sum of separate integrals over circular arcs 2 and 3. "Blind" arcs 1 and 4 are invisible to the point S because of cosine-type angular distribution of sputtered atoms. Correspondingly, the arcs KK′ and MM′ appear to be shaded from a material on the opposite side. So, if the arrival point S shifts to the arcs KK′ and MM′, integration in (5), (11) should be done over the closest material only (material I for KK′ and II for MM′)

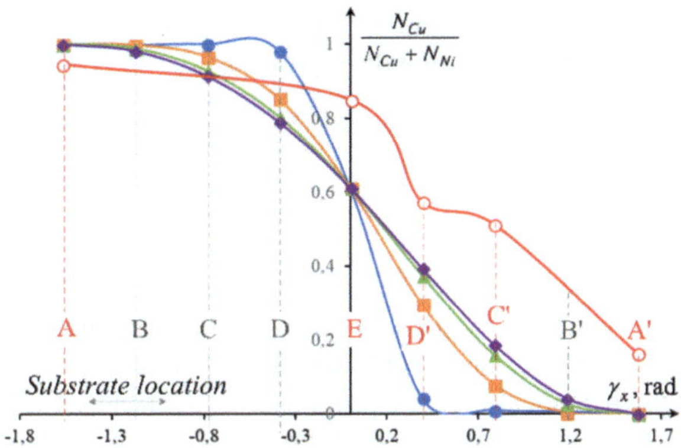

Fig. 5 Relative concentration of Cu in deposited layers versus substrate location for different values of ρ: black circle $\rho = 0.5$; black square $\rho = 0.25$; black up-pointing triangle $\rho = 0.1$; black diamond suit $\rho = 0.05$; white circle corresponds to experimental values obtained in [18]

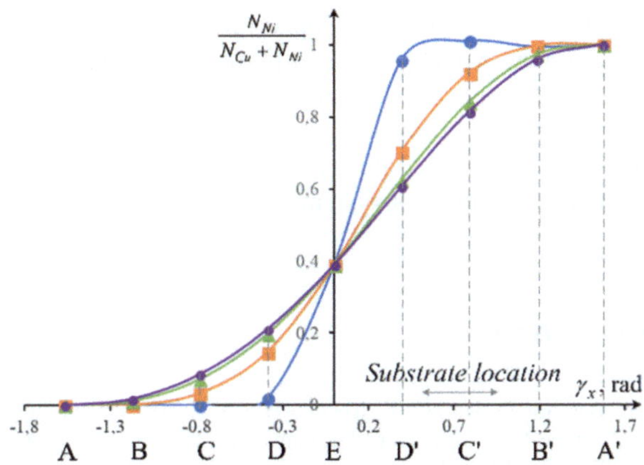

Fig. 6 Relative concentration of Ni in deposited layers versus substrate location for different values of ρ: black circle $\rho = 0.5$; black square $\rho = 0.25$; black up-pointing triangle $\rho = 0.1$; black diamond suit $\rho = 0.05$

ratio ρ: ~61% for Cu and 39% for Ni. This depends on the correlation between the sputtering rates of Cu and Ni. Indeed, sputtering coefficients for Cu and Ni are 1.1, and 0.7 and their sputtering threshold energies are 17 and 21 eV for Ar ions, so, Cu atom flux is more intense. For lower ratio ρ, that is for the higher distance between the target surface and the substrate holder, distribution of components changes more gradually.

From a comparison of calculated and experimental data in Fig. 5 it follows that the model predicts general behavior well but gives lower values of Cu concentration than measured. This can be attributed to increased values of working gas pressure. Thus, the experiment on the deposition of Cu and Ni was performed at 6.4 Pa [18]. This corresponds to the mean free path of sputtered particles in working ambient of ~3–5 cm, while the distance between target surface and substrate holder can be 1.5–5 cm. This indicates the necessity to include a scattering of sputtered atoms on ambient particles into account.

Experimentally, maximal mixing of the components near region D′ corresponded to the formation of amorphous phase relatively stable to recrystallization in absence of component segregation. The developed model also indicates that maximal mixing can be expected within the region ED′.

3 Conclusion

The proposed sputtering device can be used to deposit functional multicomponent nanocoatings onto the inner surface of low diameter pipes. To predict the properties of the coatings and to optimize the technological process, it is important to develop a mathematical model to calculate the distribution of the components on substrates. In this work, such model is proposed on the basis of cosine angular distribution of sputtered atoms and different sputtering rates of the components. It is assumed that the model will be developed further to take into account scattering of sputtered atoms on the working ambient particles depending on the pressure, the influence of sputtering relief, possible deposition on the target surface, rotation and translation of the substrate holder, etc. In its present form, the model allows predicting general behavior of sputtered fluxes and determines substrate location for maximal mixing of the two-component system.

References

1. Baranov O, Bazaka K, Kersten H et al (2017) Plasma under control: advanced solutions and perspectives for plasma flux management in material treatment and nanosynthesis. Appl Phys Rev 4:041302 (1–32)
2. Levchenko I, Keidar M, Xu S et al (2013) Low-temperature plasmas in carbon nanostructure synthesis. J Vac Sci Technol B 31:050801
3. Anders A (2010) High power impulse magnetron sputtering and related discharges: scalable plasma sources for plasma-based ion implantation and deposition. Surf Coat Technol 204:2864–2868
4. Han JG (2009) Recent progress in thin film processing by magnetron sputtering with plasma diagnostics. J Phys D Appl Phys 42:043001
5. Kelly PJ, Arnell RD (2000) Magnetron sputtering: a review of recent developments and applications. Vacuum 56:159–172
6. Musil J, Baroch P, Vlcek J et al (2005) Reactive magnetron sputtering of thin films: present status and trends. Thin Solid Films 475:208–218
7. Perekrestov VI, Mokrenko OA, Kosminska YuO (2010) Sputtering device for deposition highly porous coatings of metals or weakly volatile substances onto flat substrates in vacuum. UA patent 92525
8. Perekrestov VI, Kosminska YuO, Kornyushchenko AS (2007) Device for deposition of vacuum condensates. UA patent 80775
9. Perekrestov VI, Pogrebnyak OD, Kosminska YuO (2003) Device for deposition of condensates in vacuum. UA patent 57940A
10. Perekrestov VI, Kosminska YuO, Mokrenko OA, Dyoshin BV (2008) Device for deposition of condensates in a vacuum. UA patent 37359
11. Perekrestov VI, Pogrebnyak OD, Kosminska YuO (2003) Device for deposition of condensates in a vacuum. UA patent 57952

12. Perekrestov VI, Kosminska YuO (2004) Device for deposition of condensates in a vacuum. UA patent 69723
13. Perekrestov VI, Pogrebnyak OD, Kosminska YuO (2004) Sputtering device for deposition of condensates in a vacuum. UA patent 69974
14. Perekrestov VI, Kosminska YuO (2006) Sputtering device for deposition of condensates in a vacuum. UA patent 76257
15. Glang R (1970) Vacuum evaporation. In: Maissel L, Glang R (eds) Handbook of thin film technology, vol 1. McGraw-Hill, New York, pp 1–130
16. Depla D, Mahieu S (eds) (2008) Reactive sputter deposition. Springer-Verlag, Berlin Heidelberg
17. Wehner GK, Anderson GS (1970) Physics of sputtering by ion bombardment. In: Maissel L, Glang R (eds) Handbook of thin film technology, vol 1. McGraw-Hill, New York, pp 352–404
18. Perekrestov VI, Kravchenko SN, Kosminska YuO, Kononenko IN (2011) Structure of Ni–Cu condensates obtained at sputtering of composite rods. Metallofiz Nov Tekh+ 33:203–210

Self-organized Growth by Sputtering and Other PVD Techniques

Yu. O. Kosminska and V. I. Perekrestov

Abstract This paper presents a brief review on known self-organized growth by deposition and reports on a new approach of self-organized growth of micro- and nanostructures that can be realized by modified magnetron sputtering. Since self-organization phenomena are multidisciplinary, general ideas on self-organization and related concepts are considered first. Then, known methods of porous structure fabrication in contrast to solid layers and established self-organized growth types are discussed with an emphasis on using PVD techniques. Finally, new steps in plasma-assisted self-organized growth under quasi-equilibrium conditions are presented.

Keywords Self-organization · Sputtering · Quasi-equilibrium conditions · Growth mechanism · PVD

1 Introduction

Modern progress in nanotechnology is based on the problem of formation of various micro- and nanostructures with novel structural and functional characteristics. At present a concept of the following two synthesis approaches is developed, such as "top-down" (e.g. etching) and "bottom-up" (e.g. molecular beam epitaxy) which differ from each other considerably by structure formation mechanisms and technological solutions. Among them the former is still dominant in practical applications, often being combined with the latter in form of lithography. However, the most progressive new results, such as the growth of fullerenes, nanotubes, layered heterosystems, ordered arrays of quantum dots, etc., are achieved using the "bottom-up" techniques. Therefore, it is an atom-by-atom bottom-up assembly that serves as an attractive way to discover new knowledge in this field. It can be appropriate here to address the famous Feynman's lecture [1] and emphasize his ideas about manipulation by atoms, they're arranging the way we want and atom-by-atom assembling by physics

Yu. O. Kosminska (✉) · V. I. Perekrestov
Department of Nanoelectronics, Sumy State University, Sumy, Ukraine
e-mail: y.kosminska@phe.sumdu.edu.ua

© Springer Nature Singapore Pte Ltd. 2019
A. D. Pogrebnjak and V. Novosad (eds.), *Advances in Thin Films, Nanostructured Materials, and Coatings*, Lecture Notes in Mechanical Engineering,
https://doi.org/10.1007/978-981-13-6133-3_7

instead of traditional chemical routes. Here two different ways are possible, such as the deterministic placing of atoms like by atomic force microscopy and growth of structures based on self-organization. Importance of self-organization follows from the fact that self-organized processes in nature provide with the optimal functional effectiveness of both artificial systems and living organisms.

Today one can observe increasing attention paid by researches to study and search of self-organized physical, chemical, biological and technological processes that ensure the formation of micro- and nanosystems. But for the present very limited number of possible variants of self-organized structure growth is known for deposition of metals, semiconductors, and carbon by PVD techniques. Besides, self-organization processes at condensation of weakly volatile substances under quasi-equilibrium conditions have not almost been investigated. At the same time, the interaction of self-organized processes of dissipative and conservative character that occur away from thermodynamical equilibrium and close to it correspondingly can lead to the appearance of complex hierarchical structures with novel properties [2]. When implementing these ideas by PVD methods, one can expect both new formation mechanisms and new structural forms and characteristics of micro- and nanosystems.

Thus, this paper briefly discusses types and mechanisms of self-organized growth commonly known for the present, as well as presents new results on self-organization obtained by the authors' group for quasi-equilibrium condensation by modified magnetron sputtering.

2 General Notions of "Self"-Processes and Selectivity

2.1 Self-organization and Self-assembly

Processes of ordering or pattern formation are often observed during synthesis of solid state or molecular structures which can be described regarding self-organization and self-assembly. Both terms assume the appearance of collective ordering based on the dynamical interactions of complex system components. However, there is still a certain indistinct understanding and loose utilization of these terms due to their multidisciplinary character. Self-assembly is more often understood as molecular self-assembly and self-organization—as ordered formation of dissipative structures away from equilibrium [3]. At the same time, there are lots of examples from scientific literature that considerably extend areas of utilization of "self"-terms.

At present three general concepts of self-organization are proposed, namely dissipative, conservative and continual. conservative self-organization and self-assembly possess a lot of common features. Continual self-organization relates to evolution of individual biosystems and is not considered here.

Being a founder of synergetics, H. Haken defined self-organization as gaining by a system a spatial, temporal, combined spatial and temporal or functional structure through internal processes without specific interference from outside [4]. Specific interference can be understood as direct controlling, e.g., using templates. Basic conditions for self-organization are remoteness from equilibrium, the complexity of the system, nonlinearly interacting constituent parts and positive and negative feedback. Haken's idea has been developed further by I. Prigogine and followers as dissipative self-organization in open systems which is a nonequilibrium phase transition of "order-disorder" type. It should be noted that the meaning of the "template" becomes more blurred and includes the indirect influence of ambient conditions [3].

Conservative self-organization appears in the form of spontaneous ordering in closed systems that are near or at thermodynamical equilibrium [5]. This term is introduced by J.-M. Lehn for spontaneous formation of supramolecular assemblies from separate components. This definition is also suited for such physical processes as crystallization, recrystallization during sintering, martensite transformations, the formation of ferromagnetic and ferroelectric structures [5]. Self-assembly was firstly formulated as a more general concept that covers wider spectrum of interactions on molecular and higher levels. However, both conservative self-organization and self-assembly occur near equilibrium, are driven by minimization of Gibbs free energy, do not change their constituent parts and result in structures with new qualities. Therefore, they can be interchangeable to a certain extent.

In addition, there are processes for which one can hardly differentiate conservative and dissipative self-organization, or which are characterized by the interaction of both "self"-processes. Thus, the formation of magnetic domains and crystallization have earlier been considered as classical self-organization. However, the resulted structures remain ordered in the vicinity to equilibrium as well [6], while classical dissipative structures break down. Self-assembly can be accompanied by nonequilibrium conditions, that is constant heat fluxes and chemical reactions [2]. And finally, one can come across the ideas that both equilibrium and non-equilibrium processes can interact that increases hierarchy and complexity of a system and, as a result, provide formation of very complex hierarchical structures. Typically, this is demonstrated by self-assembly of metal nanoparticles in a polystyrene matrix or other examples from molecular biology and polymer systems [2]. Anyway, scientists agree on extreme perspectivity of self-organized "bottom-up" technologies for material science and nanotechnologies [5, 7].

2.2 General Types of Selectivity

Since supramolecular chemistry can be considered as a source of ideas concerning self-assembly or conservative self-organization, it is necessary to consider selectivity as one more central notion. Selectivity means the ability of the molecules to recognize one another. There are thermodynamic and kinetic selectivities [8]. The first variant is realized through molecule binding by the complementarity principle and the strongest possible chemical bonds. For example, it is thermodynamical selectivity that is used to bind complementary DNA parts in DNA biosensors. The second variant is realized without strong preorganization and results in a structure with the highest formation rate. Making an analogy with PVD, one can expect both types of selectivity. For instance, the effect of different growth rates of crystallographic planes forming crystal habitus or role of active nucleation centers on a growth surface is well-known. At the same time, growth selectivity in thin films based on the maximally strong chemical bonds of adatoms with the growth surface remains practically unstudied.

3 Types of Self-organized Growth by Deposition

3.1 *Stranski–Krastanov Growth*

A typical variant of self-organized growth is observed during deposition in Stranski–Krastanov mode in semiconductor heteroepitaxial and "metal–semiconductor" systems with lattice mismatch, such as Ge/Si, Si_xGe_{1-x}/Si, InAs/GaAs, (In, Ga)N/GaN, etc. Molecular beam epitaxy and metal-organic vapor phase epitaxy are commonly used in this case. Elastic stress relaxation in a stressed wetting layer ends in three-dimensional islands formation if mismatch parameter is higher 2%, instead of introducing misfit dislocations. The resulting shape of the islands is determined by a balance between elastic stress relaxation and free surface energy minimization.

3.2 *Surface Directed Self-organized Growth*

Another variant of self-organized growth of nanostructures presupposes using substrate surface as a template if the surface is structurally anisotropic and has atomic inhomogeneities of different origin [9]. For example, such inhomogeneities as monosteps and kinks are active sites for adsorption, nucleation, and growth of nanodots or nanowires. If these active cites create regular energetic relief and nucleation is suppressed on another surface cites, then one can expect self-organized growth of nanostructures [10].

Among directed growth there are:

1. growth on vicinal high index surfaces [e.g. Ge on Si(113), Si(331), Si(1 1 10); Ag and Cu on Au(7 8 8), Au(23 23 21)];
2. growth on reconstructed surfaces [e.g., metals on Si(111)-7 × 7, Au(111)-22 × $\sqrt{3}$];
3. growth at interaction with ordered adsorbate phases [e.g., Au/N/Cu(001)];
4. growth on mechanically stressed surfaces due to embedded nanostructures or subsurface hidden dislocation network [e.g., Co on Ag(001)/MgO(001), Ni on CoO(001)/Ag(001)];
5. growth on externally deformed substrates [e.g., the Volmer–Weber growth of PbTe and Sn on (111) BaF_2, (0001) GaSe].

Thus, mainly surface atomic structure, and mechanical stress fields are considered as directing forces for self-organized growth. Atoms should fix only on selected (active) nucleation cites and, thus, growth should be thermodynamically selective.

Finally, in [10] all processes of atom-by-atom structure evolution during deposition are claimed to be of self-organized character. But it is not connected to self-organized technological solutions, and the importance of direct control over atomic processes by technological parameters has been noted.

3.3 Low-Temperature Plasma-Assisted Growth

"Plasma-condensate" systems with plasma action onto growth surface have already proved their importance for nanotechnology as they allow fabricating objects considered as self-organized, among which there are semiconducting quantum dots, nanoparticle and nanowire arrays, carbon nanotubes with controlled chirality, graphene and its derivatives, nanodiamond, hybrid materials, hierarchical metamaterials, etc. This means that structure formation under plasma action is accompanied by the selective formation of condensate architecture and chemical bonds. According to [11], such variant of self-organization is determined by rates of dynamical restructuration of material fluxes in bulk and on a surface and comes when this restructuration occurs without specific external actions. As a result, self-organization is governed by surface diffusion of adatoms as a kinetic factor. Under the plasma action, the diffusion activation barrier decreases owing to heating, charge transfer and polarization in the presence of an electric field. So, adhesion energy of adatoms to the growth surface decreases. Besides, the electrical field over the substrate is also known to influence structure formation.

Plasma action onto the growth surface positively affects the ordering of metal nanoparticle arrays with a tendency to similar form and shape. In case of deposition through templates, plasma optimizes substance passing through pores, while in case of random nucleation plasma induces an electric charge of growing islands which redistributes diffusive fluxes over substrates and favors semi-spherical shaping and narrowing of size distribution [12]. In the case of post-deposition treatment by plasma self-organization can result from Ostwald ripening [13].

4 Known Approaches to Selective Fabrication of Porous Metal Layers

4.1 Porosity According to Existing Structure Zone Models

The know majority of experimentally obtained structures of PVD metallic deposits according to structure zone models (SZM) comes to thin solid films at prolonged condensation [14]. There is only one region in the modern SZMs that features porosity and relates to condensation under low substrate temperatures (below one-third of the homologous temperature) and near-thermal energy of deposited atoms. Such porous deposits consist of tapered crystallites separated by voids, are amorphous, stressed and structurally nonequilibrium. Such porosity character results from low mean free diffusion path that is very limited adatom mobility over growth surface. However, one can hardly find reports on porosity formation due to diffusion limitation by the time of adatoms being absorbed along with the realization of maximally strong chemical bonds between atoms and the growth surface. The latter approach can be realized under condensation in quasi-equilibrium conditions.

4.2 Porous Metal Layers by Traditional Techniques

Among known technologies of porous metals fabrications, dealloying and template methods are the most spread and studied. The former methods are based on the chemical or electrochemical etching of a single or several components from the alloy. The resulted porous morphology is controlled by changing the initial ratio of metals in the alloy and the etching temperature. As for using templates, first, one has to create a template with necessary pore structure, then to fill the empty spaces with metal and after that to remove the template material. Thus, the most wide-spread techniques for porous structure formation are complex and multistage processes. In this connection, it is quite attractive to replace the above techniques with the self-organized atom-by-atom assembly of porous systems within the "bottom-up" approach during condensation under quasi-equilibrium conditions.

5 Plasma-Assisted Self-organized Growth Under Quasi-Equilibrium Conditions

In this section, we present a new approach to the self-organized growth of micro- and nanostructures under plasma action onto the growth surface directly during sputter deposition. It is based on a modified magnetron sputtering combined with hollow cathode which operates under increased working gas pressure in high pure inert ambient. Self-organization, in this case, is complex [15, 16] and consists in dissipative self-organization of deposition conditions, namely low relative supersaturation of deposited substance, and conservative self-organization of structures on substrates. Due to the interdependent nature of these two constituents, it represents a system of full self-organization.

Self-organization of technological conditions results from the existence of accumulation area of a sputtered substance above the growth surface, provided by circular mass transfer and hollow cathode effect. The relative supersaturation, growth surface temperature and condensing flux are nonlinearly mutually dependent and together maintain quasi-equilibrium stead-state conditions. A macroscopic criterion of the process stationarity is time-independent stable values of the supersaturation. Furthermore, it is proposed to take into account plasma action onto the growth surface by not only decrease of diffusion barrier, but adatom desorption energy to an effective value as well.

To describe selectivity of structure growth, a concept of critical energy in the spectrum of binding energies of adatoms with growth surface has been introduced [15]. The critical energy separates higher binding energies that correspond to allowed cites of adatom fixing on the surface from lower binding energies of "forbidden" cites. The critical energy value can be shifted depending on technological conditions and supersaturation: the lower is the supersaturation, the higher is the critical energy. Thus, structural selectivity is gained when adatoms condense with the strongest chemical bonds with the growth surface. In section IIB this was mentioned as thermodynamical selectivity. Besides, the time-stable value of the critical energy serves as a microscopic criterion of stationarity.

Under the described conditions we have found formation of nanoparticle arrays during initial growth stages with a tendency to form similar shape, size and averaging distance in between due to Ostwald ripening in Volmer-Weber mode [17]. After prolonged condensation we have found the formation of highly porous metal layers for fcc, bcc and hcp lattices within the range of homologous temperatures ~0.5–0.9 (Ni, Al, Cu, Zn, Cr, Ti) [18–23]. To place the obtained structures among others in

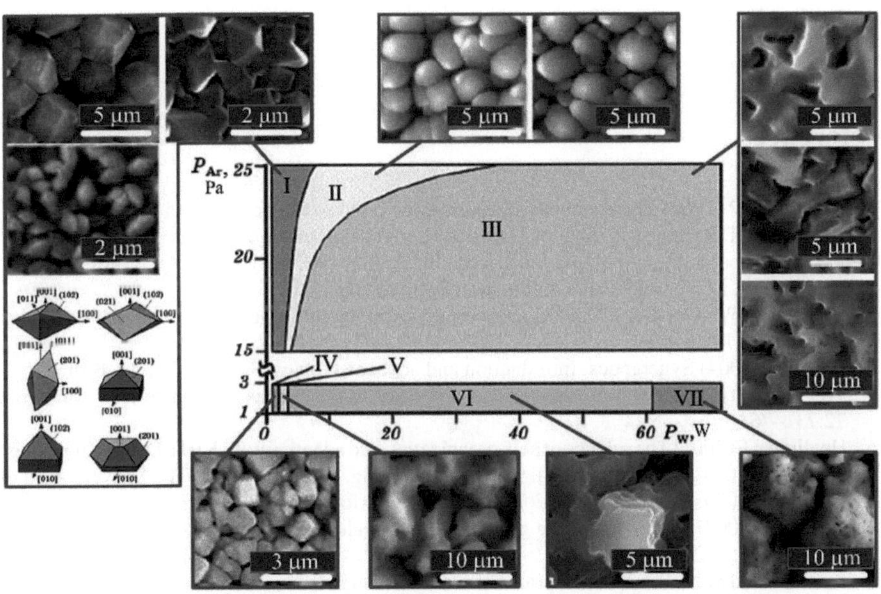

Fig. 1 Seven zones in gas pressure—discharge power coordinates which correspond to typical architectures of Al condensates obtained under quasi-equilibrium conditions under plasma action

modern SZMs, there should be used one more axis of supersaturation, that is why for now we have offered a simplified classification of our structures as shown in Fig. 1 on the example of aluminum. In the case of the strong electric field above the surface, we have observed structures resulting from focusing of weak ion fluxes onto prominent parts of the growth surface and called it as field selectivity [24]. Depending on gas pressure and applied power, structure formation occurs under the different ratio of structural and field selectivities. Among mechanisms that influence self-organized growth, in this case, the Gibbs–Thomson effect at the presence of negative surface curvature should be mentioned [24]. Thus, the structure is a competition of multiple mechanisms.

We state that proposed mechanisms of full self-organization can be more general and can be applied to describe the formation of other micro- and nanosystems, such as, for instance, multiwall carbon nanotubes by anodic arc synthesis [25], or to create new self-organized technological systems based on different physical and chemical processes.

6 Conclusion

We have briefly reviewed general notions of self-organization and underlying mechanisms of self-organized structure formation of the deposits by PVD techniques. We have shown that low-temperature plasma is a powerful active ambient to reveal self-organization of both technological deposition conditions and structure formation mechanisms.

References

1. Feynman RP (1960) There's plenty of room at the bottom. Caltech Eng Sci 23:22–36
2. Yamaguchi T, Suematsu N, Mahara H (2015) Hierarchical self-organization and self-assembly: metal nanoparticles in polymer matrices. In: Müller SC, Parisi J (eds) Bottom-up self-organization in supramolecular soft matter. Springer, pp 1–10
3. Halley JD, Winkler DA (2008) Consistent concepts of self-organization and self-assembly. Complexity 14:10–17
4. Haken H (2004) Synergetics. Introduction and advanced topics. Springer-Verlag, Berlin
5. Tret'yakov YuD (2003) Self-organisation processes in the chemistry of materials. Usp Khim 72:731–763
6. Heylighen F (2001) The science of self-organization and adaptivity. In: Kiel LD (ed) Knowledge management, organizational intelligence, and learning, and complexity. The Encyclopedia of Life Support Systems (EOLSS), Eolss Publishers, Oxford, pp 1–26
7. Wiame F (2007) Strategies for the growth of large-scale self-organized structures. Thin Solid Films 642:258–275
8. Steed JW, Atwood JL (2009) Supramolecular chemistry, vol 1. Wiley
9. Barth JV, Costantini G, Kern K (2005) Engineering atomic and molecular nanostructures at surfaces. Nature 437:671–679

10. Barna PB, Radnóczi G (2014) Structure formation during deposition of polycrystalline metallic thin films. In: Barmak K, Coffey K (eds) Metallic films for electronic, optical and magnetic applications. Woodhead Publishing Limited, Cambridge, pp 67–120
11. Ostrikov K, Neyts EC, Meyyappan M (2013) Plasma nanoscience: from nano-solids in plasmas to nano-plasmas in solids. Adv Phys 62:113–224
12. Levchenko I, Ostrikov K, Diwan K et al (2008) Plasma-driven self-organization of Ni nanodot arrays on Si(100). Appl Phys Lett 93:91–94
13. Tang J, Photopoulos P, Tserepi A, Tsoukalas D (2011) Two-dimensional nanoparticle self-assembly using plasma-induced Ostwald ripening. Nanotechnology 22:235306 (1–7)
14. Anders A (2010) A structure zone diagram including plasma-based deposition and ion etching. Thin Solid Films 518:4087–4090
15. Perekrestov VI, Olemskoi AI, Kosminska YuO, Mokrenko AA (2009) Self-organization of quasi-equilibrium steady-state condensation in accumulative ion-plasma devices. Phys Lett A 373:3386–3391
16. Perekrestov VI, Olemskoi AI, Kornyushchenko AS, Kosminskaya YuA (2009) Self-organization of plasma-condensate quasi-equilibrium systems. Phys Solid State 51:1060–1067
17. Perekrestov VI, Kosminska YuO, Kornyushchenko AS, Latyshev VM (2013) Self-organization of copper nanosystems under Volmer-Weber conditions during quasi-equilibrium condensation. Phys B 411:140–148
18. Perekrestov VI, Mokrenko AA, Kosminskaya YuA, Rubets DI (2011) Formation of nickel extended surface upon quasi-equilibrium steady-state condensation. J Surf Invest 5:667–671
19. Perekrestov VI, Kosminska YuO, Kornyushchenko AS, Mokrenko AA (2013) Impact of selective processes on Al porous structures formation during self-organized quasi-equilibrium steady-state condensation. J Porous Mater 20:967–974
20. Perekrestov VI, Kosminska YuO, Kornyushchenko AS, Latyshev VM (2014) Self-assembly of porous Cu structures during steady-state condensation of weakly supersaturated vapors. J Porous Mater 21:1159–1167
21. Perekrestov VI, Kornyushchenko AS, Latyshev VM et al (2015) Formation of porous zinc nanostructures during self-organization of critically small steady-state supersaturations. Phys Status Solidi B 252:397–403
22. Perekrestov VI, Kornyushchenko AS, Natalich VV (2014) Formation of chromium layers under Volmer-Weber conditions at critically small supersaturations. Solid State Sci 33:12–18
23. Perekrestov VI, Kosminska YuO, Mokrenko AA et al (2011) Structure formation mechanisms of low-dimensional porous titanium systems condensed under quasi-equilibrium steady-state conditions. Vacuum 86:111–118
24. Perekrestov V, Kosminska Yu, Mokrenko A, Davydenko T (2014) Self-assembly of condensates with advanced surface by means of the competing field selectivity and Gibbs–Thomson effect. Appl Surf Sci 298:171–175
25. Kosminska YuO, Perekrestov VI (2018) Regularities of self-organization of technological conditions during plasma-arc synthesis of carbon nanotubes. Diam Relat Mater 85:37–48

10. Barna PB, Radnóczi G (2014) Structure formation during deposition of polycrystalline metallic thin films. In: Barmak K, Coffey K (eds) Metallic films for electronic, optical and magnetic applications. Woodhead Publishing Limited, Cambridge, pp 67–120
11. Ostrikov K, Neyts EC, Meyyappan M (2013) Plasma nanoscience: from nano-solids in plasmas to nano-plasmas in solids. Adv Phys 62:113–224
12. Levchenko I, Ostrikov K, Diwan K et al (2008) Plasma-driven self-organization of Ni nanodot arrays on Si(100). Appl Phys Lett [illegible]
[illegible]

On the Possibility of Training Demonstration of the Giant Magnetoresistance Effect in Higher School

V. B. Loboda, M. Ya. Dovzhyk, V. O. Kravchenko, S. M. Khursenko and Yu. O. Shkurdoda

Abstract The method and technique of training demonstration of the giant magnetoresistance effect on the example of film samples (single layer Co film and three-layer film Co/Cu/Co) in CIP-geometry with the help of simple experimental equipment are presented.

Keywords Nanosize multilayer films · Anisotropic magnetoresistance · Giant magnetoresistance

1 Introduction

In the middle of the twentieth century, a new scientific and technical direction began to form in the most developed countries of the world. By the 1970s it became known as the thin films physics (FTP). The complete notion of the physics and technology of thin films of this time is given in the encyclopedic edition [1, 2]. At the end of the twentieth century, the FTP became a strong scientific basis for both applied technical directions that provide the further development of traditional microelectronics (nanoelectronics) and new material science directions in solid state physics called nanotechnology (ion-plasma technologies for creating and modifying surfaces, etc. [3–10]. And then a rather strange situation arose, when despite the determining influence of the FTP on the development of the majority of modern science-intensive technologies, in the FTP itself did not found the attainment that pretended for universal scientific recognition. And only in 2007, A. Fert and P. Grünberg for the discovery of the giant magnetoresistance effect in the FTP was awarded the Nobel Prize in Physics [11–13].

V. B. Loboda · M. Ya. Dovzhyk · V. O. Kravchenko · S. M. Khursenko
Sumy National Agrarian University, Sumy, Ukraine

Yu. O. Shkurdoda (✉)
Sumy State University, Sumy, Ukraine
e-mail: yu.shkurdoda@gmail.com

© Springer Nature Singapore Pte Ltd. 2019

A. D. Pogrebnjak and V. Novosad (eds.), *Advances in Thin Films, Nanostructured Materials, and Coatings*, Lecture Notes in Mechanical Engineering,
https://doi.org/10.1007/978-981-13-6133-3_8

The giant magnetoresistance effect (GMR) is the galvanomagnetic quantum mechanical effect observed in specially created multilayer film objects, which consist of a system of magnetic and non-magnetic conducting alternating layers. The thickness of such layers is usually about units and tens of nanometers. The effect is manifested in the fact that the electrical resistance of such objects when they are introduced into a magnetic field (magnetoresistance MR) significantly changes. The relative change in the electric resistance value $\delta = \Delta R/R_0$, where $\Delta R = R - R_0$ (R is the electrical resistance of the conductor in the magnetic field, R_0 is the electrical resistance of the conductor in the absence of a magnetic field) is one units, tens and even hundreds of percent depending on the material of the layers, number of layers and temperature. Since this is several orders of magnitude larger than the magnetoresistance of massive natural conductors, hence the name is giant magnetoresistance.

Over the past 20 years, research topics in the field of GMR have not lost their relevance and continue to be of interest to researchers [14–24]. Some practical aspects of using the GMR effect are presented in [25–28].

2 Experimental Details

In the simplest case, the GMR effect is realized in CIP geometry in three-layer film objects (FO) of type FM/NM/FM, where FM is a layer of ferromagnetic metal (Co), NM is a layer of non-magnetic metal (Cu) [14, 16]. Depending on the thickness of the layers and the heat treatment, the magnitude of the effect is $\delta = 1$–4%. Such three-layer Co/Cu/Co films with GMR are offered by us for demonstration and study of the GMR effect in higher education institutions in the field of physics, electronics or magneto-electronics. We received a certificate from the Ministry of Education and Science of Ukraine on the recognition of compliance with pedagogical requirements No 06/029 dated June 24, 2014, for a product "Film objects for demonstration and study of the giant magnetoresistance effect".

FO is obtained by methods of layer vacuum spraying of metals (Cr, Cu, Co) on glass substrates. The FO consists of (Fig. 1) from the film resistor (3) (active part of the FO) and contact pads (2) for fixing the measuring conductors. Film resistors in the substrate plane have dimensions of 2 × 10 mm and are obtained by vacuum spraying Co, Cu. The substrates are polished glass plates (1) 25 × 25 mm in size. On each plate two identical film resistors (working and backup) are sprayed.

The contact pads are made using the vacuum layer spraying Cr and Cu. The layer of chrome thickness up to 50 nm is applied to the glass first and provides adhesion to the surface of the glass of the next copper contact layer thickness up to 150 nm. The contact pads after spraying are incubated in a vacuum to a temperature of 400 °C to provide their mechanical strength. The conductors for connecting the instrument to measure the electrical resistivity of the film resistor can be connected to such contact pads or soldering or purely mechanically using spring clamps.

During the demonstration and study of the GMR phenomenon, the change in the electrical resistance R of the film resistor in an external magnetic field by induction

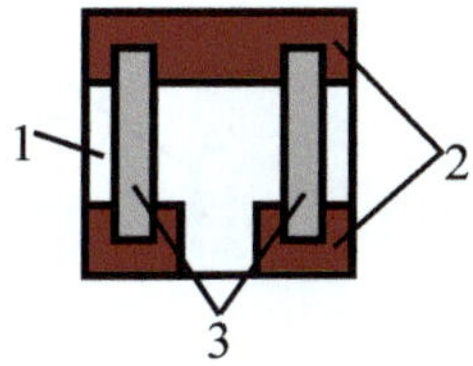

Fig. 1 Schematic image FO for demonstration and study of the giant magnetoresistance phenomenon: 1—glass substrate; 2—contact pads; 3—film resistor

B during its change is experimentally investigated, i.e. the dependence $R = R(B)$ is experimentally determined.

To do this it needed two FO—one with a film single-layer resistor with Co (thickness 40–60 nm), the second with a film three-layer Co/Cu/Co resistor. The film three-layer resistor is manufactured using a special technology by the method of layer vacuum spraying Co (up to 30 nm thick), an intermediate layer of Cu (thickness up to 1 nm) and again Co (up to 30 nm thick) in a magnetic field of a special spatial configuration. The film single layer resistor is manufactured by vacuum spraying of only Co thicknesses of about 60 nm. The resulting resistor FO is also annealing in a vacuum at a temperature of 400 °C to ensure the stability of their parameters and mechanical strength.

Using a single-layer resistor, a typical anisotropic (normal) magnetoresistance (AMR) is demonstrated and studied, and a gigantic (unusual) magnetoresistance (GMR) is demonstrated and studied using a three-layer resistor. The demonstration and study of the GMR effect with such FO is possible both at room temperature (a simple research option) and at low (cryogenic) temperatures (a more complex study option).

For demonstration and study of the GMR effect at room temperature, standard equipment of the educational laboratory of electricity and magnetism can be used: a digital ohmmeter and a magnetic system creating a homogeneous magnetic field with induction, which varies within $B = 0 \div 100$ mT (Helmholtz rings, solenoid or an electromagnet with an air gap) and the corresponding power supply (PS) of the magnetic system. The most convenient is the Helmholtz rings magnetic system. One of the possible variants of the simplest installation with Helmholtz rings is shown in Fig. 2.

The magnetic system should have a calibration curve (the graph of the dependence of the magnitude of the magnetic field induction created by it, on the magnitude of the electric current flowing through its coils).

FO with the sprayed contact pads and the film resistor under investigation is fixed in the holder, which provides the possibility of location of FO in the field of a homogeneous magnetic field of a magnetic system in the same plane in two mutually perpendicular directions in relation to the direction of the magnetic induction vector of the system (for measuring the longitudinal magnetoresistance ($R_{\parallel}$) and for measuring the transverse magnetic resistivity ($R_{\perp}$)), as shown in Fig. 3.

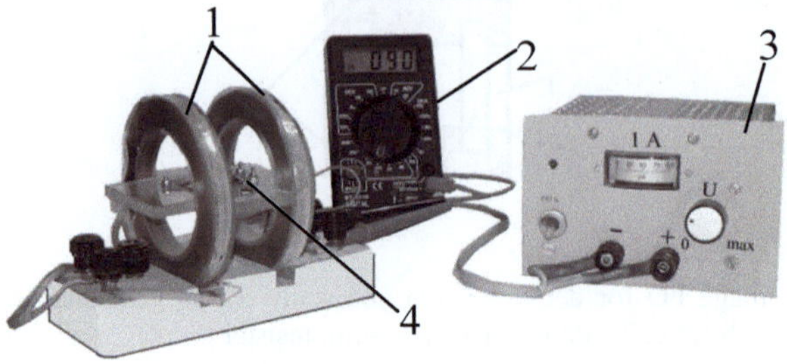

Fig. 2 Appearance of the simplest installation for the demonstration and study of the GMR phenomenon: 1—Helmholtz rings; 2—digital ohmmeter; 3—power supply of the Helmholtz rings; 4—holder with film object

Fig. 3 The location of the film object in the magnetic field: **a** when measuring the longitudinal magnetoresistance ($R_\parallel$); **b** when measuring the transverse magnetic resistor ($R_\perp$)

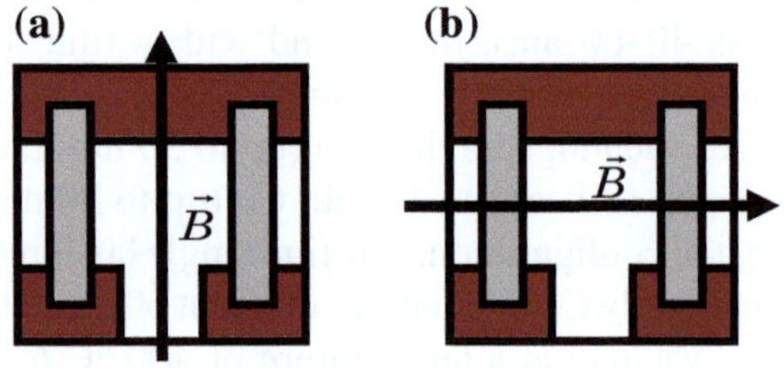

3 Method and Technology of Demonstration

3.1 Longitudinal Magnetoresistance (Dependence $R_\parallel = R_\parallel(B)$)

Before starting measurements, it is necessary to fix the FO on the holder so that the direction of the magnetic field induction vector and the direction of current through the film resistor coincide (Fig. 4a), and connect the measuring conductors of the ohmmeter to the contact pads of the FO. To get the full loop of the magnetoresistance effect (full dependence $R_{PP} = R_{PP}(\overrightarrow{B})$), you must do the following.

1. Magnetizing the film under study (film resistor) as much as possible, passing through the coils of the magnetic system the maximum electric current from the power supply. The current strength should provide creation in the location of the film object of a homogeneous magnetic field by induction not less than 100 mT.
2. Smoothly reduce the current through the coils of the magnetic system to zero and turn off the power of the magnetic system. After turning off the power of the magnetic system, record the digital ohmmeter (measure the electrical resistance of the film resistor) R_0 in the absence of a magnetic field.
3. Change the polarity of the power supply of the magnetic system (change to the opposite direction of the magnetic field induction vector).

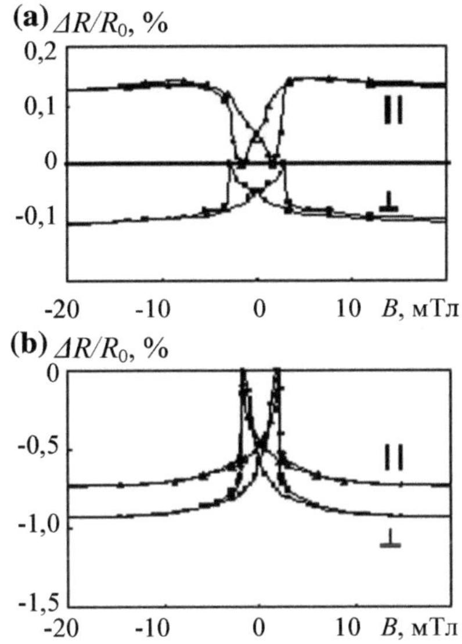

Fig. 4 The dependence of the magnetic resistance of film resistors on the induction of an external magnetic field: **a** resistor is single layer Co film (typical AMR); **b** resistor is three-layer film Co/Cu/Co (typical GMR)

4. Get the dependence of the value of the electrical resistance of the film resistor on the magnitude of the induction of the magnetic field $R_{PP} = R_{PP}(\overrightarrow{B})$. To do this, turn on the power supply of the magnetic system and, step by step, increasing the current (increasing the magnetic field induction) with a small increment, record the ohmmeter at each step.
5. Lock the ohmmeter indications after reaching the magnetic field induction value B = 100 mT and start with a small incremental step to reduce the current (to reduce the magnetic field induction) through the coils of the magnetic system to zero, again capturing at each step the ohmmeter indications.
6. Turn off the power supply of the magnetic system. Change the direction of induction of the magnetic field in the magnetic system to the opposite (change the polarity of the power supply) and again repeat the measurement of the electrical resistance of the sample in accordance with paragraphs 4 and 5 (Get the dependence $R_{PP} = R_{PP}(-\overrightarrow{B})$).
7. According to the results of the measurements, construct a complete loop of the magnetoresistive effect for the longitudinal magnetoresistance R_{PP}, that is to construct the two branches of the dependence $R_{PP} = R_{PP}(\overrightarrow{B})$ as for the current flowing through the film resistor in the direction of the vector $\vec{B}$ ($R_{PP} = R_{PP}(\overrightarrow{B})$) and for the current flowing through the film resistor in the direction opposite the vector $\vec{B}$ ($R_{PP} = R_{PP}(-\overrightarrow{B})$).

3.2 Transverse Magnitude (Dependence $R_\perp = R_\perp (B)$)

To measure the transverse magnitude resistance $R_\perp$, place the FO so that the direction of induction of the magnetic field of the magnetic system is perpendicular to the direction of flow of the electric current through the film resistor (Fig. 4b). Repeat the measurement for this position of the film object in accordance with p. 1–6 and, based on the results of measurements, construct a complete loop of the magnetoresistive effect for the transverse magnetoresistance $R_\perp$ (construct a complete graph of the dependence $R_\perp = R_\perp(\overrightarrow{B})$ similar to item 7).

Measurement carry out of the magnetic resistor for both PO with a single layer film resistor from Co and PO with a three-layer film resistor from Co/Cu/Co.

4 Results and Discussion

The obtained results of measurements of the MR must be expressed in percent using the formulas:

Longitudinal MR:

$$\delta_{\parallel} = \frac{\Delta R_{\parallel}}{R_0} = \frac{R_{\parallel} - R_0}{R_0}, \tag{1}$$

where R_{PP} is the electrical resistance of the resistor for longitudinal geometry of measurements (Fig. 3a); R_0 is the electrical resistance of the film resistor in the absence of a magnetic field;

Transverse MO:

$$\delta_{\perp} = \frac{\Delta R_{\perp}}{R_0} = \frac{R_{\perp} - R_0}{R_0}, \tag{2}$$

where $R_\perp$ is the electrical resistance of the film resistor for transverse measurement geometry (Fig. 3b).

According to the results of the measurements, construct a complete magnetic resistor loop (p. 7) (dependency graphs $\Delta R_{\parallel}/R_0 = \Delta R_{\parallel}/R_0(\vec{B})$ and $\Delta R_{\perp}/R_0 = \Delta R_{\perp}/R_0(\vec{B})$ for resistors from single layer Co film and three layer film Co/Cu/Co).

In Fig. 4 illustrates typical full-magneto-resistive loops for such films.

For film single-layer resistors from Co (Fig. 4a), there is a positive longitudinal PP (electrical resistance is increasing) and a negative transverse $\perp$ (electrical resistance decreases) magnetoresistance, which is a demonstration of AMR inherent in homogeneous ferromagnetic metals, both in massive and film conditions.

For a three-layer film resistor Co/Cu/Co (Fig. 4b), only a significant decrease in the electrical resistance (the ratio $\partial = ДR/R_0$ is less than zero) is observed regardless of the direction of the applied magnetic field, current and sample orientation (absence of anisotropy of the magnetoresistance). This is a characteristic feature of the GMR and allows a simultaneous change in the orientation of the magnetic moments of the film resistor elements (layers separated by a thin nonmagnetic layer Cu).

Compare the obtained dependencies in the case of AMR and GMR. Set the differences and determine the maximum GMR percentage.

5 Conclusions

The set of film objects of CIP-geometry created by us enables us to demonstrate and study in the courses of general physics, electronics, or magnetoelectronics, using the available simple equipment, as an ordinary anisotropic magnetoresistance (single layer film sample Co, Fig. 4a) and GMR (three-layer film sample Co/Cu/Co, Fig. 4b), which allows students to get acquainted with the physical effect of GMR.

Each sample (film object) has its own passport, which specifies its characteristics (composition electrical resistance, GMR amplitude, saturation field, coercive force, etc.) and a detailed instruction for demonstrating and studying the GMR effect.

In conclusion, the following should be noted. The GMR effect is experimentally discovered at the end of the twentieth century and the authors of the discovery in 2007 were awarded the highest scientific award—the Nobel Prize in Physics—is a fundamentally new physical phenomenon for artificially created macrosystems which structural elements have nanoscales and spin conduction electrons in which plays a main role.

Its practical use is one of the components of nanotechnology and spintronics, and the proposed set of film objects and the above-described method of their research, in our opinion, allow them to acquaint with them students of physical and technical specialties of universities.

References

1. Meisell L, Gleng R (1977) Tehnologiya tonkih plenok: spravochnik, vol 1. Sow. Radio, Moskow
2. Meisell L, Gleng R (1977) Tehnologiya tonkih plenok: spravochnik, vol 2. Sow. Radio, Moskow
3. Pogrebnjak AD, Lebed AG, Ivanov YuF (2001) Modifcation of single crystal stainless steel structure (Fe–Cr–Ni–Mn) by high-power ion beam. Vacuum 63(4):483–486. https://doi.org/10.1016/S0042-207X(01)00225-1
4. Pogrebnjak AD, Bazyl EA (2001) Modification of wear and fatigue characteristics of Ti–V–Al alloy by Cu and Ni ion implantation and high-current electron beam treatment. Vacuum 64(1):1–7. https://doi.org/10.1016/S0042-207X(01)00160-9
5. Pogrebnjak AD, Isakov IF, Opekunov MS et al (1987) Increased wear resistance and positron annihilation in Cu exposed to high power ion beam. Phys Lett A 123(8):410–412. https://doi.org/10.1016/0375-9601(87)90043-0
6. Goncharov AA, Yunda AN, Komsta H et al (2017) Effect of structure on physicomechanical properties of transition metals diboride films. Acta Phys Pol A 132(2):270–273
7. Yakovin S, Zykov A, Dudin S et al (2017) Plasma assisted deposition of TaB_2 coatings by magnetron sputtering system. Probl At Sci Technol 107(1):187–190

8. Bazhin AI, Goncharov AA, Pogrebnyak AD et al (2016) Superhardness effect in transition metal diborides films. Phys Met Metall 117(6):594–601
9. Goncharov AA, Yunda AN, Shelest IV et al (2017) Effect of the magnetron sputtering parameters on the structure and substructural characteristics of tantalum diboride films. J Nano-Electron Phys 9(4):04014
10. Pogrebnjak AD, Bondar OV, Abadias G et al (2016) Structural and mechanical properties of NbN and Nb–Si–N films: experiment and molecular dynamics simmulations. Ceram Int 42(10):11743–11756. https://doi.org/10.1016/j.ceramint.2016.04.095
11. Baibich MN, Broto JM, Fert A et al (1988) Giant magnetoresistance of (001)Fe/(001)Cr magnetic superlattices. Phys Rev Lett 61(21):2472–2475
12. Binasch G, Grünberg P, Saurenbach F et al (1989) Enhanced magnetoresistance in layered magnetic structures with antiferromagnetic interlayer exchange. Phys Rev B 39(7):4828–4830
13. Fert A (2008) Proischojdenie, razwitie i perspektiwi spintroniki. Usp Phys Nauk 178(12):1336–1348
14. Shkurdoda YuO, Chornous AM, Shabelnyk YuM et al (2017) The influence of the concentration of components in magnetic layers on the magnetoresistive properties of three-layer film systems based on FexNi1−x and Cu. J Magn Magn Mater 443:190–194. https://doi.org/10.1016/j.jmmm.2017.07.078
15. Loboda VB, Protsenko IE (1981) Structure and electrical resistance of thin scandium films (III). Study on electrical properties. Kristall Tech 16(4):489–494
16. Loboda VB, Shkurdoda YuO, Kravchenko VO et al (2011) Structure and magnetoresistive properties of polycrystalline Co/Cu/Co films. Metallofiz Noveishie Tekhnol 33(2):161–169
17. Loboda VB, Kolomiets VM, Shkurdoda YuO et al (2012) Structure and magnetoresistive properties of nanocrystalline film systems based on Co, Fe, Ag, and Cu. Metallofiz Noveishie Tekhnol 34(8):1043–1055
18. Loboda VB, Khursenko SN (2006) Structure and electrical conductivity of ultrathin Ni–Cu films. JETP 103(5):790–794
19. Shkurdoda YO, Chornous AM, Loboda VB et al (2016) Structure and magnetoresistive properties of three-layer film systems based on permalloy and copper. J Nano- Electron Phys 8(2):02056
20. Loboda VB, Kolomiets VM, Khursenko SM et al (2014) The electrical conductivity of the three-layer polycrystalline films Co/Ag(Cu)/Fe in the conditions of atoms interdiffusion. J Nano- Electron Phys 6(1):04032
21. Protsenko IY, Mehta PK, Odnodvorets LV et al (2014) Magnetoresistive properties of quasi granular film alloys FexPt1−x at the low concentrations of Pt atoms. J Nano- Electron Phys 6:01031
22. Synashenko OV, Tkach OP, Buryk IP et al (2009) Magnetoresistive properties of multilayer nanodimensional film systems. Probl At Sci Technol 18:169
23. Protsenko I, Odnodvoretz L, Chornous A (1998) Electroconductivity and tensosensibility of multilayer films. Metallofiz Noveishie Tekhnol 20:36
24. Lytvynenko IM, Pazukha IM, Pylypenko OV et al (2015) Structural, magnetic and magnetoresistive properties of ternary film Ni–Fe–Co alloy. Metallofiz Noveishie Tekhnol 37(10):1377
25. Pogorily AM, Ryabchenko SM, Tovstolytkin AI (2010) Spintronika. Osnovni javisza. Tendenzii rozvitku. Ukr Phys J 6(1):37–97 [In Ukrainian]
26. Lukashevich MG (2003) Vvedenie v magnitoelektroniku. BGU, Minsk [In Russian]
27. Sisoeva S (2005) Avtomobilnie datchiki polozenija. Komponenty technol 4(5):60–68
28. Romanova I (2014) Magnitoresistivnaya pamyat' MRAM. Elektron-nauka-technol-biznes 8(00140):72–77

Multifractal Analysis of the Surfaces of Protective (TiAlSiY)N, $Me_{1-x}N/CrN$ and $Me_{1-x}N/ZrN$ Coatings

Ya. Kravchenko, B. Natalich, M. Opielak and V. Borysiuk

Abstract In the present paper, a technique for the preparation of protective (TiAl-SiY)N, Me_xN/CrN and Me_x/ZrN coatings is shown. The algorithm of multifractal fluctuation analysis is described, and the results of the numerical investigation of microroughness of surfaces of investigated samples are presented. As follows from the calculated numerical parameters, the surface of (TiAlSiY)N/CrN coating is the smoothest and the surface of (TiAlSiY)N is the most non-uniform.

Keywords Surface morphology · Multifractal analysis · Self-similarity · Vacuum deposition

1 Introduction

Numerical methods for digital image analysis can be a useful technique in materials science when studying of the microstructure of sample's surface by its electron microscopic images. As it is known, most of the practicable surfaces can have a self-similar structure, which makes to be possible the application of fractal analysis technique to study their structures. Nowadays several techniques and computer algorithms to calculate the quantitative characteristics of self-similar structures have been developed. Generalized Hurst exponent is one of the most used parameters of self-similarity of the surface, which allows describing surface structure quantitatively [1–6]. Thus, the application of fractal analysis in materials science makes it possible to calculate numerical parameters for describing the surface structure of material using mathematical algorithms.

Ya. Kravchenko (✉) · B. Natalich · V. Borysiuk
Sumy State University, Sumy, Ukraine
e-mail: y.kravchenko@phe.sumdu.edu.ua

M. Opielak
Lublin University of Technology, Lublin, Poland

© Springer Nature Singapore Pte Ltd. 2019
A. D. Pogrebnjak and V. Novosad (eds.), *Advances in Thin Films, Nanostructured Materials, and Coatings*, Lecture Notes in Mechanical Engineering,
https://doi.org/10.1007/978-981-13-6133-3_9

In this paper, the surfaces protective nitride coatings have been investigated by methods of numerical studies of digital images, in particular, in the context of multifractal fluctuation analysis (MFFA).

Much attention is paid to the development of functional nitride coatings for tribotechnical purposes [7–10]. The modification of the material surface by vacuum condensates deposition makes it possible to increase the period of service, in the case of cutting tools and rendering new properties such as corrosion and oxidation resistance, thermal stability, etc. The surface condition (the degree of roughness, the size of the droplet fraction and the presence of microcracks) has a significant influence on the properties of the condensable coating.

The improvement of nitride coatings, often binary ones, can occur by adding doping additives and producing multicomponent coatings, as well as by fabrication complex layered systems by combining of multicomponent and binary layers in a repeated bilayer period.

The as-deposited multicomponent (TiAlSiY)N coating and a series of layered coatings with a variable composition of bilayer Me_xN/CrN and Me_x/ZrN were obtained by vacuum arc deposition technique.

2 Models and Methods

A two-dimensional method of multifractal fluctuation analysis (MFFA) was used to estimate the scaling characteristics of the surface. It allows obtaining a complete description of the structure of the investigated surface in the context of described below numerical procedure [11–16].

According to Refs. [17–19], the surface is presented by a two-dimensional array $X(i,j)$, where i, j are arguments that take the value $i = 1, 2 \ldots M$ and $j = 1, 2 \ldots N$. Then the surface is divided into $M_s \times N_s$ square sections, which do not intersect with each other and have the size of $s \times s$. The numbers $M_s = [M/s]$ and $N_s = [N/s]$ represent the integer parts obtained after dividing the intervals of changing the arguments i, j into segments. The values of the investigated functions in each segment is designated by the indices υ, ω and represent them by the sequence: $X_{\upsilon,\omega}\,(i, j) = X(l_1 + i, l_2 + j)$. Arguments $1 \leq i, j \leq s$, are changed in the segment that determined by $l_1 = (\upsilon - 1)s$ and $l_2 = (\omega - 1)s$.

The cumulative sum for each segment calculated from the following:

$$u_{\upsilon\omega}(i, j) = \sum_{k_1=1}^{i} \sum_{k_2=1}^{j} X_{\upsilon\omega}(k_1, k_2) \tag{1}$$

The dependence $u_{\upsilon\omega}(i, j)$ determines the fractal surface geometrically.

As in the case of time series, the irregular dependence of $U_{\upsilon,\omega}(i, j)$ should be calculated from the smooth surface $\widetilde{U}_{\upsilon,w}(i, j)$. It takes into account the trend in the variation of the initial function $U_{\upsilon,\omega}(i, j)$ and is called the trend.

$$\tilde{u}_{\upsilon\omega}(i, j) = ai + bj + c \tag{2}$$

where the coefficients *a* and *b* are determined by the method of least squares. During calculation of intermediate values, the functions with much higher accuracy can be used, but this will noticeably increase the time of computations [17–22]. The residual matrix is calculated from the formula:

$$\varepsilon_{\upsilon\omega}(i, j) = u_{\upsilon\omega}(i, j) - \tilde{u}_{\upsilon\omega}(i, j) \tag{3}$$

Its use gives the segment-specific variances

$$F^2(\upsilon, \omega, s) = \frac{1}{s^2}\sum_{i=1}^{s}\sum_{j=1}^{s}\varepsilon^2_{\upsilon\omega}(i, j) \tag{4}$$

The averaging over all segments leads to complete dispersion

$$F_q(s) = \left\{\frac{1}{M_s N_s}\sum_{\upsilon=1}^{M_s}\sum_{\omega=1}^{N_s}[F(\upsilon, \omega, s)]^q\right\}^{1/q} \tag{5}$$

which is deformed by the parameter *q*. On that basis, negative values of q increase the contribution of segments corresponding to small fluctuations, and positive ones allocate large values of $F^2(\upsilon, \omega, s)$. For $q = 0$, the above formula should be replaced by the expression:

$$F_0(s) = \exp\left\{\frac{1}{M_s N_s}\sum_{\upsilon=1}^{M_s}\sum_{\omega=1}^{N_s}\ln[F(\upsilon, \omega, s)]\right\} \tag{6}$$

The value of s varied from $s_{min} \approx 6$ to $s_{max} \approx \min(M, N)/4$. For self-similar sets, this leads to a scaling relation between the variance $F_q(s)$ and the scale *s*:

$$F_q(s) \sim s^{h(q)} \tag{7}$$

where *h(q)* is the generalized Hurst exponent. In a log-log plot, the latter dependence represents a straight line, the slope of which gives the exponent *h(q)* at different values of *q* parameter.

The production of the coatings TiAlSiY-series was made by the vacuum-arc deposition technique in a Bulat-6 [23]. The coatings were deposited onto 12 × 18H9T steel substrates (analog of SUS321 and 321S51 steels) with a sample size of 18 × 20 × 2 mm^3. The constant rotation speed of the substrate holder was 8 rpm, the distance from the evaporator to the substrate was fixed at 250 mm.

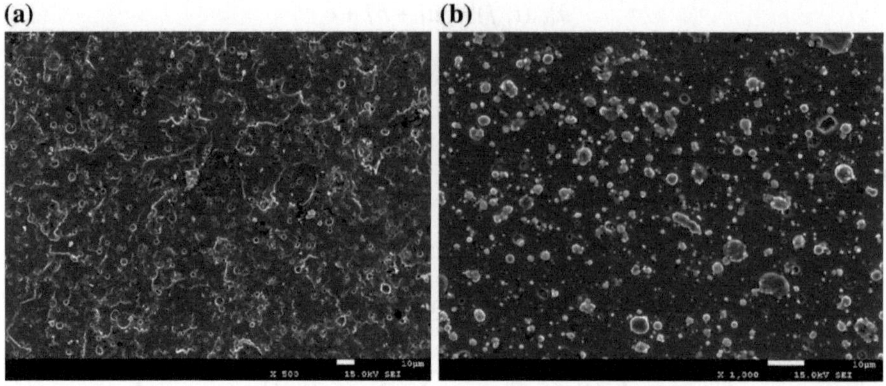

Fig. 1 SEM images of the surface of multilayered coatings: (TiAlSiY)N (**a**), (TiAlSiY)N/ZrN (**b**)

The deposition of the coatings was performed from two metal sources with an atomic ratio of the composite cathode material as follows: 58 at.% Ti, 38 at. % Al, 3 at.% Si, 1 at.% Y and Cr and Zr cathodes with a material purity of 99.8% (in the case of layered condensates). Sintering of the composite cathode (TiAlSiY) was performed in a SPS 25-10.

Figure 1a, b shows surface SEM micrographs of multielement (TiAlSiY)N and multilayered (TiAlSiY)N/ZrN coatings.

Layered Me_xN/CrN and Me_x/ZrN condensates were obtained under the following deposition conditions: arc current of 100/90 A, coil focus of 0.5/0.4 A, negative potential applied to the substrate was −110 V, the substrate temperature was 250 °C. For multielement (TiAlSiY)N coating deposition conditions were changes as arc current and coil focus were 100 and 0.5 A, respectively, negative potential applied to the substrate was increased to −150 V, the substrate temperature was 300 °C. The pressure of the working atmosphere for all coatings was fixed at $P_N = 4 \times 10^{-3}$ Torr.

The surface morphology was analyzed using a scanning electron microscope JEM-7001TTLS (JEOL) with an Oxford X-Max detector. The sample holder in the microscope chamber was modified, which had led to a significant increase in the slope of the sample and obtain a more detailed picture of the surface microrelief. Numerical models based on the corresponding electron microscopic images were constructed for all surfaces investigated in this work. Thus, each surface was represented as a two-dimensional array of data, whose element indices were set to the pixel indices of the corresponding images, and the elements themselves were set as the brightness values of these pixels. Graphical views of 3D models obtained in this way are shown in Fig. 2.

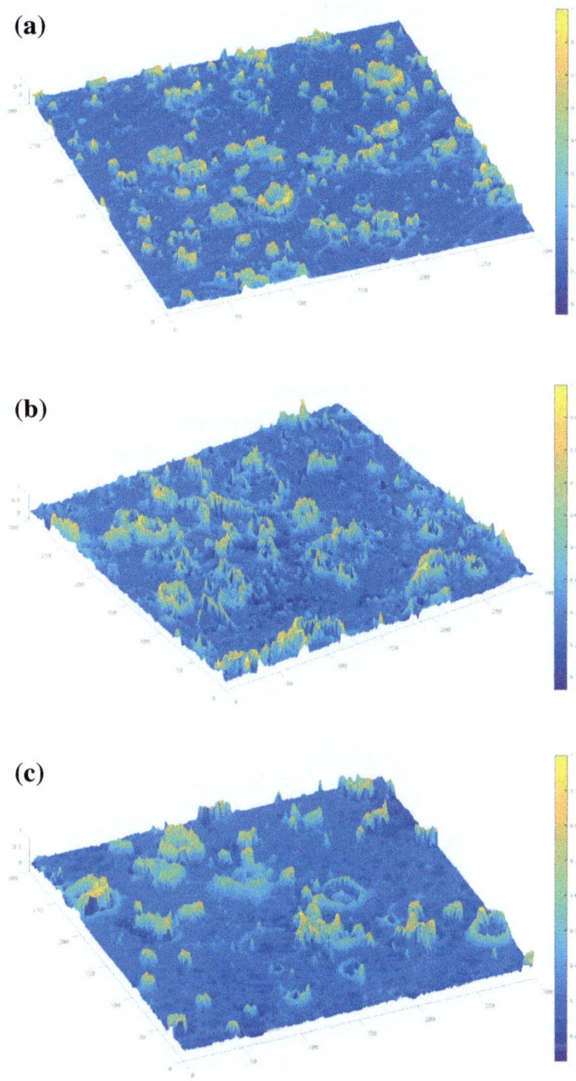

Fig. 2 3D models of the selected coverage areas: (TiAlSiY)N (**a**), (TiAlSiY)N/ZrN (**b**), (TiAlSiY)N/CrN (**c**)

3 Results and Discussion

Vacuum arc deposition method is characterized by the presence of high-ionized plasma flows of the evaporated material. Meanwhile, the presence of droplets from the evaporated target in the plasma flows is an important feature of this method. The microdroplets with the size 0.1 ... 20 μm were observed on the surface of all samples except Zr-containing sample. The formation of multiple craters was registered for

Me_x/ZrN coating, which was probably caused by an increase of the negative bias potential $-U_s$.

The presence of a macro-droplet components affects the uniformity of the surface structure of the condensates, introducing additional distortions, boundaries, and pores into it, which leads to an increase in the coefficient of friction, a decrease in wear resistance and corrosion resistance of the coatings from tribotechnical designation.

MFFA-analysis allows making a quantitative assessment of the degree of fluctuation of the digital surface height map from a smooth surface, according to which it is possible to conclude the roughness of the material surface, which makes this method a promising technique in analyzing the structure of the investigated nitride coatings [24–28].

The results of the carried fluctuation analysis are shown in Fig. 3.

Figure 3 demonstrates the calculated dependences of the fluctuation function (5) on the scale s, constructed in a two-dimensional graph of numerical data that uses logarithmic scales on both the horizontal and vertical axes. As the figure shows, the dependencies obtained for investigated samples have a pronounced linear character of the form, especially when $q > 0$. This type of dependence is typical for self-similar objects, and therefore the Hurst exponent $h(q)$ can be calculated for all samples by linear regression of the obtained dependences.

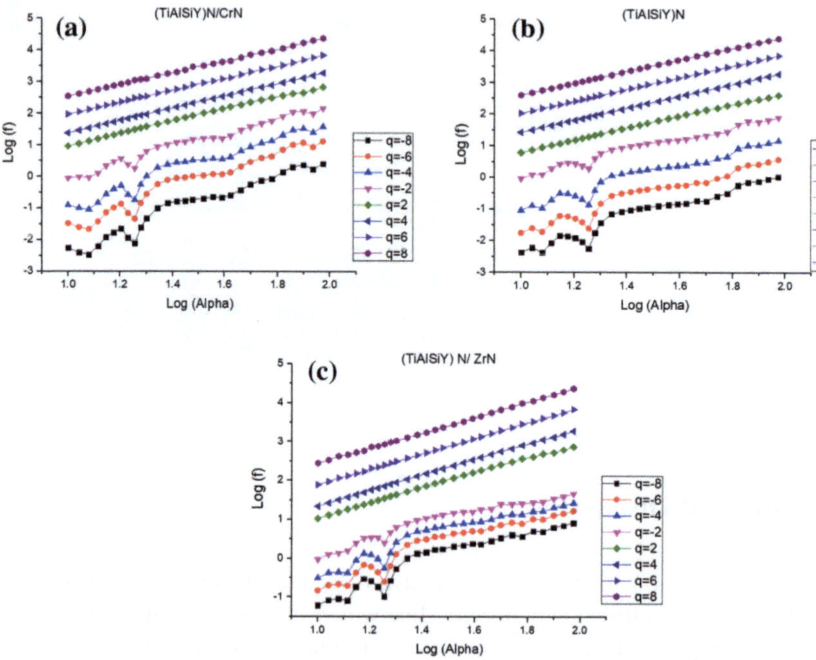

Fig. 3 Log-log dependencies of the calculated fluctuation functions Eq for TiAlSiY-series Curves shifted vertically for clearance

Figure 3 shows the dependencies of the generalized Hurst exponent calculated for all studied samples. As can be seen from the presented dependences, (TiAlSiY)N coating surface is characterized by the largest range of variation of $h(q)$ values, after which follows (TiAlSiY)N/ZrN and further (TiAlSiY)N/CrN. Mathematically this means that for the surface image of (TiAlSiY)N sample, the altitude fluctuations (4) relatively to the smooth surface (2) have the highest values in comparison with the other samples. This may mean that the surface of this coating has a large value of the relative surface roughness.

It should also be noted that the results of carried fluctuation analysis of SEM images and microstructure of surfaces as well as calculated numerical parameters can be used to parameterization the roughness of the surfaces of the investigated protective coatings [29–31].

Acknowledgements The authors gratefully to Ministry of Education and Science of Ukraine for financial support (Project No. 0117U003923).

References

1. Mandelbrot BB (1982) The fractal geometry of nature. Freeman, San Francisco
2. Family F, Vicsek T (1991) Dynamics of fractal surfaces. World Scientific, Singapore
3. Paladin G, Vulpiani A (1987) Anomalous scaling laws in multifractal objects. Phys Rep 156(4):147–225. https://doi.org/10.1016/0370-1573(87)90110-4
4. Feder J (1988) Fractals. Plenum Publishers, New York
5. Talibuddin S, Runt JP (1998) Reliability test of popular fractal techniques applied to small two-dimensional self-affine data sets. J Appl Phys 76(9):5070. https://doi.org/10.1063/1.358490
6. Sornette D (2001) Critical phenomena in natural science. Springer, New York
7. Pogrebnjak AD, Bor'ba SO, YaO Kravchenko et al (2016) Effect of the high doze of N + (1018 cm^{-2}) ions implantation into the (TiHfZrVNbTa)N nanostructured coating on its microstructure, elemental and phase compositions, and physico-mechanical properties. J Superhard Mater 38(6):393–401. https://doi.org/10.3103/S1063457616060034
8. Berladir KV, Budnik AO, Dyadyura KA et al (2016) Physicochemical principles of the technology of formation of polymer composite materials based on polytetrafluoroethylene—a review. High Temp Mater Process 20(2):157–184. https://doi.org/10.1615/HighTempMatProc.2016017875
9. Pogrebnjak A, Maksakova O, Kozak C et al (2016) Physical and mechanical properties of nanostructured (Ti–Zr–Nb)N coatings obtained by vacuum-arc deposition method. Prz Elektrotechniczny 2016(8):180–183. https://doi.org/10.15199/48.2016.08.49
10. Pogrebnjak AD, Beresnev VM, Kolesnikov DA et al (2013) Multicomponent (Ti-Zr-Hf-V-Nb)N nanostructure coatings fabrication, high hardness and wear resistance. Acta Phys Pol A 123(5):816–818. https://doi.org/10.12693/APhysPolA.123.816
11. Makowiec D, Galaska R, Dudkowska A et al (2006) Long range dependencies in heart rate signal—revisited. Phys A 369(2):632–644. https://doi.org/10.1016/j.physa.2006.02.038
12. Kravchenko Ya O, Coy LE, Peplińska B et al (2018) Nano-multilayered coatings of (TiAlSiY)N/MeN (Me = Mo, Cr and Zr): influence of composition of the alternating layer on their structural and mechanical properties. J Alloy Compd 767:483–495. https://doi.org/10.1016/j.jallcom.2018.07.090
13. Pogrebnjak AD, Borisyuk VN, Bagdasaryan AA et al (2014) The multifractal investigation of surface microgeometry of (Ti-Hf-Zr-V-Nb)N nitride coatings. JNEP 6(4):04018

14. Lyashenko IA, Borysiuk VN, Manko NN (2014) Statistical analysis of self-similar behaviour in the shear induced melting model. Cond Matt Phys 17(2):23003. https://doi.org/10.5488/CMP.17.23003
15. Kravchenko Y, Maksakova O, Drodziel P et al (2016) Effect of thermal annealing and deposition conditions on the structure and mechanical properties of a multilayer nitride coating based on Ta. High Temp Mater Process 20(1):85–92
16. Olemskoi OI, Danyl'chenko SM, Borysyuk VM et al (2009) Multifractal analysis of X-ray patterns of complex condensed media. Metallofiz Noveishie Tekhnol 31(6):777–789
17. Gu GF, Zhou WX (2006) Detrended uctuation analysis for fractals and multifractals in higher dimensions. Phys Rev E 74:061104. https://doi.org/10.1103/PhysRevE.74.061104
18. Lee JW, Lee KE, Rikvold PA (2006) Multifractal behavior of the Korean stock-market index KOSPI. Phys A 364:355–361. https://doi.org/10.1016/j.physa.2005.08.082
19. Halsey TC, Jensen MH, Kadano LP et al (1986) Fractal measures and their singularities: the characterization of strange sets. Phys Rev A 33:1141–1151. https://doi.org/10.1103/PhysRevA.33.1141
20. Kantelhardt JW, Zschiegner SA, Koscielny-Bunde E et al (2002) Multifractal detrended fluctuation analysis of nonstationary time series. Phys A 316(1–4):87–114. https://doi.org/10.1016/S0378-4371(02)01383-3
21. Chhabra AB, Sreenivasa KR (1991) Negative dimensions: theory, computation, and experiment. Phys Rev A 43:1114–1117. https://doi.org/10.1103/PhysRevA.43.1114
22. Kimiagar S, Sadegh Movahed M, Khorram S et al (2009) Fractal analysis of discharge current uctuations. J Stat Mech 3:03020. https://doi.org/10.1088/1742-5468/2009/03/P03020
23. Biksa A, Yamamoto K, Dosbaeva G et al (2010) Wear behavior of adaptive nano-multilayered AlTiN/MexN PVD coatings during machining of aerospace alloys. Tribol Int 43(8):1491–1499. https://doi.org/10.1016/j.triboint.2010.02.008
24. Olemskoi AI, Yushchenko OV, Borisyuk VN, Zhilenko TI, Kosminska YuO, Perekrestov VI (2012) Hierarchical condensation near phase equilibrium. Phys A 391:3277–3284. https://doi.org/10.1016/j.physa.2011.10.027
25. Kavasseri RG, Nagarajan R (2005) A multifractal description of wind speed records. Chaos Soliton Fract 24(1):165–173. https://doi.org/10.1016/j.chaos.2004.09.004
26. Olemskoi A, Shuda I, Borisyuk V (2010) Generalization of multifractal theory within quantum calculus. EPL 89(5):1–12. https://doi.org/10.1209/0295-5075/89/50007
27. Pogrebnyak AD, Shpak AP, Azarenkov NA et al (2009) Structures and properties of hard and superhard nanocomposite coatings. Phys-Usp 52:29–54. https://doi.org/10.3367/UFNe.0179.200901b.0035
28. Pogrebnjak AD, Bondar OV, Abadias G et al (2016) Structural and mechanical properties of NbN and Nb-Si-N films: experiment and molecular dynamics simulations. Ceram Int 42(10):11743–11756. https://doi.org/10.1016/j.ceramint.2016.04.095
29. Pogrebnjak AD, Bazyl EA (2001) Modification of wear and fatigue characteristics of Ti-V-Al alloy by Cu and Ni ion implantation and high-current electron beam treatment. Vacuum 64(1):1–7. https://doi.org/10.1016/S0042-207X(01)00160-9
30. Pogrebnjak AD, Rogoz VM, Bondar OV et al (2016) Structure and physicomechanical properties of NbN-based protective nanocomposite coatings: a re-view. Prot Met Phys Chem Surf 52:802–813. https://doi.org/10.1134/S2070205116050191
31. Kadyrzhanov DB, Zdorovets MV, Kozlovskiy AL et al (2018) Influence of ionizing irradiation on the parameters of Zn nanotubes arrays for design of flexible electronics elements. Dev Methods Measur 9(1):66–73. https://doi.org/10.21122/2220-9506-2018-9-1-66-73

Multilayer Nickel–Copper Metal Hydroxide Coating as Cathode Material for Hydrogen Evolution Reaction

A. Maizelis and B. Bairachniy

Abstract The comparison of the electrochemical behavior of electrodes with Ni–Cu alloy coating and multilayer coating consisting of alternating layers of alloy and metal hydroxide layers in alkaline water electrolysis is presented. The coatings were electrodeposited in polyligand pyrophosphate-ammonia electrolyte. It is shown that the electrode with multilayer coating is more corrosion resistant, has higher activity in the hydrogen evolution reaction due to both more developed surface and higher intrinsic catalytic activity. The value of the exchange current of hydrogen evolution reaction on the multilayer coating is comparable to the value of the exchange current on the Ni–Cu–Mo ternary alloy and nickel-containing electrodes with the highly developed 3D nanostructured surface.

Keywords Alkaline water electrolysis · Multilayer coating · Nickel · Copper · Electrode · Hydrogen evolution reaction · Overvoltage · Catalytic activity · Exchange current

1 Introduction

Hydrogen is a viable alternative to fossil fuel, and alkaline water electrolysis is a promising method of hydrogen production. The most important task of electrolysis is the correct choice of the electrode material. The ideal properties of an electrode used for water electrolysis are a large surface area, good electrical conductivity, good stability, low overpotential, low cost and ease of fabrication [1].

Among the non-noble metals, Ni is a promising candidate for low-cost HER catalysts. However, the hydrogen evolution reaction (HER) rate on Ni is hindered by the hydrogen desorption step, due to the large value of hydrogen binding energy on that metal [2].

A. Maizelis (✉) · B. Bairachniy
Department of Technical Electrochemistry, National Technical University "Kharkiv Polytechnic Institute", Kharkiv, Ukraine
e-mail: a.maizelis@gmail.com

© Springer Nature Singapore Pte Ltd. 2019
A. D. Pogrebnjak and V. Novosad (eds.), *Advances in Thin Films, Nanostructured Materials, and Coatings*, Lecture Notes in Mechanical Engineering,
https://doi.org/10.1007/978-981-13-6133-3_10

Solmaz et al. [3] found that the electrodeposited NiCu coating has a higher electrocatalytic activity for the HER in the alkaline medium than the Cu and Ni electrodes due to the porosity as well as the synergistic interaction of the Ni and Cu.

It was shown that the NiCu coating containing 14.6% Ni has a compact and heterogeneous structure with good time stability over electrolysis [4]. The combination of Ni with Cu leads to a decrease of the hydrogen binding energy and, consequently, to an enhanced HER activity of the Ni–Cu alloy when compared to Ni [2].

It was found that Ni–10Cu alloy exhibited considerably higher intrinsic activity than Ni plate and other Ni–Cu bimetallic alloys with Cu content of 5, 10, 20, 30 and 50 wt.% prepared by powder metallurgy and polished [5]. Among the Ni–Cu alloys containing 6–81 mol.% Ni electrodeposited on a Cu substrate electrodes having low Ni content gave high electrocatalytic activities [6]. According to [7] among the electrodeposited NiCu alloys, containing from 16 to 59% Ni the $Ni_{51}Cu_{49}$ catalyst achieved the highest current density for HER based on the geometrical area.

Nanostructured alloys have recently attracted great attention for hydrogen evolution due to their unique electron structure and large surface area as, for example, three-dimensional (3D) CoNiCu nano-network structures [8] and Cu/Ni–Co 3D hierarchical nanostructures [9].

Nickel–copper alloy is deposited with subsequent etching for obtaining the nanoporous nickel [10]. In [2] 3D nickel–copper nanostructured foams were prepared by galvanostatic electrodeposition. Cu–Ni MMFs consisting of Ni-rich and Cu-rich phases have been successfully synthesized by the hydrogen bubble template electrodeposition method [11]. A way to fabricate the NiCu/Al_2O_3/nano-carbon network (NCN) composite electrode by NiCu particles coelectrodeposition, using a novel conductive alumina/NCN composite material as the support was presented in [12].

A hierarchical porous $Ni(OH)_2$/NiCu electrode was fabricated by cyclic voltammetry (CV) treatment, electrochemical dealloying after brush plating or replacing reaction. It was shown that modifying NiCu electrode by $Ni(OH)_2$ leads to the highest HER activity due to the synergy between the $Ni(OH)_2$ and NiCu surfaces [1, 13].

Therefore, the nickel–copper nanostructured materials are promising for use in alkaline water electrolysis, and their properties depend on the production conditions.

The authors have shown in previous works that nanostructured multilayer coatings deposited from polyligand electrolytes have enhanced catalytic activity, improved mechanical and anticorrosive properties [14–16]. As the known from literature nanostructured nickel–copper coatings are deposited from sulfate and citrate electrolytes only, the study of other methods for their electrodeposition and from other electrolytes is of interest.

This study aims to compare the behavior of nanostructured nickel–copper metal hydroxide multilayer coating and a nickel–copper alloy coating deposited from a polyligand pyrophosphate-ammonia electrolyte as a cathode material in the alkaline water electrolysis.

2 Materials and Methods

Electrochemical studies were carried out using potentiostat P-45X. Nickel–copper alloy coatings and multilayer coatings consisting of alternating layers of a nickel—copper alloy and a mixture of metals with their hydroxides were deposited on a copper wire of 0.5 mm diameter and 10 mm height using a combined nickel–copper anode. A polyligand pyrophosphate-ammonia electrolyte was used for the deposition. Such electrolyte allows electrodeposition of well-adhered coatings on an electronegative substrate [17, 18]. Electrolyte composition is: 0.055 mol L^{-1} $CuSO_4$, 0.55 mol L^{-1} $NiSO_4$, 1.0 mol L^{-1} $K_4P_2O_7$, 2.4 mol L^{-1} $NH_3(NH_4^+)$.

The obtained multilayer coatings were tested in the hydrogen evolution reaction (HER) in the electrolyte containing 5 mol L^{-1} NaOH without of inert gas purging. The counter electrode in this study was made of platinum. A saturated silver/silver chloride electrode was used as the reference electrode. The potential value was corrected with the ohmic drop by the electrolyte resistance calculated of 100 mV impulse for 20 ms.

3 Results and Discussion

Cyclic volt-ampere dependencies (CVA) obtained on platinum in a pyrophosphate-ammonia solution reveal areas of potential corresponding to deposition of various types of films (Fig. 1). The CVA were obtained by (i) scanning the potential from the open-circuit value in the cathode direction, (ii) holding the value of working electrode potential at the chosen value for accumulating the electrolysis products, and (iii) scanning potentials in the anode direction until the electrolysis products were dissolved. At the same time, anodic peaks can give information about the phase composition of thin films obtained at the potentials of the cathode branch of the CVA [19, 20].

After the accumulation of electrolysis products at the potential of −0.4 V an anode peak appears at the anode branch of the CVA at approximately −0.2 V (curve 1). This is the peak of the dissolution of copper deposited at the initial stage of cathode branch of CVA. The NiCu alloy begins to deposit actively at potential of −0.9 V. Two peaks are observed at the anode branch of the CVA after the potential delay at −1.175 V (curve 2). The potential region of the first peak (from −0.35 to −0.1 V) corresponds to the copper and the copper-enriched phase dissolution. At the potentials of the second peak (from 0.2 to 0.5 V) nickel and/or nickel-enriched phase dissolves. During potential delay at the current density that slightly excesses the limiting current (−1.2 V, curve 4), the current increases (unlike curve 2), which indicates the formation of larger crystals and developed surface.

The copper content in the film virtually does not change since the area of the first anodic peak remains virtually unchanged. The second anode peak shifts towards larger values of the potential and expands due to the appearance of the shoulder on

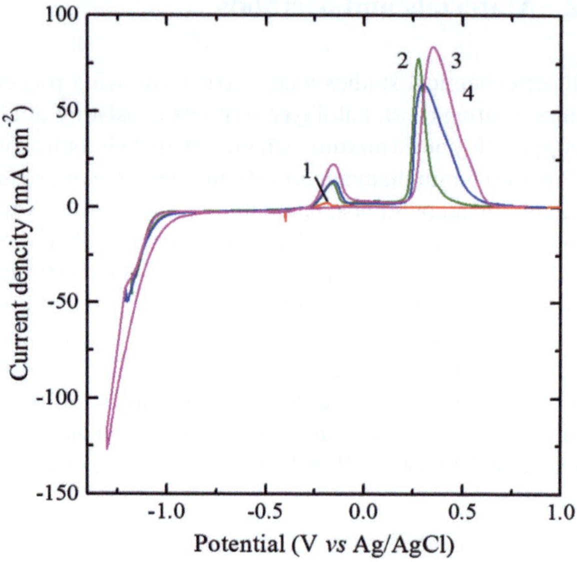

Fig. 1 CVA on Pt in pyrophosphate-ammonia electrolyte containing 0.055 mol L^{-1} $CuSO_4$, 0.55 mol L^{-1} $NiSO_4$, 1.0 mol L^{-1} $K_4P_2O_7$, 2.4 mol L^{-1} $NH_3(NH_4^+)$. The potential scan rate is 50 mVs^{-1}. The potential delay at −0.4 V (1), −1.175 V (2), −1.2 V (4), −1.3 V (3) was for 5 s

the downward branch. This indicates both the increase in the nickel content in the film and the qualitative changes associated with inclusion of metal hydroxides in the film. With further cathodic polarization, when the potential is delayed at −1.3 V (curve 3), the current has time to increase even more. Moreover, on the developed surface of the electrode, the cathodic part of backward branch proceeds at more positive potentials. The amount of deposit that dissolves in the anode peaks increases, but the ratio of the peak area is practically unchanged, which indicates a small change in the composition of films obtained on the upward brunch after limiting cathodic current.

Two types of coatings were formed in the pyrophosphate-ammonia electrolyte: (i) Ni–Cu alloy, which was deposited at −1.175 V, and (ii) multilayer coating consisting of alternating layers of alloy and metal hydroxide layers. The alloy layers of the multilayer coating were deposited at −1.175 V, the metal hydroxide layers were obtained at −1.35 V. The characteristics of the obtained coatings are given in Table 1. In contrast to the short-time delay of the potential under conditions of CVA, with a longer potentiostatic electrodeposition of alloy layer (see curve 2 in Fig. 2), the current increases that lead to increasing in the development of film surface. The surface develops faster during multilayer coating formation due to the higher values of deposition currents of metal hydroxide layers deposition (curve 1).

The coatings electrodeposition was stopped when the current of the alloy deposition was 24 mA in both cases, i.e., when the electrochemically accessible for deposition surface areas of both coatings were equal. The geometric surface area of the electrodes with coating increased to 0.322 cm^2 in the case of the alloy coating and to 0.204 cm^2 in the case of a multilayer coating, which is proportional to the amount of electricity Q spent for coating deposition (see Table 1).

Table 1 Parameters of the coatings

Parameters	Coating alloy	Multilayer
Number of layers	1	9 + 8
Amount of electricity (C)	16.27	10.85
Geometrical surface area (cm^2)	0.322	0.204
Real surface area (cm^2)	0.95	1.61

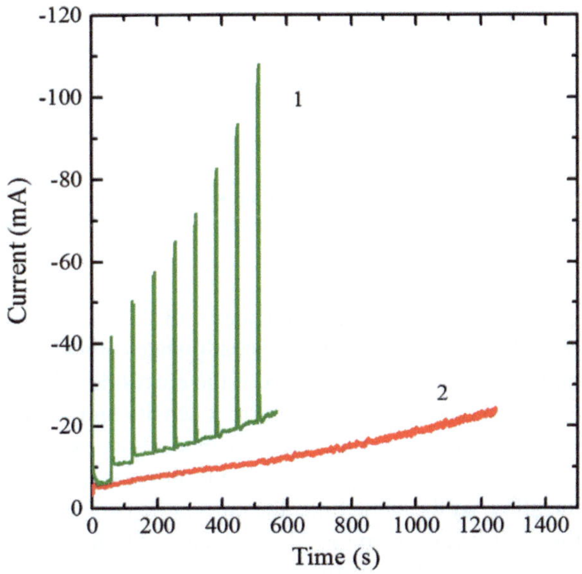

Fig. 2 Chronoamperograms of Ni–Cu alloy deposition at −1.175 V (2) and multilayer coating deposition at periodical potential change from −1.175 V to −1.35 V (1)

The surface properties of the obtained coatings were stabilized by cyclic voltammetry in a 5 M solution of NaOH in the range of potentials from open-circuit potential to −1.4 V (Fig. 3). The hydroxy compounds of the coatings are reduced at the initial part of the curves, mainly in the first cycle. It is more significant on the surface of the multilayer coating (Fig. 3a). With an increase in the number of cycles, the rate of hydrogen evolution decreases due to an increase in the degree of screening of the surface by bubbles.

The change in current value also decreases. By the 50th cycle, surface properties are stabilized. The current at the boundary of the cycling at −1.4 V in the case of multilayer coating (Fig. 3b) is approximately twice higher as compared to the case of alloy coating (Fig. 3a), despite the fact that its geometric area of sample with alloy coating is less by 1.6 times (see Table 1). Additionally, as a result of cycling the current is reduced by 40.3 and 34.8% on the surface alloy and multilayer coating, respectively.

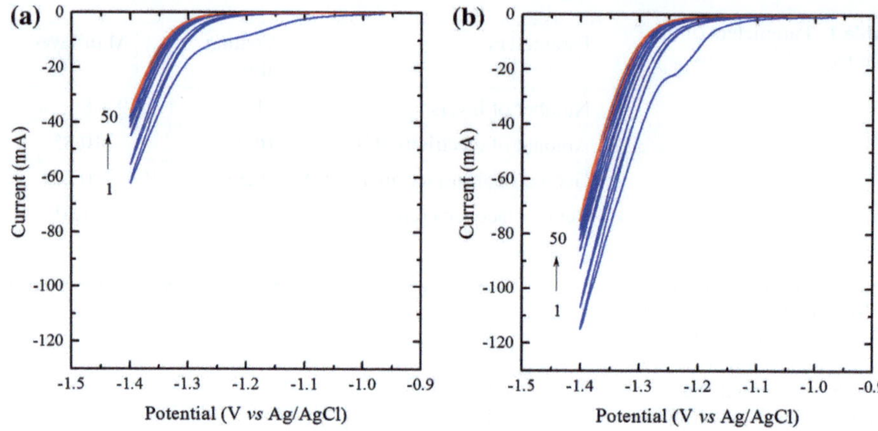

Fig. 3 The 1st, 2nd, 10th, 20th, 30th, 40th and 50th cycles of CVA in 5 M solution of NaOH on electrodes with alloy (**a**) and multilayer (**b**) coatings. The potential scan rate is 50 mV s^{-1}

The electrocatalytic performance of the electrodes was determined using a 5 M NaOH solution based on their steady-state cathodic polarization curves (Fig. 4), normalized to the geometric area of the electrodes. The overvoltage value of hydrogen evolution reaction at a fixed current density and current density for fixed overvoltage value are the important characteristics of the electrode. Figure 4 illustrates the higher rate of HER on the surface of the multilayer coating as compared to the surface of the alloy. At the same time, the current density of HER at overvoltage of 200 mV on the multilayer coating is 23.8 mA cm^{-2} (Table 2), which is slightly higher than for metallurgical alloys (from 9.1 mA cm^{-2} to 20.6 mV cm^{-2} in 6 M KOH, depending on the alloy composition [5]), as well as in oxygen-free 1 M KOH solution on coatings obtained in sulfate-citrate electrolytes by galvanostatic deposition (21.87 mA cm^{-2} [3]) and on alloys obtained by brush plating (12.66–19.13 mA cm^{-2} [13]). The overvoltage of HER at 100 mA cm^{-2} on the alloy (301.3 mV, Table 2) corresponds to the value of HER overvoltage, for example, on a CuNiCo ternary alloy electrodeposited in a citrate electrolyte (from 288 mV to 317 mV depending on the alloy composition [21]), but on a multilayer coating it is significantly lower (233 mV, Table 2).

Improved characteristics of electrodes can be caused by their chemical composition and structure, the degree of increase in the real surface area during electrodeposition. To obtain comparable kinetic parameters of HER, the polarization curves of Fig. 4 were normalized to the real surface area and shown in semilogarithmic coordinates (Fig. 5). The real surface area was determined by comparing the value of the double layer capacitance calculated from cyclic voltammetry obtained on the studied electrodes with the capacity of the double layer on a smooth metal surface [22, 23]. The values of the real surface area of the electrodes (Table 1) indicate a higher degree of development of the multilayer coating surface, which explains the higher currents on the cyclic curves in Fig. 3, and higher current densities on the polarization curves in Fig. 4.

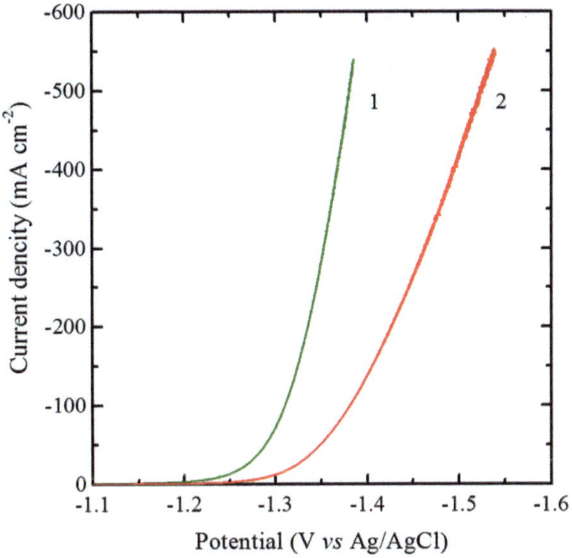

Fig. 4 Polarization curves in 5 M NaOH solution on electrodes with multilayer (1) and alloy (2) coating. The potential scan rate is 1 mV s^{-1}. Cathodic currents are normalized to the geometric area

Table 2 Parameters of HER on electrodes

Parameters	Coating alloy	Multilayer
Current density of HER at overvoltage of 200 mV (mA cm^{-2})	4.2	23.8
Overvoltage of HER at 100 mA cm^{-2} (mV)	301.3	230.6
Tafel coefficient b (mV)	72.8	67.6
Tafel coefficient a (mV)	497	370
j_0 (int) (mA cm^{-2})	0.147×10^{-3}	3.34×10^{-3}
j_0 (apper) (mA cm^{-2})	0.46×10^{-3}	16.37×10^{-3}

Kinetic parameters of HER on the studied electrodes are given in Table 2. The obtained values of Tafel coefficient b are close to those given in [5] for sintered copper-nickel alloys.

The intrinsic exchange current density j_0(int) for the electrode with multilayer coating is 3.34×10^{-3} mA cm^{-2}, which is 22.7 times higher as compared to the electrode with Ni–Cu alloy coating (see Table 2). This indicates that the activity of the multilayer coating in HER is higher than that of the alloy coating, not only due to a larger real surface area but also due to a higher intrinsic catalytic activity. It is comparable to the sintered NiCuMo alloys (j_0(int) from 2.49×10^{-3} to 4.03×10^{-3} mA cm^{-2} depending on the alloy composition [22]).

The activity of cathode j_0 (apper) with multilayer coating is about 16.37×10^{-3} mA cm^{-2}, which is 35.6 times higher than that for an electrode with Ni–Cu alloy coating (see Table 2). This is lower than for 3D hierarchical Cu/(Ni–Co) coating

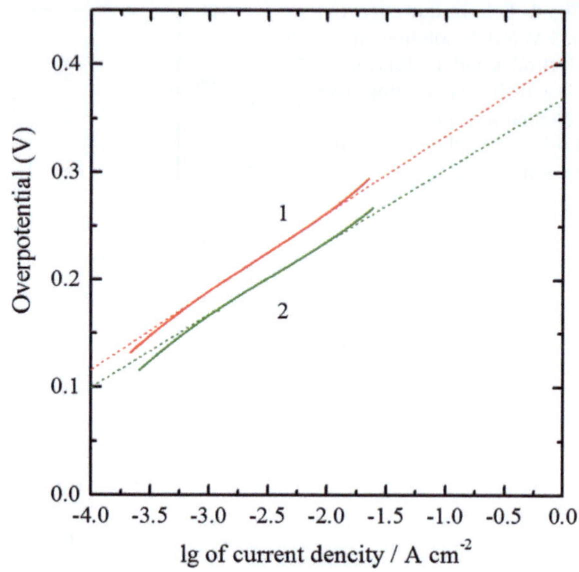

Fig. 5 Semilogarithmic polarization curves in 5 M NaOH solution on electrodes with alloy (1) and multilayer (2) coatings. Cathodic currents are normalized to the real area

(25–46 mA cm^{-2} [9]), but this is higher than those of 3D hierarchical Cu/Ni coating (14 mA cm^{-2} [9]) and Ni–Cu composite electrode with a hierarchical structure (1.47–2.69 mA cm^{-2} [1]).

Since the electrodes corrode and lose their activity and lifetime during shutdown electrolysis, the determination of their corrosion behavior is important for their practical applications. The corrosion potential of the unoxidized multilayer coating (curve 2 in Fig. 6) is approximately 50 mV more positive than the corrosion potential of the unoxidized alloy surface (curve 1 in Fig. 6). The increase in the contribution of the cathodic component to the corrosion process is compensated by a decrease in the activity of the anodic component of coating dissolution.

When the current is turned off, the potential of the studied cathodes in the alkaline non-deaerated electrolyte is slowly shifted to the positive side (the inset to Fig. 6). Its value stabilizes within half an hour in the potential region of the copper from alloy oxidation (see an anodic branch of curve 1). The corrosion potential of the oxidized surface of the multilayer coating (curve 2′) is by 65 mV more positive than the corrosion potential of the oxidized alloy surface curve 1′), and the corrosion current is almost an order of magnitude lower. This indicates the higher corrosion resistance of the multilayer coating in an alkaline solution as compared to the alloy coating.

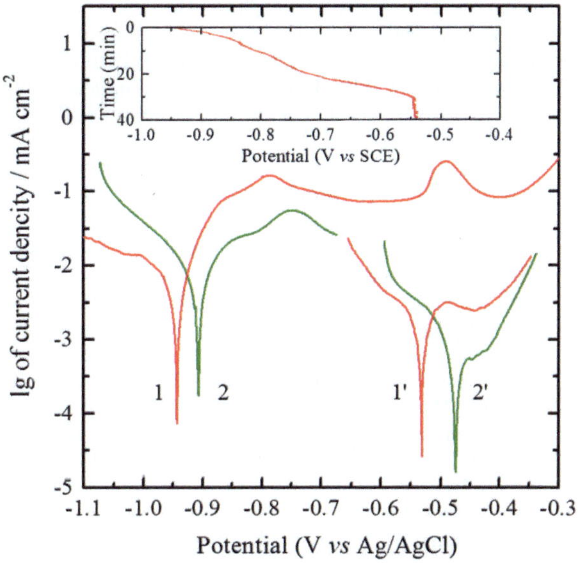

Fig. 6 Corrosion diagrams of electrodes with alloy (1, 1′) and multilayer (2, 2′) coatings in 5 M NaOH solution after current switching off (1, 2) and after staying in electrolyte without current for 40 min (1′, 2′). Insert—changing of open-current potential of electrode with alloy coating after current switching off

4 Conclusions

An electrode with a multilayer coating consisting of alternating layers of Ni–Cu alloy and metal hydroxide layers with a smaller geometrical surface area has a more developed real area as compared to the Ni–Cu alloy coating. Polarization studies have shown that this electrode also has higher corrosion resistance in an alkaline medium, higher catalytic activity in the hydrogen evolution reaction. Comparison of the characteristics of the multilayer coating obtained in the polyligand pyrophosphate-ammonia electrolyte with other nickel-containing coatings (including 3D structures) shown the perspectivity of use of electrodes with multilayer coating as a cathode in the alkaline water electrolysis.

References

1. Jiao P, Duan N, Zhang C et al (2012) The effect of cobalt ion on the hydrogen evolution reaction in sulfate solution. Int J Hydrogen Energy 31(40):17793–17800. https://doi.org/10.1016/j.ijhydene.2016.07.111
2. Cardoso DSP, Eugénio S, Silva TM et al (2015) Hydrogen evolution on nanostructured Ni–Cu foams. RSC Adv 5(54):43456–43461. https://doi.org/10.1039/C5RA06517H
3. Solmaz R, Döner A, Kardaş G (2008) Electrochemical deposition and characterization of NiCu coatings as cathode materials for hydrogen evolution reaction. Electrochem Commun 10(12):1909–1911. https://doi.org/10.1016/j.elecom.2008.10.011

4. Solmaz R, Döner A, Kardaş G (2009) The stability of hydrogen evolution activity and corrosion behavior of NiCu coatings with long-term electrolysis in alkaline solution. Int J Hydrogen Energy 34(5):2089–2094. https://doi.org/10.1016/j.ijhydene.2009.01.007
5. Wang K, Xia M, Xiao T et al (2017) Metallurgically prepared NiCu alloys as cathode materials for hydrogen evolution reaction. Mater Chem Phys 186:61–66. https://doi.org/10.1016/j.matchemphys.2016.10.029
6. Ngamlerdpokin K, Tantavichet N (2014) Electrodeposition of nickel-copper alloys to use as a cathode for hydrogen evolution in an alkaline media. Int J Hydrogen Energy 39(6):2505–2515. https://doi.org/10.1016/j.ijhydene.2013.12.013
7. Ahn SH, Park HY, Choi I et al (2013) Electrochemically fabricated NiCu alloy catalysts for hydrogen production in alkaline water electrolysis. Int J Hydrogen Energy 38(31):13493–13501. https://doi.org/10.1016/j.ijhydene.2013.07.103
8. Wang C, Li W, Lu X et al (2012) Facile synthesis of porous 3D CoNiCu nano-network structure and their activity towards hydrogen evolution reaction. Int J Hydrogen Energy 37(24):18688–18693. https://doi.org/10.1016/j.ijhydene.2012.09.149
9. Wang N, Hang T, Chu D, Li M (2015) Three-dimensional hierarchical nanostructured Cu/Ni-Co coating electrode for hydrogen evolution reaction in alkaline media. Nano-Micro Lett 7(4):347–352. https://doi.org/10.1007/s40820-015-0049-1
10. Chang JK, Hsu SH, Sun IW, Tsai WT (2008) Formation of nanoporous nickel by selective anodic etching of the nobler copper component from electrodeposited nickel-copper alloys. J Phys Chem C 112(5):1371–1376. https://doi.org/10.1021/jp0772474
11. Zhang J, Baró MD, Pellicer E, Sort J (2014) Electrodeposition of magnetic, superhydrophobic, non-stick, two-phase Cu–Ni foam films and their enhanced performance for hydrogen evolution reaction in alkaline water media. Nanoscale 6(21):12490–12499. https://doi.org/10.1039/C4NR03200D
12. Zhao X, Fuji M, Shirai T et al (2011) Electrocatalytic evolution of hydrogen on the NiCu/Al_2O_3/nano-carbon network composite electrode. J Mater Sci 46(13):4630–4637. https://doi.org/10.1007/s10853-011-5365-3
13. Yin Z, Chen F (2014) Electrochemically fabricated hierarchical porous Ni $(OH)_2$/NiCu electrodes for hydrogen evolution reaction. Electrochim Acta 117:84–91. https://doi.org/10.1016/j.electacta.2013.11.102
14. Maizelis A, Bairachniy B (2017) Electrochemical formation of multilayer SnO_2–Sb_xO_y coating in complex electrolyte. Nanoscale Res Lett 12(1):119. https://doi.org/10.1186/s11671-017-1902-6
15. Maizelis A, Bairachniy B (2017) Electrochemical formation of multilayer metal and metal oxide coatings in complex electrolytes. In: International conference on nanotechnology and nanomaterials. Springer, Cham, pp 557–572. https://doi.org/10.1007/978-3-319-56422-7_41
16. Maizelis AA, Bairachniy BI, Trubnikova LV, Savitsky BA (2012) The effect of architecture of the Cu/(Ni–Cu) multilayer coatings on their microhardness. Funct Mater 19(2):238–244
17. Maizelis AA, Bairachnyi BI, Tul'skii GG (2018) Contact displacement of copper at copper plating of carbon steel parts. Surf Eng Appl Electrochem 54(1):12–19. https://doi.org/10.3103/S1068375518010106
18. Maizelis AA, Tul'skii GG, Bairachnyi VB, Trubnikova LV (2017) The effect of ligands on contact exchange in the NdFeB–Cu^{2+}–$P_2O_7^{4-}$–NH_4^+ system. Russ J Electrochem 53(4):417–423. https://doi.org/10.1134/S1023193517040085
19. Maizelis AA (2017) Voltammetric analysis of phase composition of Zn-Ni alloy thin films electrodeposited under different electrolyze modes. In: 2017 IEEE 7th international conference nanomaterials: application properties (NAP-2017). IEEE, Piscataway, pp 02NTF13-1-02NTF13-5. https://doi.org/10.1109/NAP.2017.8190373
20. Maizelis A, Bairachny B (2017) Voltammetric analysis of phase composition of Zn–Ni alloy thin films electrodeposited from weak alkaline polyligand electrolyte. J Nano- Electron Phys 9(5):1–7. https://doi.org/10.21272/jnep.9(5).05010

21. Goranova D, Lefterova E, Rashkov R (2017) Electrocatalytic activity of Ni–Mo–Cu and Ni—Co–Cu alloys for hydrogen evolution reaction in alkaline medium. Int J Hydrogen Energy 42(48):28777–28785. https://doi.org/10.1016/j.ijhydene.2017.10.002
22. Yu L, Lei T, Nan B et al (2015) Mo doped porous Ni–Cu alloy as cathode for hydrogen evolution reaction in alkaline solution. RSC Adv 5(100):82078–82086. https://doi.org/10.1039/C5RA15612B
23. Xia M, Lei T, Lv N, Li N (2014) Synthesis and electrocatalytic hydrogen evolution performance of Ni–Mo–Cu alloy coating electrode. Int J Hydrogen Energy 39(10):4794–4802. https://doi.org/10.1016/j.ijhydene.2014.01.091

Effect of Nano-Structured Factors on the Properties of the Coatings Produced by Detonation Spraying Method

L. Markashova, Yu. Tyurin, O. Berdnikova, O. Kolisnichenko, I. Polovetskyi and Ye. Titkov

Abstract Surface treatment by detonation spraying of composite coatings significantly increases the service life of structures. The demanded direction of application of this method is spraying of powders of various systems. The purpose of this paper is to study the structure (phase composition, microhardness, grain and subgrain structure, dislocation density, dispersed micro- and nano-particles, their composition, and distribution), and the properties of metal-ceramic coatings: Al_2O_3–Ti; Al_2O_3–Al; $ZrSiO_4$ sprayed on various substrates materials (aluminum, titanium). The investigations of structural-phase state of coatings were carried out at all the structural levels using comprehensive methodological approach including optical metallography, analytical scanning electron microscopy, X-ray structural phase analysis, as well as transmission micro diffraction electron microscopy. As a result of the carried out work the experimental data on the full complex of structural and phase parameters of coatings, sprayed at different modes of cumulative-detonation spraying, were obtained. The obtained results showed the prospects of using multi-chamber detonation spraying, which ensures the necessary complex of mechanical properties of coatings by forming the most favorable structure. The surface treatment of metals by the detonation spraying of coatings makes it possible to obtain composite surface layers with high operational properties. The matrix of such coatings is characterized by a significant dispersion of the substructure, the formation of strengthening phases of nanoscale dimensions and a uniform distribution of dislocation density.

Keywords Coatings · Detonation spraying · Structure · Phase composition · Nano-particles · Dislocation density · Hardening · Local internal stresses · Crack resistance

---

L. Markashova · Yu. Tyurin · O. Berdnikova (✉) · O. Kolisnichenko · I. Polovetskyi · Ye. Titkov
E.O. Paton Electric Welding Institute of the NAS of Ukraine, Kiev, Ukraine
e-mail: omberdnikova@gmail.com

© Springer Nature Singapore Pte Ltd. 2019

A. D. Pogrebnjak and V. Novosad (eds.), *Advances in Thin Films, Nanostructured Materials, and Coatings*, Lecture Notes in Mechanical Engineering,
https://doi.org/10.1007/978-981-13-6133-3_11

1 Introduction

Surface treatment by detonation spraying of composite coating significantly increases the service life of structures, especially, due to its wear, heat and corrosion resistances. This makes them indispensable in the manufacture of structures of responsible use in heavy machine building, shipbuilding, transport, the chemical and aviation industries. The demanded direction of application of the method is spraying of powders of various systems. At the E.O. Paton Electric Welding Institute of the NAS of Ukraine, the technology and equipment for detonation spraying were developed by which high-quality coatings are formed with high materials utilization factor and productivity [1, 2].

Multi-chamber detonation device is designed for the formation of high-quality powder materials on the part surface (Fig. 1a). This device realizes a mode of detonation burning of the gas mixture in specially profiled chambers. Accumulation of burning energy from chambers forms an area of compression of combustion products at the input in a nozzle—cylindrical shank. Expansion and outflow of combustion products through the nozzle is carried out in the form of a high-speed jet, which is used for acceleration and heating of powder material (Fig. 1b, c). Modeling of combustion processes in double-chamber structure of the device showed that velocity of combustion products from the open part of the nozzle reaches 1.8 km/s as a result of the accumulation of energy of detonation burning from two chambers. This allows accelerating ceramic powder material to 1.2–1.5 km/s rate [3].

The peculiarity of multi-chamber detonation device is a continuous feed of components of combustion mixture and powder by transporting gas, and gas-dynamic dosing of powder, and synchronizing the feed of powder dose in combustion products jet. Standard powders, used in geothermal technologies, as well as powder mixtures, can be feed into the device. Gas-dynamic systems for dosing and powder feed provide high accuracy and synchronizing of Multi-chamber detonation device work and powder feeder. This guarantees material use ratio for metals up to 90%, metal-ceramic to 80%, ceramics to 70%. The investigation show [4] that a cumulative-detonation technology provides formation of dense coating materials having the next porosity for metals $<0.3\%$, metal-ceramic $<0.5\%$, ceramic $<0.1\%$. The efficiency of coating formation for metal up to 3 kg/h, metal-ceramic to 2 kg/h, ceramics to 1 kg/h.

2 General Information

2.1 Objectives

The purpose of this paper is to study the structure (phase composition, microhardness, grain and subgrain structure, dislocation density, dispersed micro- and nano-particles, their composition, and distribution), and the properties of metal-ceramic coatings:

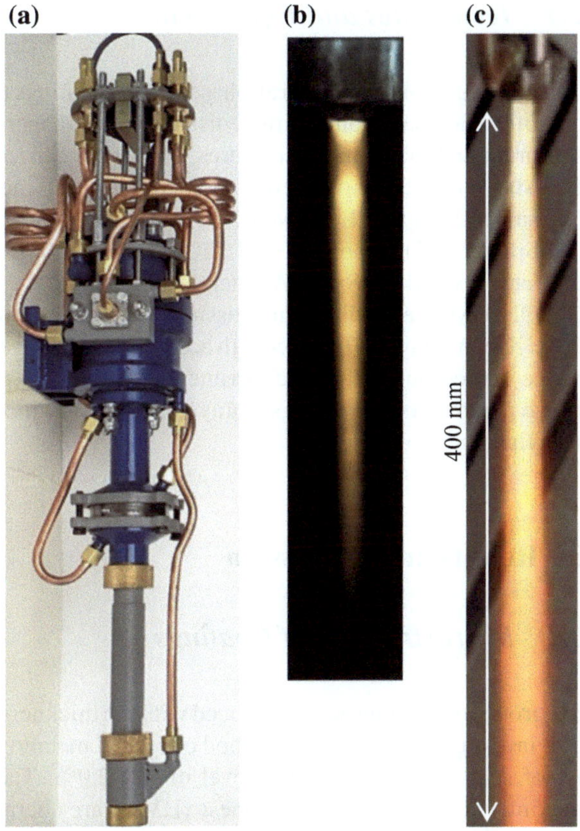

Fig. 1 Multi-chamber detonation device (**a**). Outflow of the supersonic pulse jet from the nozzle of the detonation device: without powder (**b**); with powder (**c**)

Al_2O_3–Ti; Al_2O_3–Al; $ZrSiO_4$ sprayed on various substrates materials (aluminum, titanium).

2.2 Materials and Conditions for Obtaining the Coatings

The coatings were sprayed using powders of $ZrSiO_4$ (No. 1; 2) with fractional composition 7–60 μm with Co–Cr–Al–Y (No. 2) underlayer 7–60 μm and mechanical mixture of initial Al_2O_3 (No. 4; 5; 6) powder (H.C. Starck Company: AMPERIT® 740.0), fraction composition 5–22 μm with additives (5%) of pure powders of Al or Ti (5–22 μm).

Detonation spraying mode: detonation frequency 20 Hz; distance to the sample 55 mm; travel speed 1500 mm/min with the same number of passes; the length/diameter ratio of the gun barrel is 500/16 mm and the combustible gas to the oxidizer 5.0; 5.4 ($ZrSiO_4$) and 5.0; 5.8 (Al_2O_3). The sample size is 15 × 10^3 mm.

2.3 *Techniques and Equipment*

The investigations of structural-phase state of coatings (micro-hardness, volume fraction of pores, phase composition, distribution of dispersed phases, character of grain, sub-grain and dislocation structures, etc.) were carried out at all the structural levels using comprehensive methodological approach including optical metallography (Versamet-2, Japan; Leco-M400, USA), analytical scanning electron microscopy (Philips SEM-515, Holland and JEOL JAMP 9500F with energy dispersion spectrometer Oxford EDS INCA Energy 350, Japan), X-ray structural phase analysis (DRON-UM1), as well as transmission micro diffraction electron microscopy (JEM-200CX, company <<JEOL>> with accelerating voltage of 200 kV, Japan). As a result of the carried out work the experimental data on the full complex of structural and phase parameters of coatings, sprayed at different modes of cumulative-detonation spraying, were obtained.

3 Results and Discussion

3.1 *Microstructure of Coatings*

Six groups of coatings were produced with the thickness (δ) of up to 250 μm (Table 1). The investigations using the method of optical metallography showed that the porosity of such coatings is at the level of 0.7–2.9%. The minimum porosity and the maximum integrated microhardness ($HV_{0.3}$) are characteristic for coatings No. 3, 4 produced using the powders Amperit 740.0 with the additives of 3–5% Ti.

3.2 *Phase Analysis of Coatings*

Using X-ray diffraction phase analysis of the produced coatings No. 1, 2 it was revealed the formation of coatings of identical phase composition (ZrO_2, SiO_2, and residual $ZrSiO_4$) at approximately equal in the content of the forming phase components (61–64, 33, and 3–6%).

Analogous diffraction phase analysis for coatings No. 3 and 4, it was revealed that the use of powders Amperit 740.0 + 3% Ti (No. 3) and Amperit 740.0 + 5% Ti (No. 4) promotes the formation of coatings of identical phase composition (γ-Al_2O_3; α-Al_2O_3; $AlTi_3$) at approximately equivalent content of the forming phase components. However, in the coatings Amperit 740.0 + 5% Ti (No. 4) integrated microhardness ($HV_{0.3}$) is by 17% increased (from 8900–10,990 to 9660–13,770 MPa) as compared to coating Amperit 740.0 + 3% Ti (No. 3), Table 1.

The spraying of powders Amperit 740.0 + 3% Al (No. 5) and Amperit 740.0 + 5% Al (No. 6) promotes the formation of coatings also of similar phase composition (γ-

Table 1 Results of investigations of coating structure

No.	Powder substrate	δ (μm)	$HV_{0.3}$ (MPa)	Phase composition
1	$ZrSiO_4$ titanium	300	$\frac{5270}{7360}$	53.44% $ZrO_2^{(T)}$ 10.12% $ZrO_2^{(M)}$ 33.0% $SiO_2^{(A)}$ 3.35% $ZrSiO_4^{(T)}$
2	$ZrSiO_4$ (Co–Cr–Al–Y*) aluminium	160 120*	$\frac{7100}{8560}$ $\frac{5760*}{6180}$	49.3% $ZrO_2^{(T)}$ 11.7% $ZrO_2^{(M)}$ 33.0% $SiO_2^{(A)}$ 5.8% $ZrSiO_4^{(T)}$
3	Al_2O_3 + 3% Ti titanium	200	$\frac{8900}{10{,}990}$	69.1% $Al_{2.66}O_4^{(K)}$ 17.8% α-$Al_2O_3^{(R)}$ 13.0% $AlTi_3^{(H)}$
4	Al_2O_3 + 5% Ti titanium	230	$\frac{9660}{13{,}770}$	67.0% $Al_{2.66}O_4^{(K)}$ 18.0% α-$Al_2O_3^{(R)}$ 15.0% $AlTi_3^{(H)}$
5	Al_2O_3 + 3% Al aluminium	250	$\frac{8900}{10{,}250}$	69.0% $Al_{2.66}O_4^{(K)}$ 15.2% α-$Al_2O_3^{(R)}$ 15.8% $Al^{(K)}$
6	Al_2O_3 + 5% Al aluminium	225	$\frac{7900}{10{,}520}$	69.0% $Al_{2.66}O_4^{(K)}$ 15.2% α-$Al_2O_3^{(R)}$ 15.8% $Al^{(K)}$

*—underlayer; $^{(K)}$—cubic lattice; $^{(R)}$—rhombohedral lattice; $^{(H)}$—hexagonal lattice $^{(T)}$—tetragonal lattice; $^{(M)}$—monoclinic lattice; $^{(A)}$—amorphous-nanocrystalline

Al_2O_3; α-Al_2O_3; Al) at approximately the same content of forming phase components and levels of integrated microhardness, Table 1.

3.3 *Fine Structure of Coatings*

Results of transition electron microscopic examinations provided the opportunity to study features of fine structure of coatings and its parameters: changes in density and character of dislocation density (ρ) in different structural components (in the inner volumes and along the structural boundaries); character of the forming substructure, its parameters; size of subgrain (d_S); diameter of particles (d_P) of phase precipitations; effective distances between the forming phases (λ), etc. (Table 2).

In the case of using $ZrSiO_4$ powder, the particle size of nanosized phase precipitates (10–60 nm) in the surface layers of coatings No. 2 (Fig. 2b) decreases by a factor of 1.3 compared to No. 1 (Fig. 2a), and in the underlayer material it is the same (20–100 nm) (Fig. 2c). The distance between the forming

Table 2 Parameters of fine structure of coatings

Structural parameters	Coatings			
	No. 1	No. 2	No. 4	No. 6
d_s (μm)	0.2/0.5	0.3/0.5	0.1/0.4	0.1/0.6
d_p (nm)	20/100	10/60	20/100	10/80
λ (nm)	10/50	10/50	10/30	10/50
ρ (s m^{-2})	(5–9) × 10^9 coating	6 × 10^9–10^{10} coating	(3–5) × 10^9 coating	(2–3) × 10^9 coating
	(2–4) × 10^{10} substrate	(4–6) × 10^9 substrate	(6–7) × 10^{10} substrate	(5–6) × 10^9 substrate

dispersed phases (ZrO_2, SiO_2) is almost the same (10–50 nm), which characterizes the uniform distribution of the phases formed in the coating matrix. The density of dislocations on the outer surface of the coatings: (5–9) × 10^9 cm^{-2} (No. 1) and 6 × 10^9–10^{10} cm^{-2} (No. 2). While at the boundary of the base material (2–4) × 10^{10} cm^{-2} (No. 1) and ρ = (4–6) × 10^9 cm^{-2} (No. 2).

For the coatings Al_2O_3 with the most favorable structural and phase changes (high microhardness, minimum porosity, etc.), namely—Amperit 740.0 + 5% Ti (No. 4, Fig. 2d) and for comparison—Amperit 740.0 + 5% Al (No. 6, Fig. 2e) the following regularities were revealed.

In the case of using Amperit 740.0 + 5% Ti powder (No. 4), the particle size of the phase precipitates (10–100 nm) in the surface layers of coatings decreases by a factor of 2 compared to the coatings Amperit 740.0 + 5% Al (No. 6). Also, the distance between the emerging dispersed phases (up to 10–30 nm) decreases in 2.0–2.3 times, which characterizes the increase in the volume fraction in the matrix of the phases formed. There is a refinement (by 1.4 times) of the substructure (0.1–0.4 μm) with increasing dislocation density on the outer surface of the coatings: from (2–3) × 10^9 cm^{-2} (No. 6) to (3–5) × 10^9 cm^{-2} (No. 4). While the dislocation densities in the coatings at the contact zone are (5–6) × 10^9 and (6–7) × 10^{10} cm^{-2}, respectively.

4 Analytical Evaluations of Coating Operational Properties

The executed complex of experimental investigations at all the structural levels allowed to carry out analytical evaluations of the (differentiated) contribution of different structural and phase factors and parameters, formed in the coatings, in change of strength characteristics and to determine the structural factors influencing the on character and distribution of local inner stresses, which are the potential sources of incipience of cracks [5–7].

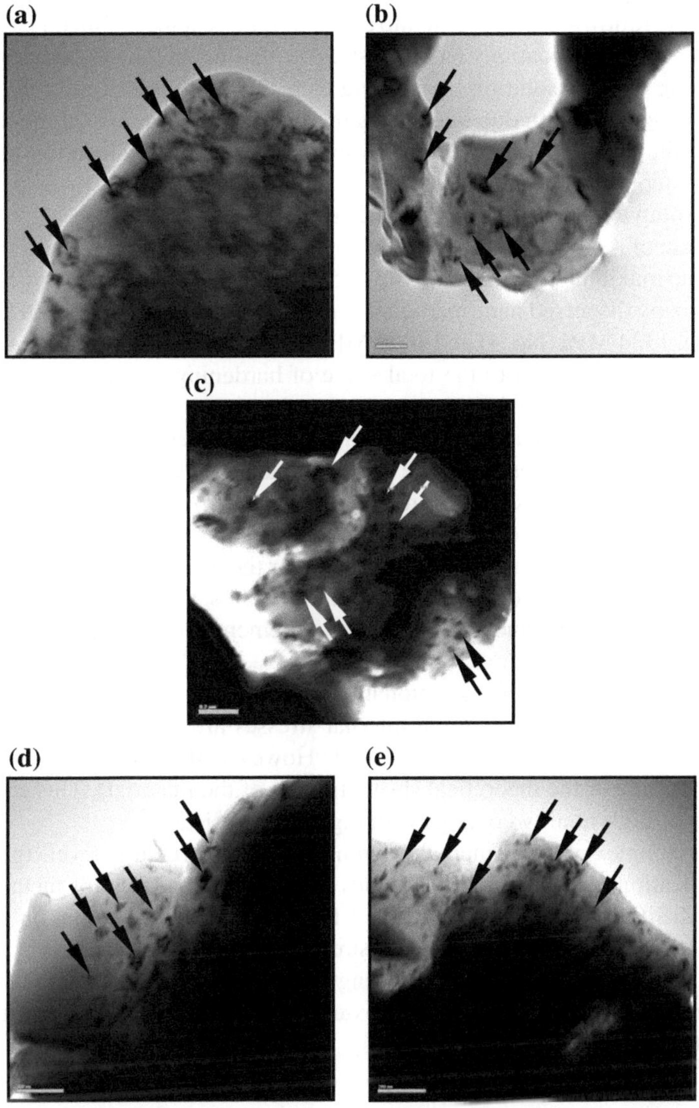

Fig. 2 Fine structure: coating No. 1 (**a**); coating No. 2 (**b**, **c**—underlayer); coating No. 4 (**d**); coating No. 6 (**e**). Arrows denote particles of nanoscale phases

At that, the following were taken into account: the resistance of metal lattice to the movement of free dislocations (friction stress of lattice or Peierls-Nabarro stress); the hardening of solid solution by alloying elements and impurities (solid solution hardening); the hardening due to changes in grain and sub-grain size (dependences of Hall-Petch, grain boundary and substructure hardening); dislocation hardening, caused by interaction between dislocations; the hardening, caused by dispersed particles according to Orowan (dispersion hardening) [8–17].

In the case of Amperit 740.0 + 5% Ti (No. 4) and Amperit 740.0 + 5% Al (No. 6) powders, the maximum contribution (up to 56%) to hardening is made by hardening of the matrix by dispersed nano-particles of phase precipitates (dispersion hardening by Orovan): 1334 MPa (No. 4) and 1070 MPa (No. 6). In the case of $ZrSiO_4$ powder, the maximum contribution to the total value of hardening of phase separations by dispersed particles is up to 44%.

Calculating-analytical methods for determining the level of local internal stresses have made it possible to estimate the crack resistance of coatings. The dislocation structure was taken into account in studies of the fine structure. From the analysis of various approaches to the determination of the mechanisms of crack initiation and destruction of materials, an estimate was made for the level of local internal stresses on the basis of the dislocation theory of crystalline solids, which relates the formation of internal stresses to the nucleation and rearrangement of a dislocation structure [5, 6, 18].

In the investigated options of detonation spraying, a low level of local internal stresses is observed. The maximum internal stresses are formed in the area of the coating-substrate interface in coating No. 1. However, their level does not exceed 960 MPa (or 0.22 of the theoretical shear strength of the material). This ensures the production of high-quality coatings with high crack resistance.

Carried complex of experimental examinations at all structural levels (grain, sub-grain, dislocation, dispersed nano-particles of phase precipitates) of metal-ceramic coatings ($ZrSiO_4$, Al_2O_3–Ti/Al) obtained by multi-chamber detonation spraying, shown that the greatest influence on their strength and crack resistance is provided by uniform distribution of the emerging strengthening dispersed phases, the formation of a subgrain structure, the absence of extended and dense dislocation clusters—concentrators of local internal stresses

5 Summary

Based on experimental data obtained at various structural levels, analytical assessments of the influence of specific structural parameters (phase composition, grain, subgrain, and dislocation structures) on the mechanical properties and crack resistance of the coatings under investigation were made.

Structural factors influencing the character and distribution of local internal stresses, which are potential sources of crack initiation in structural microregions, are determined.

Analytical estimates show that the most significant contribution to the operational properties (strength, crack resistance) of the coatings under investigation is made by: uniform distribution of the forming strengthening dispersed phases, refinement of grain and subgrain structures in the absence of extended and dense dislocation accumulation-concentrators of local internal stress.

The obtained results showed the prospects of using multi-chamber detonation spraying, which ensures the necessary complex of mechanical properties of coatings by forming the most favorable structure.

References

1. Tyurin Y, Kolisnichenko O, Duda I (2009) Comparative analysis of cumulative-detonation and HVOF thermal spray installations efficiency. Hardening Technol Coat 5:27–33
2. Tyurin Y, Kolisnichenko O, Poleshuk M (2009) Cumulatively-detonation device for thermal spray coatings. In: Proceedings of the 9th international conference films and coatings, St. Petersburg, 26–29 May 2016, pp 44–46
3. Kolisnichenko OV, Tyurin YN, Tovbin R (2017) Efficiency of the coating deposition process using multi-chamber detonation device. Paton Weld J 10:18–23. https://doi.org/10.15407/tpwj2017.10.03
4. Tyurin Y, Kolisnichenko O, Jia J et al (2016) Characterization and properties of cermet coatings formed by a new multi-chamber detonation sprayer. In: International thermal spray conference (ITSC 2016), Shanghai/P.R. China, 10–12 May 2016, 324, DVS-Berichte, pp 862–865
5. Markashova LI, Tyurin YN, Kolisnichenko OV et al (2017) Effect of structure on properties of Al_2O_3 and Al (or Ti) mechanical mixture coatings produced by multichamber detonation spraying method. Paton Weld J 9:27–32. https://doi.org/10.15407/tpwj2017.09.05
6. Markashova LI, Poznyakov VD, Gaivoronskii AA et al (2012) Estimation of the strength and crack resistance of the metal of railway wheels after long-term operation. Mat Sci 47(6):799–806. https://doi.org/10.1007/s11003-012-9458-1
7. Markashova L, Berdnikova O, Bernatskyi A et al (2017) Physical and mechanical properties of high-strength steel joints produced by laser welding. In: Young scientists forum on applied physics and engineering (YSF), IEEE International, p 88–91. https://doi.org/10.1109/YSF.2017.8126596
8. Suzuki HO (1967) On the yield strength of polycrystalline metals and alloys. In: Structure and mechanical properties of metals. Metallurgiya, Moscow, pp 255–260 [In Russian]
9. Ashby MF (1972) On the Orowan Stress. In: Physics of strength and plasticity. Metallurgiya, Moscow, pp 88–107 [In Russian]
10. Gol'dshtein MI, Litvinov VS, Bronfin BM (1986) Metallophysics of high-strength alloys. Metallurgiya, Moscow [In Russian]
11. Conrad H (1973) Model of strain hardening for explaining the influence of grain size on the stress of metal flow. In: Gordienko LK (ed) Ultrafine grain in metals. Metallurgiya, Moscow, pp 206–219 [In Russian]
12. Armstrong RW (1973) Strength properties of metals with ultrafine grain. In: Gordienko LK (ed) Ultrafine grain in metals. Metallurgiya, Moscow, pp 11–40 [In Russian]
13. Petch NJ (1953) The cleavage strength of polycrystalline. J Iron Steel Inst 173(5):25–28
14. Orowan E (1954) Dislocation in metals. AIME, New York
15. Ashby MF (1983) Mechanisms of deformation and fracture. Adv Appl Mech 23:118–177
16. Kelly A, Nicholson RB (1963) Precipitation hardening. Pergamon, New York
17. Ebelling R, Ashby MF (1966) Yielding and flow of two phase copper alloys. Phil Mag 13(7):805–809
18. Panin VE, Likhachev VA, Grinyaeva YV (1985) Structural levels of deformation of solids. Nauka, Novosibirsk [In Russian]

Nanostructures in Welded Joints and Their Interconnection with Operation Properties

Liudmyla Markashova, Olena Berdnikova, Tetiana Alekseienko, Artemii Bernatskyi and Volodymyr Sydorets

Abstract The role of structural factors in ensuring optimal properties of materials and their operational reliability has been shown. Studying the phase composition and nanostructures in the welded joints (heat-affected zone, weld metal) of high-strength structural steel produced by advanced high-speed technologies hybrid laser-arc welding. Structural paramcters (dimension of grains and subgrains, dislocation density, nanoparticles) and phase changes in the welded joints were studied by using of analytical scanning electron microscopy, optical metallography, and for fine study the transmission electron microscopy was used as well. The most influential structural factors are the dispersing of martensite structure and bainite substructure, equable distribution of particles of structural phases and the absence of extended dislocation clusters—zones of crack incipience and propagation. Such substructure of welding joints of high-strength structural steel produced by laser-arc welding provides the high complex of strength properties and crack resistance.

Keywords Hybrid laser-arc welding · High-strength structural steel · Welded joint · Microstructure · Nanoparticles · Dislocation density · Mechanical properties · Local inner stresses

1 Introduction

Progressive modern technologies of laser [1–5] and hybrid laser-arc welding [6–10] are used today in current engineering for production of materials with specific service properties. Such technologies provide joints with minimum size of welds and heat affected zone in welding of dissimilar materials [11–13] as well as high-strength steels [4, 10, 14–19]. First experimental works on hybrid technologies, in which TIG welding was carried out with additional influence of laser beam, (Great Britain, USA,

---

L. Markashova · O. Berdnikova (✉) · T. Alekseienko · A. Bernatskyi · V. Sydorets
E.O. Paton Electric Welding Institute of the National Academy of Sciences of Ukraine, Kiev, Ukraine
e-mail: omberdnikova@gmail.com

© Springer Nature Singapore Pte Ltd. 2019

A. D. Pogrebnjak and V. Novosad (eds.), *Advances in Thin Films, Nanostructured Materials, and Coatings*, Lecture Notes in Mechanical Engineering,
https://doi.org/10.1007/978-981-13-6133-3_12

Germany, including together with Ukrainian scientists etc.) appeared in the 80th of the XX century. Today, specialists from Germany (Fraunhofer Institute, Chemnitz, Institute of applied beam technologies), Finland (Technological University, Lappeenranta), USA, China as well as Ukrainian scientists (E.O. Paton Electric Welding Institute) carry works on hybrid technologies and research of these processes as well as issue series of patents in the different countries [3, 6–10, 18].

To optimize the parameters of welding process most of technological experiments on hybrid welding is dedicated to interaction between arc and laser radiation [3, 7, 20–22], process of spatial stabilization of welding arc and localization of its effect by laser radiation [3, 7, 20], determination of effect of shielding gas [3, 20], arc power [3, 7, 20–22], welding speed and welding current on change of penetration depth [3, 7, 14–18, 20–22]. Investigation of peculiarities of hybrid process, their effect on weld formation, getting the advantages and eliminating the disadvantages of laser and arc constituents is now the main problem of technologists and researches. In comparisons with laser welding, which characterizes by high thermal locality of metal effect and high cooling rates, hybrid laser-arc welding allows reducing cooing rate and, respectively, effect the process of solidification in order to form in weld metal and heat affected zone the structures providing strength as well as crack resistance of welded joints [3, 7, 10, 20–22].

2 General Information

2.1 Objectives

The aim of the present work is determination of technological mechanisms of nanostructure formation in welded joints and detection of their interaction with a set of mechanical characteristics, i.e. ductility, strength and crack resistance, which define the service properties of welded joint. Peculiarities of structural changes in weld metal of high-strength steel joints produced by hybrid laser-arc welding were investigated at all structural level (from grain to dislocation) that allowed obtaining comprehensive and reliable result.

2.2 Welded Joints Materials and Their Fabrication Conditions

The investigations were carried out at butt-welded joints of high-strength structural steel with plate thickness up to 10 mm. Chemical elements composition of the structural high-strength steel: 0.183% C, 0.98% Mn, 2.07% Ni, 1.19% Cr, 0.08% V, 0.22% Mo, 0.33% Si.

Welded joints were manufactured at various speeds Vw = 20; 25; 30 mm/s of laser-arc welding. Powerful CO_2-laser (wavelength of laser radiation is 10.6 μm, focal spot diameter of laser beam is 0.4 mm) was a technological tool. No filler materials were use. In all modes, the laser radiation power was 4.4 kW. Shielding gas flow rate of the mixture 18% CO_2 + 82% Ar was 14 l/min. Such modes allow providing heat-affected zone (HAZ) metal cooling in 600–500 °C temperature interval with $W_{6/5}$ = 58; 61; 63 °C/s rate, respectively.

2.3 *Techniques and Equipment for Research*

Chemical elements composition and structural-phase state in the weld metal and in the heat-affected zone were studied optical metallography (OM), analytical scanning electron microscopy (SEM), and transmission micro diffraction electron microscopy (TEM) as well. Microscopes: JEM-200CX (TЭM, Japan, company JEOL); SEM-515 (Netherlands, company Philips); Versamet-2 (OM, USA, company Unitron).

3 Results and Discussion

3.1 *Results of Mechanical Tests*

Based on the results of mechanical tests it was determined that the highest indices of yield limit (σ_y) and tensile strength (σ_t) has the welds produced at V_W = 25 mm/s ($\sigma_{0.2}$ = 1138–1156 MPa; σ_t = 1319–1326 MPa). However, in this case there is a significant (1.6 times) reduction of ductility indices (ψ = 35.6–40.8%) of weld metal and impact toughness (KCV^{-40} = 50.4–65.7 J/cm^2) in HAZ metal zone in comparison with c V_W = 20 mm/s (KCV^{-40} = 79.9–93.5 J/cm^2) and V_W = 30 mm/s (KCV^{-40} = 66.5–96.7 J/cm^2), Table 1.

3.2 *Mircostructures of Welded Metal and HAZ*

As a result of the study of the structural-phase components (martensite M, upper bainite B_U, lower bainite B_L, grain size, microhardness) formed in the weld metal and the HAZ the following is revealed (Table 2).

It is determined that increase of welding speed from V_W = 20 to 25 mm/s changes phase composition of weld metal and HAZ of welded joints from bainite-martensite to martensite as well as varies relationship of structural constituents in HAZ metal of welded joints, namely volume fraction of B_L decreases (2–3 times) at rise (3 times) of M portion, that as a result provides the maximum indices of strength, however, leads to significant reduction (1.6 times) of ductility (ψ) in weld.

B_U (60–70%) structures are mainly formed in the case of increase of welding rate to $V_W = 30$ mm/s. Such structural changes can result in non-uniform level of mechanical properties in welding zone and decrease of welded joint crack resistance.

3.3 *Fine Structure of Welded Metal and HAZ*

Comprehensive TEM studies of the fine structure (the dislocation density in structural components, substructures in internal volumes of grains, etc.) have given such results. When welding speed is $V_W = 20$ mm/s, in the weld metal and the overheating area of HAZ the width of laths of the B-structures is 0.5–1.2 μm (in the weld, Fig. 1a,

Table 1 Mechanical properties of welded joints

V_W (mm/s)	$\sigma_{0.2}$ (MPa)	σ_t (MPa)	δ_5 (%)	Ψ (%)	KCV, J/cm² at t (°C)		
					Weld	Weld	HAZ
					+20 °C	−40 °C	−40 °C
20	851.8	1068.3	12.7	59.6	61.4	35.7	93.5
	963.6	1189.8	9.3	63.7	90.0	40.9	79.9
25	1138.1	1326.6	10.0	40.8	64.5	59.1	50.4
	1156.3	1319.7		35.6	58.6	45.2	61.7
30	991.5	1078.1	9.7	59.6	58.7	41.9	90.1
	982.1	1088.9	14.0		81.2	29.7	66.5

Table 2 Structure parameters (D_G—grain size; HV—microhardness; V_{fr}—volume fraction of phase components) of the welded joints

Zone	D_G (μm)	HV (MPa)	V_{fr} (%)		
			B_L	B_U	M
$V_W = 20$ mm/s					
Weld	30/120 × 170/350	3800–4010	60	20	20
HAZ	30/60	3540–3900	80	5	15
$V_W = 25$ mm/s					
Weld	30/80 × 150/300	3800–4010	20	15	65
HAZ	25/50	3540–3900	30	20	50
$V_W = 30$ mm/s					
Weld	20/80 × 150/250	3360–3940	10	60	30
HAZ	20/40	3360–4010	20	70	50

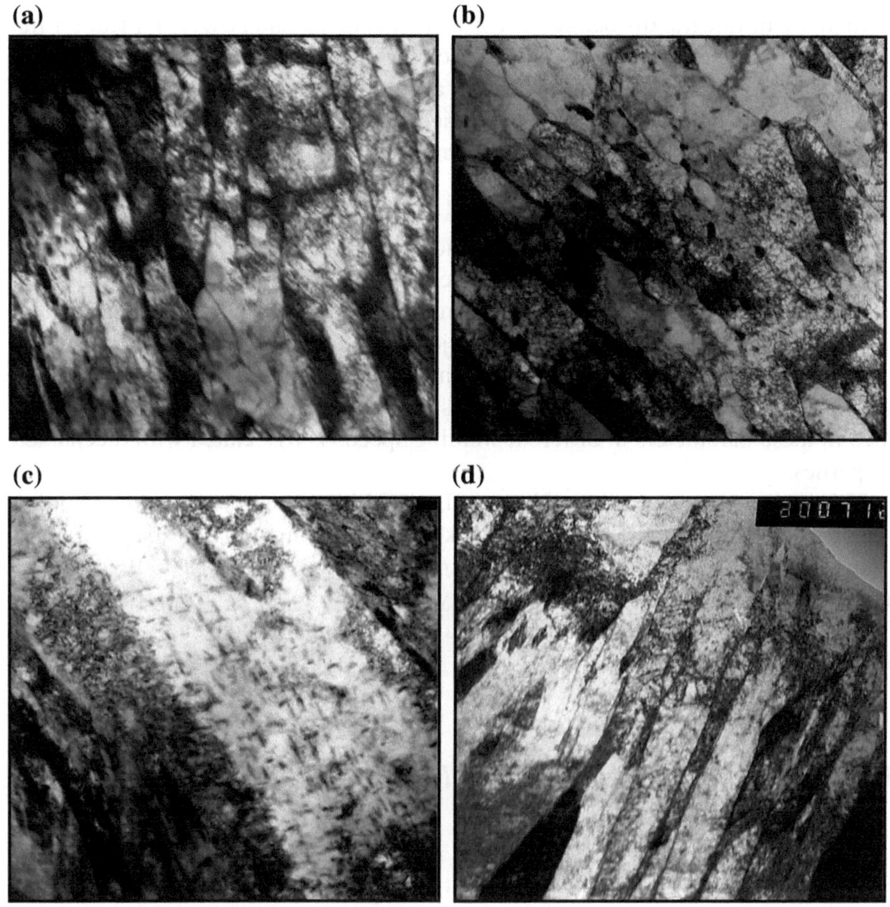

Fig. 1 The microstructure of lower bainite (**a**, **b**, ×30,000), martensite (**c**, ×30,000) and upper bainite (**d**, ×20,000) at welding speeds: **a–c**—20 mm/s; **d**—30 mm/s

b) and 0.4–0.9 μm (in the HAZ), respectively. Width of laths of the M-structures of metal in weld zone is 0.7–1.7 μm, Fig. 1c.

Formation of inner substructure of $d_S = 0.1–0.3$ μm size with uniform distribution of dislocation volume density up to $\rho = (4–6) \times 10^{10}$ cm^{-2} (Fig. 1a, b) is observed in weld metal at $V_W = 20$ mm/s in B_L lathes. In M structure dislocation density makes $\rho = (8 \times 10^{10}–1 \times 10^{11})$ cm^{-2}. B_L fragmentation is typical for joint HAZ metal, refinement of lath structure of B_L and M is observed (10–20%).

Investigation of nanostructures in the welding zone at laser-arc welding show that we have nanoparticles of carbide phases Fe_3C with a size from 30 up to 100 nm in the substructure of the laths of tempered martensite and lower bainite (Fig. 1a–c). The distance between the forming dispersed phases is $\lambda_P = 40–100$ nm.

Increase of integral dislocation density in weld metal and HAZ is typical for fine structure of welding zone metal in the case of $V_W = 30$ mm/s, mainly B_U structure is formed (Fig. 1d). Extended dislocation accumulations up to $\rho = (1–1.5) \times 10^{11}$ cm^{-2} is mainly formed along B_U boundaries in weld metal along grain boundaries. This creates a gradient of dislocation density in such element of structures and, respectively, gradient of inner stresses.

If we consider this from the point of view of uniform refinement of structural parameters, absence of gradients along the grain structure and microhardness, the most favorable structure of the studied welded joints is obtained at welding speed of 20 mm/s. Such speed provides and guarantees the uniform level of mechanical properties and crack resistance of welded joints [4].

The study of the fine structure determined the quantitative parameters of the nanostructures: upper and lower bainite, as well as tempered martensite. The difference in these parameters at different welding speeds is associated with mechanical properties.

This is proved by evaluation of input of B_L and M structures in total strengthening. Nanoparticles of carbide phases Fe_3C make significant input in dispersion strengthening. Besides, their uniform distribution in B_L structure (in contrast to B_U) provides uniform distribution of dislocation density according to Orowan mechanism [23]. There are no extended dislocation accumulations, which can develop concentrators of internal stresses, in B_L structure.

3.4 The Experimental-Analytical Technique of Estimation of Strengthening

The executed complex of experimental investigations at all the structural levels allowed carrying out analytical evaluations of the specific contribution of different structural and phase factors and parameters, formed in the investigated weld beads, in change of strength characteristics σ_T and determining the structural factors cardinal influencing on the character and distribution of local inner stresses ($\tau_{L/IS}$), which are the potential sources of incipience and propagation of cracks in the investigated structural microregions [23–27].

In the application of technique the experimental data were used that obtained by optical microscopy, SEM and TEM: grain size; the width of the laths of B structures; effective spacing between carbide phases, that is with taking into account the parameters of the fine structure—the dislocation density.

The following was determined as a result of carried evaluations of change integral hardening along the welding zone. Under conditions of $V_W = 20$ mm/s a calculation value of yield limit $\Sigma\Delta\sigma_T$ in weld metal and HAZ makes 917 and 1077 MPa, respectively, and maximum input in this value is made by substructure ($\Delta\sigma_{Sg} = 318–356$ MPa), dispersion ($\Delta\sigma_{D.H.} = 253–295$ MPa) and dislocation ($\Delta\sigma_D$ $= 157–180$ MPa) hardening. Input of B_L is maximum one (Fig. 2).

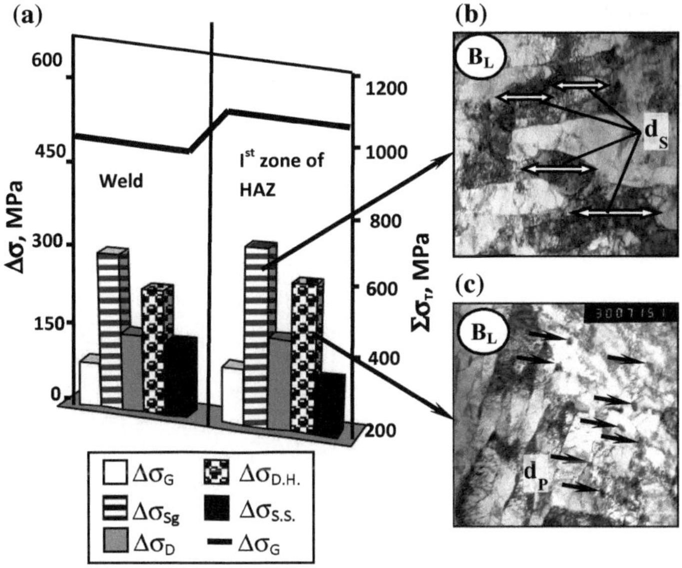

Fig. 2 The contribution of various components $\Delta\sigma$ of structural hardening to the calculated value of hardening $\Sigma\sigma_T$ of the metal of welded joints (**a**) and the fine structure of the main structural components: **b**, **c**—lower bainite at a welding speed of 20 mm/s, ×30,000 ($\Delta\sigma_G$—grain hardening; $\Delta\sigma_{D.H.}$—dispersion hardening; $\Delta\sigma_{Sg}$—subgrain hardening; $\Delta\sigma_{S.S.}$— solid solution hardening; $\Delta\sigma_D$—dislocation hardening)

It have been revealed that if the speed of hybrid laser-arc welding increases the strengthening effect of the structures increases. This is due to an increase in the contribution of the substructure hardening, dispersion hardening and grain hardening (Fig. 2) what was shown by comparison of the strengthening effect of the formed structures in the metal of the welded joints of structural steel.

Calculation estimations of local internal stresses ($\tau_{L/IS}$, [27]) given on diagrams of Fig. 3 show the following. The lowest values are observed at comparatively uniform their distribution and make $\tau_{L/IS}$ = 1470–1890 MPa or 0.22 from the theoretical level of material shear strength (V_W = 20 mm/s), that is assisted by formation of fine grain structures of B_L in uniform distribution of dislocation density. Such a result is first of all achieved in formation of joint weld metal of nanoparticles of carbide phases, which is uniformly distributed on internal volumes of substructure of lower bainite and tempered martensite.

4 Conclusions

The use of high-speed welding technologies allows obtaining phase composition and nanostructural components that provide high operational properties of welded joints.

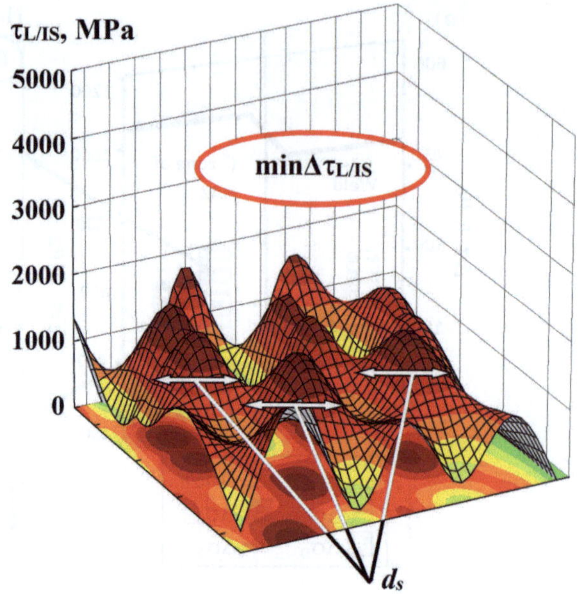

Fig. 3 Local internal stress distribution ($\tau_{L/IS}$) in the weld metal in structural zones in the lower bainite (B_L) at the welding speed of 20 mm/s

As a result of the action of a high-speed thermal-deformation cycle of hybrid laser-arc welding on the weld metal and the heat affected zone of high-strength steel, a homogeneous dispersed fine-grained lower bainite structure and tempered martensite is formed with a uniform distribution of carbide-type nano-phases in them.

The most favorable structure (formation of fine grain lower bainite structures, absence of extended dislocation accumulations, uniform dislocation distribution) in the investigated joints is formed at welding speed 20 mm/s.

References

1. Kovacs T (2018) Laser welding process specification base on welding theories. Proc Manufact 22:147–153. https://doi.org/10.1016/j.promfg.2018.03.023
2. Auwal ST, Ramesh S, Yusof F et al (2018) A review on laser beam welding of copper alloys. Int J Adv Manuf Tech 96(1–4):475–490. https://doi.org/10.1007/s00170-017-1566-5
3. Katayama S (2013) Handbook of laser welding technologies. Woodhead Publishing Limited, Cambridge. https://doi.org/10.1533/9780857098771.backmatter
4. Markashova L, Berdnikova O, Bernatskyi A et al (2017) Physical and mechanical properties of high-strength steel joints produced by laser welding. In: Young scientists forum on applied physics and engineering (YSF), IEEE International, p 88–91. https://doi.org/10.1109/YSF.2017.8126596
5. Unt A, Poutiainen I, Grünenwald S et al (2017) High power fiber laser welding of single sided T-joint on shipbuilding steel with different processing setups. Appl Sci 7(12):1276. https://doi.org/10.3390/app7121276

6. Reisgen U, Krivtsun I, Gerhards BA et al (2016) Experimental research of hybrid welding processes in combination of gas tungsten arc with CO_2- or Yb:YAG-laser beam. J Laser Appl 28(2):022402. https://doi.org/10.2351/1.4944096
7. Shelyagin VD, Krivtsun IV, Borisov YuS et al (2005) Laser-arc and laser-plasma welding and coating technologies. Avtom Svarka 8:49–54
8. Kesse MA, Gyasi EA, Kah P (2017) Usability of laser-TIG hybrid welding processes. In: The 27th international ocean and polar engineering conference, International Society of Offshore and Polar Engineers, pp 42–49
9. Bunaziv I, Akselsen OM, Salminen A, Unt A (2016) Fiber laser-MIG hybrid welding of 5 mm 5083 aluminum alloy. J Mat Proc Techn 233:107–114. https://doi.org/10.1016/j.jmatprotec.2016.02.018
10. Berdnikova O, Poznyakov V, Bushma O (2016) Laser and hybrid laser-arc welding of high strength steel N-A-XTRA-70. Mat Sci Forum 870:630–635. https://doi.org/10.4028/www.scientific.net/MSF.870.630
11. Murzin SP, Liedl G (2017) Laser welding of dissimilar metallic materials with use of diffractive optical elements. Comput Opt 41(6):848–855. https://doi.org/10.18287/2412-6179-2017-41-6-848-855
12. Cherepanov AN, Mali VI, Maliutina IN et al (2017) Laser welding of stainless steel to titanium using explosively welded composite inserts. Int J Adv Manuf Technol 90(9–12):3037–3043. https://doi.org/10.1007/s00170-016-9657-2
13. Siora OV, Bernatsky AV (2011) Development of basic processing methods of laser welding of joints of dissimilar metals. Metallofiz Noveishie Tekhnol 33:569–576
14. Farrokhi F, Siltanen J, Salminen A (2015) Fiber laser welding of direct-quenched ultrahigh strength steels: evaluation of hardness, tensile strength, and toughness properties at subzero temperatures. J Manuf Sci Eng 137(6):061012. https://doi.org/10.1115/1.4030177
15. Berdnikova O, Sydorets V, Alekseienko T (2014) Structure and properties of laser-welded joints from high-strength steels. Appl Mech Mat 682:240–245. https://doi.org/10.4028/www.scientific.net/AMM.682.240
16. Sokolov M, Salminen A, Khlusova E et al (2015) Testing of new materials and computer aided optimization of laser beam welding of high-strength steels. Phys Proc 78:255–264. https://doi.org/10.1016/j.phpro.2015.11.036
17. Kurc-Lisiecka A, Lisiecki A (2017) Laser welding of the new grade of advanced high-strength steel DOMEX 960. Mater Tehnol 51(7):199–204. https://doi.org/10.17222/mit.2015.158
18. Cao X, Wanjara P, Huang J et al (2011) Hybrid fiber laser-arc welding of thick section high strength low alloy steel. Mater Des 32(6):3399–3413. https://doi.org/10.1016/j.matdes.2011.02.002
19. Rossini M, Spena P, Cortese L et al (2015) Investigation on dissimilar laser welding of advanced high strength steel sheets for the automotive industry. Mat Sci Eng: A 628:288–296. https://doi.org/10.1016/j.msea.2015.01.037
20. Krivtsun I, Reisgen U, Semenov O et al (2016) Modeling of weld pool phenomena in tungsten inert gas, CO_2-laser and hybrid (TIG + CO_2-laser) welding. J Laser Appl 28(2):022406. https://doi.org/10.2351/1.4943994
21. Rethmeier M, Gook S, Lammers M et al (2009) Laser-hybrid welding of thick plates up to 32 mm using a 20 kW fibre laser. Trans JWRI 27(2):74–79. https://doi.org/10.2207/qjjws.27.74s
22. Semenov I, Krivtsun I, Reisgen U (2016) Numerical study of the anode boundary layer in atmospheric pressure arc discharges. J Phys D: Appl Phys 49(10):105204. https://doi.org/10.1088/0022-3727/49/10/105204
23. Orowan E (1954) Dislocation in Metals. AIME, New York
24. Gol'dshtein MI, Litvinov VS, Bronfin BM (1986) Metallophysics of high-strength alloys. Metallurgiya, Moscow
25. Conrad H (1973) Model of strain hardening for explaining the influence of grain size on the stress of metal flow. In: Gordienko LK (ed) Ultrafine grain in metals. Metallurgiya, Moscow, pp 206–219

26. Petch NJ (1953) The cleavage strength of polycrystalline. J Iron Steel Inst 173(5):25–28
27. Panin VE, Likhachev VA, Grinyaeva YuV (1985) Structural levels of deformation of solids. Nauka, Novosibirsk

Development of a Controlled In Situ Thin-Film Technology for Porous Anodic Alumina-Based Nanostructures

T. Lebyedyeva, I. Frolov, M. Skoryk and P. Shpylovyy

Abstract This work is aimed at development of the technology of thin-film porous anodic alumina/aluminum (PAA/Al) structures for optical sensors based on metal-clad waveguides and Al nanomesh films for transparent conductive layers. The method of optical control directly in the process of the structures formation which consists in monitoring of the angular dependence of monochromatic light reflection was used both for the formation of the PAA/Al structures by anodic oxidation, for the widening of pores by wet etching, as well to control the etching of the PAA and monitoring the remaining Al nanomesh.

Keywords Aluminum thin films · Magnetron sputtering · Anodic oxidation · Porous anodic alumina · Al nanomesh · In-situ monitoring · Reflective curves

1 Introduction

To the date, there are several directions, in which the works on creation of porous anodic alumina (PAA) films and PAA based nanostructures are being carried out [1]. An advantage of PAA for practical application is the simplicity and the cheapness of its production and the possibility of forming arrays of homogeneous pore diameters ranging from a few to hundreds of nanometers.

Ordered nanostructures on substrates of various materials, including semiconductors, metals, dielectrics, and polymers can be produced in the pores of the nanopatterns from anodic alumina. Current works on the formation of arrays of various kinds of nanoelements—nanodots, nanowires, nanorings, etc. are quite successful.

T. Lebyedyeva (✉) · I. Frolov · P. Shpylovyy
Department of Sensory Devices, Systems and Technologies of Noncontact Diagnostics, V.M. Glushkov Institute of Cybernetics, NAS of Ukraine, Kiev, Ukraine
e-mail: lebtetyana@gmail.com

M. Skoryk
NanoMedTech LLC, Kiev, Ukraine

© Springer Nature Singapore Pte Ltd. 2019
A. D. Pogrebnjak and V. Novosad (eds.), *Advances in Thin Films, Nanostructured Materials, and Coatings*, Lecture Notes in Mechanical Engineering,
https://doi.org/10.1007/978-981-13-6133-3_13

A two-stage anodizing [2] or pre-texturing of the surface of the aluminum [3–5] is commonly used for well-ordered arrays.

Porous anodic alumina thin films in themselves are the subject of development for use as sensory layers. A wide range of impedance, capacitive, acoustic, optical and other sensors and biosensors are being developed [6]. The devices on the surface plasmon resonance (SPR) and waveguide sensors, including sensors on a metal clad waveguide (MCWG) with PAA/Al thin films structure, where the remaining unanodized Al (thickness of 20–10 nm) acted as the metal layer to excite the plasmon for waveguide coupling, are among the most sensitive optical sensors [7–9].

Aluminum nanomesh thin films (nano-sized networks non-anodized Al on a substrate) made by electrochemical anodizing and wet etching of porous anodic alumina have special importance as one of the promising candidates for transparent conductive layers (TCLs) which possess both high conductivity and high optical transparency [10, 11].

The researches on the formation of porous anodic alumina (PAA) films for the use of their nanoscale properties can be divided into two main areas: growing oxide on a bulk aluminum, foils, plates, and on films deposited on the substrate. The development of technological processes of PAA creation on bulk aluminum usually presumes the improvement of methods of ordered structures creation with the implementation of a two-stage process of anodizing. This is where significant success has already been achieved [1]. Nanostructured PAA films growth on thin Al layers deposited on hard and flexible substrates, which are already widely used in microelectronics, seems to bring more promising results. This allows us to combine microelectronic processes with the process of growing of nanostructured PAA layer. However, the method of growing of ordered PAA on deposited thin aluminum films is yet to be discovered [12, 13]. The structural difference in PAA films, grown on the bulk aluminum (plates or folds) and the deposited films of aluminum is usually attributed to the fact that on the contrary to the annealed and subjected to the electrochemical polishing surface of bulk aluminum, the surface of the deposited film has high surface roughness, and films themselves may be disordered small-grain structure with a considerable difference in grain sizes and a large number of randomly distributed inter-grain boundaries. Aluminum films produced under various conditions also differ from each other—according to structural parameters, the density of grains packing and average values of surface roughness [14].

This work was aimed at the development of the technology of thin-film PAA/Al structures for optical sensors based on metal-clad waveguides (MCWG) and Al nanomesh films for transparent conductive layers. The method of optical control directly in the process of the formation of PAA/Al structures, which consists in the monitoring of the angular dependence of reflection on monochromatic light, was used by us earlier for MCWG PAA/Al production [9]. It was used both for the formation of the PAA/Al structures by anodic oxidation and for the widening of pores by wet etching. Since the angular dependences of the reflection curves of the PAA/Al structures vary both with a change in the dielectric layer and with the slightest change in the thickness of Al, the method can be applied to control the etching of the PAA, as well as to control the remaining Al nanomesh.

2 Experiment

The plates of optically polished glass with $n_D = 1.609$, 20 × 25 mm in size, 1 mm thick were used as substrates. The substrates were previously cleaned in chromic acid solution for 6 h. After washing in deionized water, they were treated in the high-frequency oxygen plasma. Immediately before the deposition, the final cleaning of the substrate was carried out in the vacuum chamber by bombarding their surface with argon ions with an energy of 400 eV for 10 min. The adhesive layer of niobium 2–5 nm thick and aluminum films were deposited in a vacuum chamber equipped with a turbomolecular pump by magnetron sputtering at a constant current. Optimization of the deposition regime of aluminum films was carried out by us earlier [13].

The parameters related to the deposition of aluminum films are:

Target aluminum	(99.999% purity)
Gas argon	(99.95% purity)
Base pressure	1×10^{-6} Torr
Argon pressure	1.8×10^{-3} Torr
Substrate-to-target distance	45 mm
Deposition rate	1.1 nm/s

Anodic oxidation of aluminum films was carried out in a two-electrode cell. Anodization modes and control of chronometric dependencies were provided by means of a self-made computerized device. A vacuum deposited gold film on a glass substrate served as the cathode. The formation of PAA was carried out in the potentiostatic mode under 40 V, but in the beginning, the oxide was grown in a constant current regime. The area for the formation of PAA was limited by a mask of photoresist. To avoid etching of the aluminum surface in a photoresist developer, the mask was stamped. 0.3 M solution of oxalic acid was used as the electrolyte. The electrolyte was stirred with a magnetic stirrer; its temperature was 17 °C. Widening of PAA pores by etching was carried out in a 5% aqueous solution of H_3PO_4 at room temperature. An angular reflection curves at a wavelength of 670 nm was monitored using the "Plazmontest" device [9].

To form Al nanomesh selective etching of PAA in an etchant composition of 10% H_3PO_4 + 2% CrO_3 at a temperature of 18 °C was carried out. The change in the reflection curves during etching of PAA at a wavelength of 635 nm was monitored with a self-made instrument providing a large range of light angles incidence.

Samples of aluminum films, PAA and Al nanomesh have been studied by SEM Tescan Mira 3 LMU. Before the SEM investigations, the Au layer thickness of 10 nm was thermally evaporated in a vacuum on the surface of the specimens to form a conducting film to avoid the charging problem.

3 Results

Fabrication of nanostructured PAA/Al coatings for MCWG sensors and transparent conductive Al nanomesh films consist of such main steps as deposition of Al film, anodic oxidation, and chemical etching. As the characteristics of devices based on those structures change drastically at nanometer scale varying of PAA and Al layers thickness, in situ method of their control should be applied. The sensitive method is the method of control of reflection curves. Since the angular dependences of the reflection curves of the PAA/Al structures vary both with a change in the dielectric layer and with the slightest change in the thickness of Al, the method can be applied to control the etching of the PAOA, as well as to control the remaining Al nanomesh.

This method was chosen for in situ control of anodic oxidation and pore widening at manufacturing of MCWG and elimination of PAA at manufacturing of Al nanomesh films.

The chronometric dependences I(t) obtained during single-stage anodic oxidation of aluminum films with a thickness of 230 and 340 nm with Nb adhesive layer 1–2 nm in 0.3 M oxalic acid solution at a voltage of 40 V are shown in Fig. 1.

When the structure PAA/Al for MCWG was formed, the anodization was stopped at the reflection curve of the desired shape was obtained [9]. It corresponds to the onset of the current drop on chronoampermetric curve (Fig. 1, curve A). At Al nanomesh formation (Fig. 1, curve B), approaching of the anodization front to substrate always accompanies by a smooth current drop (section 1 on the insert in Fig. 1) with a subsequent sharp current jump down (section 2) and a further gradual decrease (section 3). We believe that this jump is because the anodization front has reached an adhesive layer. This adhesive layer is probably a mixture of niobium and aluminum atoms due to metal interdiffusion [15], and it also may be partially oxidized prior to the deposition of aluminum because of the getter properties of niobium. The jump on the curve I(t) we consider as a signal that continuous aluminum has been converted to the nanoscale. We stopped the anodizing, as soon as section 2 appeared on curve I(t).

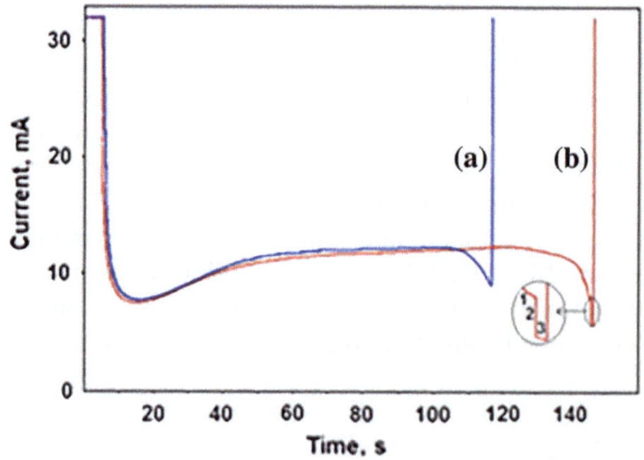

Fig. 1 Chronoampermetric curves obtained during anodic oxidation of the Al film. **a** Al thickness of 230 nm, PAA/Al MCWG formation, **b** Al thickness of 340 nm, Al nanomesh formation

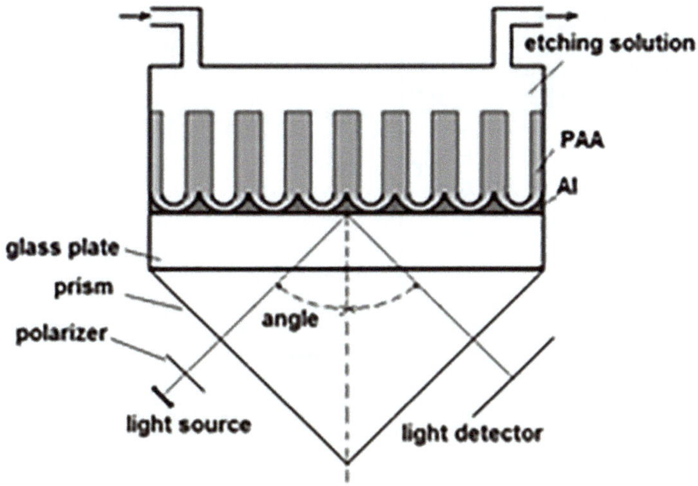

Fig. 2 Scheme of reflective curves measurement

The decrease in current as the anodization front approaches the substrate occurs very rapidly, and it is difficult to control the time of anodization stoppage reproducibly. To ensure the steady state growth of the PAA and to ensure the reproducibility of the stopping time, it seems that the anodizing temperature must be lowered.

After thorough washing and drying of the samples, we carried out the etching of PAA in an aqueous solution of 10% H_3PO_4 + 2% CrO_3 at room temperature (18 °C). The control of PAA etching was made by a self-made computerized device with a *p*-polarized light source of 635 nm wavelength. The range of light incidence angles of 40–68° was provided by a stepping motor. Figure 2 shows the scheme of reflective curves measurement.

Figure 3 shows the sequence of reflective curves obtained at PAA etching. The time interval between the curves was 20 min. The initial reflective curve 1 in Fig. 3, B is a first-order waveguide mode of a thin film structure of PAA/Al. In the etching process, the porous oxide layer is gradually etched and converted to a combination of a thin alumina barrier layer and/or PAA residues lying on the surface of the Al nanomesh. The waveguide minimum shifts to the left (curves 2, 3, 4) and gradually disappears, and in the viewing angle region, the left part of the SPR minimum comes visible (curve 5). When etching proceeds curve 6 is formed from curve 5. Curve 6 corresponds to surface plasmon resonance on a nanoscale Al film. Due to the limited range of the device angles, we can see only the left side of the SPR curve. The invariance of the SPR curve with further etching (curve 7 coincides with curve 6) indicates that PAA is completely etched away. There remained only Al nanomesh, which remained unchanged under this etching regime.

The SEM images are shown in Fig. 4a, b correspond to the reflection curves 1 and 7 in Fig. 3, respectively. Figure 4a is the SEM of the PAA surface after the anodic oxidation. The inset in Fig. 4a shows the surface of the initial aluminum. Figure 4b,

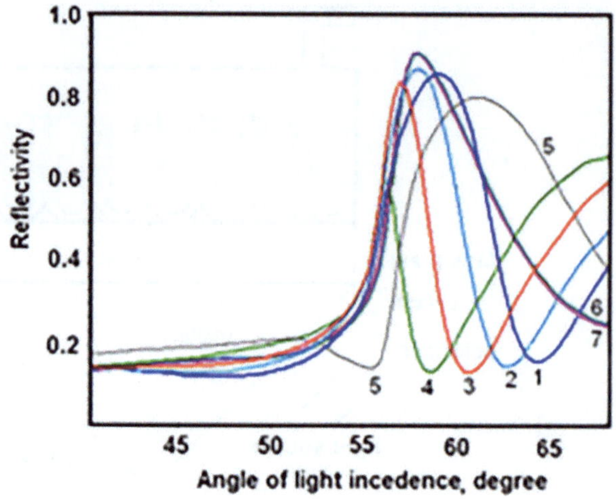

Fig. 3 The reflection curves obtained during the etching of PAA/Al structure

c are the Al nanomesh after etching of the PAA. Most of the cells have a hexagonal shape, but there are also cells with 5 and 4 vertices. Size of the cells ranges from 90 to 120 nm. The nanostructure is heterogeneous, the discontinuities of Al lines are visible in some places. Figure 4c shows a relatively homogeneous area of the Al nanomesh. The width of the aluminum lines is generally about 20 nm. A bright spot measuring 20–25 nm is at the center of almost every cell, apparently a gold-coated bump from an anodic niobium oxide (or a mixture of niobium and aluminum oxides [16]).

From the SEM data, we can see that the Al nanomesh reflects a certain extent the structure of the aluminum film from which it was formed. AFM analysis of the Al film relief showed that the roughness of substrate surface (RMS) is 3–4 nm, a difference of the film thickness on areas between protruding crystallites is 10–15 nm, the height of big crystallites is 50–150 nm [13]. The difference in the thickness of the Al film leads to the fact that different sections of the anodization front reach the substrate not simultaneously. At first, at thinner places, the formation of nanoscale from aluminum begins earlier. The longer formation time of the nanoscale after touching the bottom of the pore of the cover leads to a thinning of the aluminum lines in places with a lower thickness of the aluminum film. In these places, the breaks in the aluminum lines may appear until the anodization stops. Second, it is likely that the local increase in temperature caused by the current flow in places where the Al nanoset is thinner can further accelerate the oxidation. Third, the question of the relationship between the purity of aluminum into grains and on grain boundaries remains open, and the purity affects the formation of PAA and, consequently, Al nanoset.

In our opinion, to increase the homogeneity of the Al nanomesh, it is necessary to ensure the uniformity of the thickness of Al films, possibly by precision electrochemical polishing. For the reproducibility of the moment when the formation of the PAA stops, work should be carried out to optimize the composition of the electrolyte and to decrease of the anodic oxidation temperature.

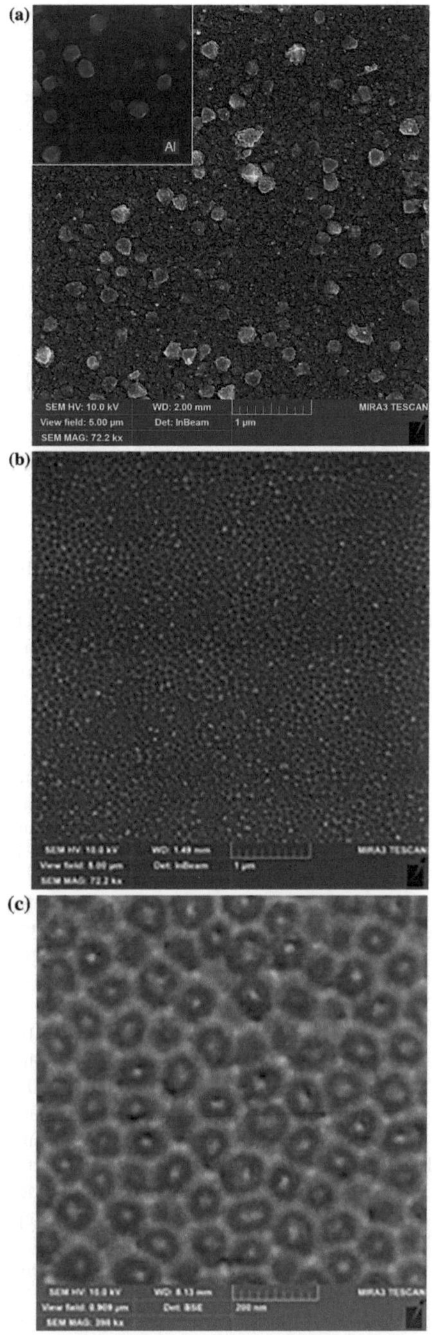

Fig. 4 SEM of the PAA: **a** surface after anodic oxidation, **b, c** Al nanomesh after etching of the PAA

4 Conclusions

An in situ controlled technology for the formation of metal-clad waveguides sensors PAA/Al and the formation of transparent Al nanomesh from thin aluminum films was developed. It includes deposition of Al film by a direct current magnetron on glass substrates with an adhesive layer of niobium, anodic oxidation for PAA/Al structure formation and chemical etching for PAA pores widening or PAA elimination. It was shown that the method of monitoring of the angular reflection curves could be successfully applied in situ both for monitoring of the anodic oxidation process in the formation of the structure of PAA/Al nanomesh and for controlling the etching of PAA. The application of this method makes it possible to obtain reproducible results in the formation of nanostructures with layers of nanometer thickness.

References

1. Losic D, Santos A (2015) Nanoporous alumina: fabrication, structure, properties, and applications. Springer, Heidelberg
2. Masuda H, Fukuda K (1995) Ordered metal nanohole arrays made by a two-step replication of honeycomb structures of anodic alumina. Science 268(5216):1466–1468
3. Lei Y, Cai W, Wilde G (2007) Highly ordered nanostructures with tunable size, shape, and properties: a new way to surface nano-patterning using ultra-thin alumina masks. Prog Mater Sci 52:465–539
4. Wang C, Tanaka S, Saito K, Shimizu T, Shingubara S (2014) Fabrication of ordered arrays of anodic aluminum oxide pores with interpore distance smaller than the pitch of nano-pits formed by ion beam etching. J Mater Sci Nanotech 1(1):1–6
5. Yasui K, Nishio K, Nunokawa H, Masuda H (2005) Ideally ordered anodic porous alumina with sub-50 nm hole intervals based on imprinting using metal molds. J Vac Sci Technol B. https://doi.org/10.1116/1.1941247
6. Santos A, Kumeria T, Losic D (2014) Nanoporous anodic alumina: a versatile platform for optical biosensors. Materials 7:4297–4320
7. Biednov M, Lebyedyeva T, Shpylovyy P (2015) Gold and aluminum based surface plasmon resonance biosensors: sensitivity enhancement. Proceedings of the SPIE 9506 optical sensors. https://doi.org/10.1117/12.2177992
8. Hotta K, Yamaguchi A, Teramae N (2010) Properties of a metal clad waveguide sensor based on a nanoporous-metal-oxide/metal multilayer film. Anal Chem 82(14):6066–6073
9. Voitovich I, Lebyedyeva T, Rachkov O, Gorbatiuk O, Shpylovyy P. Anodic alumina-based nanoporous coatings for sensory applications. In: Fesenko O, Yatsenko L (eds) Nanoplasmonics, nano-optics, nanocomposites, and surface studies. Springer, Heidelberg, p 423
10. Li Y, Chen Y, Qiu M, Yu H, Zhang X, Sun X, Chen R (2016) Preparation of aluminum nanomesh thin films from an anodic aluminum oxide template as transparent conductive electrodes. Sci Rep 6:20114. https://doi.org/10.1038/srep20114
11. Hubarevich A, Marus M, Stsiapanau A, Smirnov A, Zhao J, Fan W, Wang H, Sun X (2015) Transparent conductive nanoporous aluminum mesh prepared by electrochemical anodizing. Phys Status Solidi A Appl Mater Sci 212(10):2174–2178
12. Ottone C, Laurenti M, Bejtka K, Cauda V (2014) Effects of the Film thickness and roughness in the anodization process of very thin aluminum films. J Mater Sci Nanotech 1(1):1–9
13. Lebyedyeva T, Kryvyi S, Lytvyn P, Skoryk M, Shpylovyy P (2016) Formation of nanoporous anodic alumina by anodization of aluminum films on glass substrates. Nanoscale Res Lett 11(203):1–11

14. Lita A, Sanchez JE (1999) Characterization of surface structure in sputtered Al films: correlation to microstructure evolution. J Appl Phys 85(2):876–882
15. Lubenchencko A, Afanas'ev V, Lukashevsky M, Norell M, Pavolotsky A, Fedorovich S (2008) Study of Nb/Al interface combining spectroscopy of reflected electrons with ion sputtering. J Phys Conf Ser 97(1):1–6
16. Mozalev A, Sakairi M, Saeki I, Takahashi H (2003) Nucleation and growth of the nanostructured anodic oxides on tantalum and niobium under the porous alumina film. Electrochem Acta 48(20):3155–3170

The Structure and Tribological Properties of Ni/MoS_2 Composite Layers Formed on Aluminum and Its Alloys

B. Kucharska, J. Mizera, M. Szumiata, A. Zagórski and J. R. Sobiecki

Abstract This work investigates the influence of process parameters on microstructure, adhesion of layers to the substrate and wear resistance of Ni/MoS_2 micro- and nanocrystalline composite layers. Pure nickel coatings were also fabricated and tested for comparison. The layers were deposited by electrochemical reduction on pure A2 aluminum and on PA6 alloy. The layers were produced at a current density of 3 A/dm^2 and a mixing speed of 400 rpm. The bath composition was also modified by the addition of saccharin and the disperse phase of molybdenum disulfide in the bath (2.5 and 5 g/l). Materials prepared in this way underwent structural investigations (using scanning electron microscopy and optical microscopy), a scratch test and tribological tests (using a ball-on-disc method). The results of the tests allowed to determine the dependence of the structure and properties of the produced coatings on the production parameters. The composition of the electrolyte affects the size of crystallites as well as the morphology and topography of the surface. The scratch test confirmed the good adhesion of the nickel and composite layers to the substrate. The best lubricating properties were shown by nanocrystalline coatings. The too high content of molybdenum disulfide in the bath caused an undesirable increase in the coefficient of friction.

Keywords Ni-MoS_2 composite layers · An electrochemical method · Molybdenum disulfide · Aluminum · Tribology

1 Introduction

An extremely important area in the dynamic development of industry is surface engineering, which gives great opportunities in obtaining favorable properties of materials. By using its techniques, it is possible to obtain materials with better technical parameters, such as a lower coefficient of friction, greater resistance to abrasion and

B. Kucharska (✉) · J. Mizera · M. Szumiata · A. Zagórski · J. R. Sobiecki
Faculty of Materials Science and Engineering, Warsaw Technical University, Warsaw, Poland
e-mail: bhkucharska@gmail.com

© Springer Nature Singapore Pte Ltd. 2019
A. D. Pogrebnjak and V. Novosad (eds.), *Advances in Thin Films, Nanostructured Materials, and Coatings*, Lecture Notes in Mechanical Engineering,
https://doi.org/10.1007/978-981-13-6133-3_14

wear, or good corrosion resistance [1]. Appropriate coatings protect the part against the aggressive environment and difficult weather and operational conditions. Nowadays, aesthetic values are also important, and here also surface engineering has a huge scope—the coatings not only protect but also can improve the appearance of the object and give it aesthetic qualities. A very common method of applying coatings is the electrochemical method, the advantages of which include low process temperature, simple equipment as well as relatively low costs [2]. The method has been used in this work due to the advantages above, as well as due to the possibility of obtaining composite coatings. The production of such composite layers is not technologically difficult, and the process is very similar to traditional electrochemical nickel plating. The main difference is the bath composition, to which the required additional ingredient is added (usually in the form of a powder). In the case of the tests presented in work, these were particles of molybdenum disulfide, which due to their specific chemical structure are characterized by excellent mechanical properties. Their presence reduces the coefficient of friction and acts as a lubricant, which may result in good tribological properties of coatings with embedded MoS_2 particles [3, 4].

Aluminum is a very reactive metal—it is instantly oxidized in air and in water. On its surface, a thin Al_2O_3 layer is formed, which protects it against corrosion, but at the same time reduces the adhesion of the applied layers, and may even prevent their coating [5, 6]. Therefore, special surface preparation is required for the planned coatings, which include mechanical surface preparation (removal of impurities, other corrosion products) and degreasing, but also additional galvanizing or contact tinning. After preliminary mechanical removal of impurities from the surface, it is degreased and activated (in alkaline or acidic solution) to remove the oxide layer. Then it is galvanized twice, using lightning in nitric acid between the two processes. Other acids may also be used. However, nitric acid has optimal properties when applied to aluminum alloys. Iron, nickel or copper salts are sometimes added to the zinc bath, as well as complexing compounds to improve the quality of the zinc layer [6]. Between all the above processes, the surface is rinsed to remove any residue from the operation. After completing these steps, one can proceed to the proper application of the metallic coating.

2 Experimental Method

The coatings were produced on pure A2 aluminum and on PA6 alloy. Microcrystalline and nanocrystalline layers were produced with two different thicknesses: 30 and 4 μm at a current density (3 A/dm^2) and mixing speed (400 rpm). Bath composition was controlled: the content of molybdenum disulphide in the bath (2.5 and 5 g/l) and the addition of saccharin. Materials prepared in this way have been subjected to microstructure tests, scratch test and tribological tests. In the case of aluminum and its alloys, it is very important to properly prepare their surface to obtain a coating with adequate adhesion and high purity. In the case of aluminum and PA6 alloy, the substrate was of good quality, therefore its surface was not polished, but it was galvanized to remove the oxide layer and to protect the surface from further oxidation. Oxides are formed as a result of the immediate oxidation of aluminum in the air and make it difficult to apply a protective coating, as well as reduce its adhesion to the substrate. The galvanizing process took place in several stages. The first step was to degrease the surface, which aims to get rid of all impurities and ensure better adhesion of the coating. Galvanizing was performed in a bath containing ZnO, NaOH, $FeCl_3 \cdot 6H_2O$, $NiSO_4 \cdot 6H_2O$, $NaKC_4H_4O_6 \cdot 4H_2O$ and $C_6H_5Na_3O_7$. Galvanizing was performed at a temperature of 32 °C. The composite coating was produced by electrochemical nickel plating. The main components of the electrolyte (in this case the so-called Watts bath) are nickel salts (nickel (II) sulphate (VI), nickel (II) chloride) and boric acid. Microcrystalline Ni/MoS_2 composite coatings were prepared in a Watts base bath modified with the addition of molybdenum disulphide. Baths with two different MoS_2 content were used: 5 and 2.5 g/l. In the case of nanocrystalline coatings, saccharin was added to the electrolyte to obtain a more fine crystallite structure and a higher gloss. In order to compare, pure nickel coatings were also produced with the same producing parameters.

The morphology of nickel and composite Ni/MoS_2 coatings was examined by scanning electron microscopy using a HITACHI SU-70 microscope, equipped with an EDS adapter, allowing analysis of the chemical composition of the produced coatings. The microstructure of Ni/MoS_2 nickel and composite coatings was examined using the NIKON Eclipse LU150 N optical microscope. Equipped with a diamond indenter from Rockwell Scratch tester CMS Instruments, it was used to determine the adhesion and mark friction coefficients of nickel and composite layers by scratch technique. The test was carried out under a load of 10 N, and the scratch length was 3 mm. Tribological tests were carried out on a T-21 ball-on-disc tribotester (ITEE Radom). It was used to test the wear resistance and coefficient of friction in the association of disc—ball working in sliding motion. The ball made of GB GCr15 Bearing Steel was pressed with 1 N force to rotate at 143 rpm. Shield. The T-21 device was equipped with a control and measurement system that allows measuring the friction force, total linear wear of friction junction elements, temperature, rotational speed, and friction path.

3 Results

Figure 1 shows the particle size distribution of molybdenum disulfide. 60% of the volume of MoS_2 powder used are particles with dimensions of 5–17 μm, and 40% of particles with a diameter of 18–70 μm, and the morphology of molybdenum disulfide is shown in Fig. 2.

In the first stage of the process (Fig. 3) in the case of coatings deposited in the Watts base bath with higher MoS_2 content (5 g/l) nickel crystallites are visible, whereas the MoS_2 is low and the layer is compact. At the later stage of the process (Fig. 4), the matrix is not visible; nickel builds mainly on the MoS_2 particles, which results in the formation of porous structures. The structure is not very compact, porous, which is visible in the images of metallographic lateral defects (Fig. 5).

For both microcrystalline layers (Fig. 6) and nanocrystals (Fig. 7), nickel is deposited in the further stage on the MoS_2 particles, which occur in the form of agglomerates, which results from their poor dispersion. In the case of nanocrystalline layers, nickel is deposited in the form of spherical structures, whereas in the case of microcrystalline coatings in the form of rice grains. It is also visible the graining of the nickel matrix grains deposited in a bath modified with the addition of saccharin

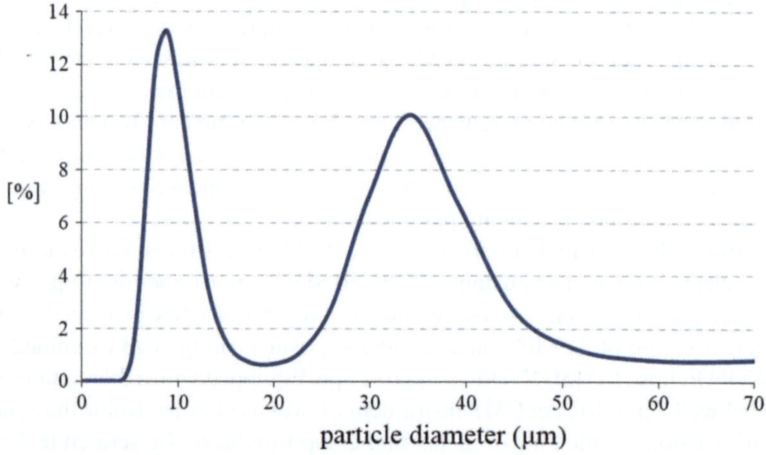

Fig. 1 The particle size distribution of molybdenum disulphide

Fig. 2 Morphology of molybdenum disulphide particles

Fig. 3 Microstructure of microcrystalline Ni/MoS_2 coatings after 6 min deposited in a bath containing 5 g/l MoS_2 (coating thickness 4-μm)

Fig. 4 Morphology of microcrystalline Ni/MoS_2 coatings after 30 min deposited in a bath containing 5 g/l of MoS_2 (coating thickness 30 μm)

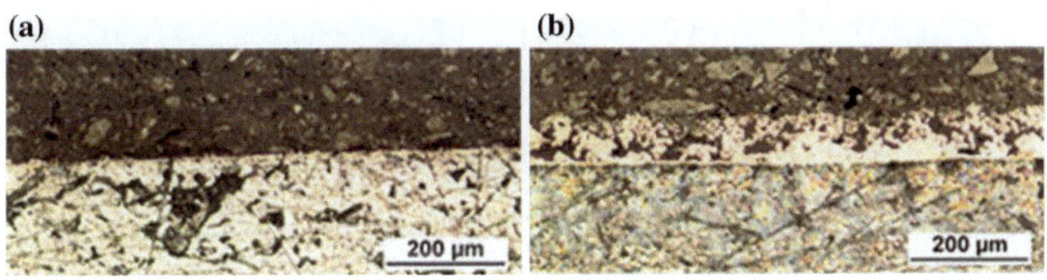

Fig. 5 Microstructure in cross-section of microcrystalline Ni/MoS_2 coatings deposited in a bath containing 5 g/l of MoS_2 (4 μm (**a**) and 30 μm (**b**) coatings)

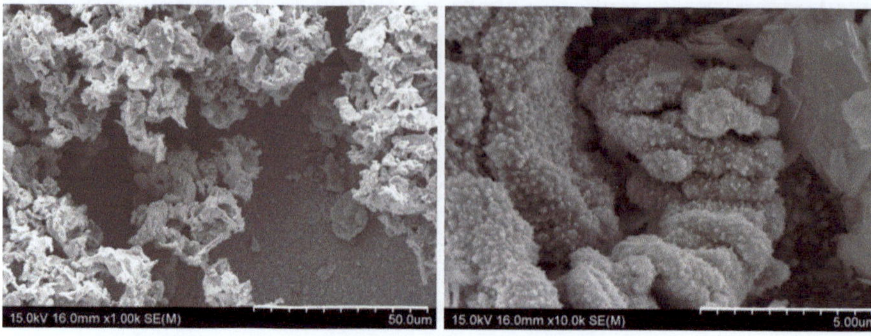

Fig. 6 Microstructure of microcrystalline Ni/MoS_2 coatings deposited in a bath containing 2.5 g/l MoS_2 (30-μm coating)

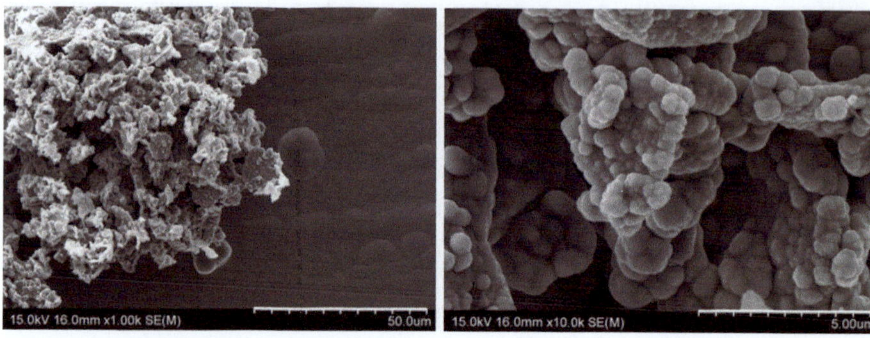

Fig. 7 Microstructure of nanocrystalline Ni/MoS_2 coatings deposited in a bath containing 2.5 g/l MoS_2 (30-μm coating)

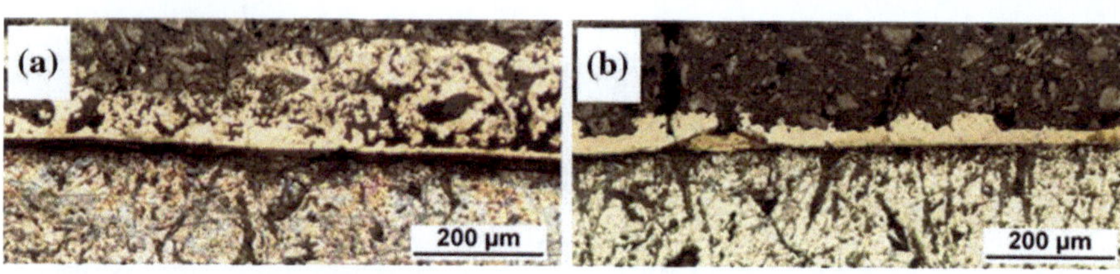

Fig. 8 Microstructure in cross-section of microcrystalline (**a**) and nanocrystalline (**b**) Ni/MoS_2 composite coatings deposited in a bath containing 2.5 g/l MoS_2 (30-μm coating)

in relation to those deposited in the Watts base bath. Composite microcrystalline coatings are characterized by higher porosity in relation to nanocrystalline coatings and higher content of embedded MoS_2 particles, which is visible in cross sections (Fig. 8).

During the scratch test of microcrystalline nickel and composite Ni/2.5MoS_2 coatings, surface behavior can be observed when interacting with the Rockwell diamond indenter on the sample surface. In the case of a pure nickel coating, there are visible cracks after a short test (with a load of 10 N), while composite coatings show a greater scratch resistance—no delamination and decohesion of the layer are observed on their surfaces.

In the assessment of mechanical properties in both the microcrystalline and nanocrystalline layers, the coefficient of friction remains relatively constant (Fig. 9). There are fluctuations which, together with the increase in pressure, stabilize and the graph is smoother. The nanometric layer shows better lubricating properties because it has a lower coefficient of friction at the beginning, and also the variations are not as large as in the micrometer layer.

The coefficient of friction (determined during the ball-on-disc test) of a pure microcrystalline nickel coating is kept at a relatively constant level of just over 1 (Fig. 10). The addition of molybdenum disulfide to the bath in the amount of 2.5 g/l significantly reduces the coefficient of friction (at the beginning up to six times) up to a certain working time of the sample (about 280 s), and later (about 670 s) reaches the value of 1.1, which corresponds to the coefficient of friction of the nickel coating without the addition of MoS_2. It is different in the case of a composite coating with a higher concentration of molybdenum disulfide (5 g/l): instead of a decrease in the coefficient of friction, its very large increase is observed—more than twice after the time of 1000 s. This is most likely due to the fact that the coating largely consists of a soft MoS_2 and a small amount of nickel matrix. Then the ball breaks the coating very quickly and easily (consisting of too much of the dispersive phase) and reaches the aluminum substrate.

Similar observations can be made with nanometric nickel and composite coatings: the coefficient of friction of a pure coating remains at a similar (constant) level of around 1.3. In turn, the addition of molybdenum disulfide to the Watts bath in the amount of 2.5 g/l causes a decrease in the coefficient of friction by about half.

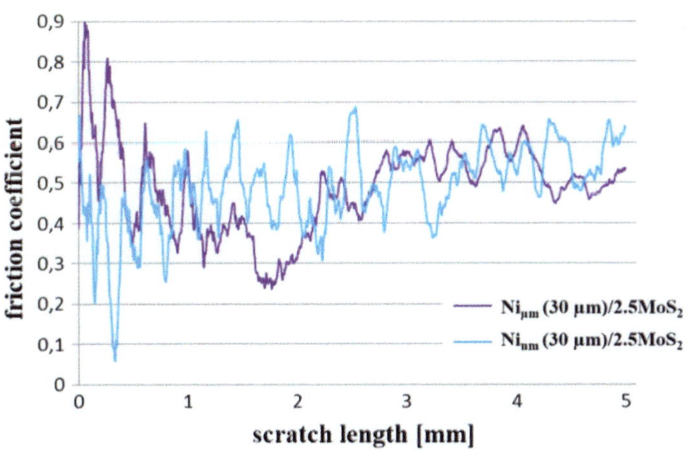

Fig. 9 Scratch test of composite coatings: micrometric and nanometric

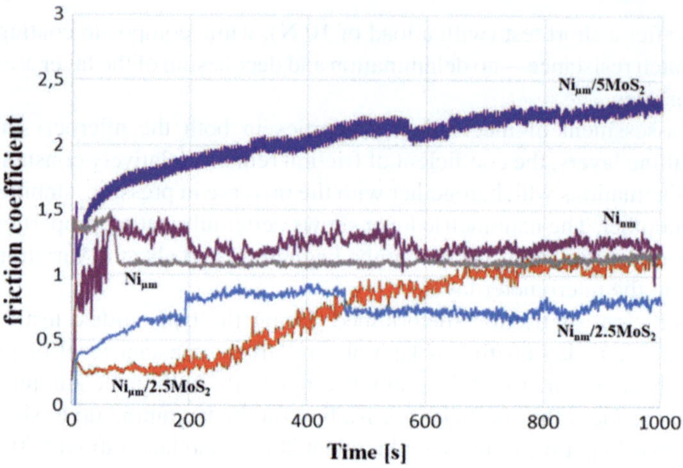

Fig. 10 Graph of the coefficient of friction as a function of time of microcrystalline and nanocrystalline nickel and composite coatings Ni/2.5MoS_2 and Ni/5MoS_2

4 Conclusions

1. Pure nickel coating consists of smooth, fine, spherical nickel crystallites in the case of nanocrystalline layers (nickel is deposited in the form of spherical structures), whereas in the case of microcrystalline coatings—in the form of rice grains.
2. Composite microcrystalline coatings are characterized by higher porosity compared to nanocrystalline coatings and higher content of embedded MoS_2 particles.
3. Composite coatings show greater scratch resistance compared to pure nickel coating.
4. All tested layers show good adhesion to the substrate.
5. The best tribological properties are shown by the nanocrystalline layer with the addition of molybdenum disulphide in the amount of 2.5 g/l for Watts bath (the lowest coefficient of friction).

References

1. Dykas BD, Stringer DB, LaBerge KE (2012) Evaluation methodology for surface engineering techniques to improve powertrain efficiency in military vehicles. Army Research Lab
2. Kucharska B (2017) Micro- and nano-crystalline Ni/Al_2O_3 and Ni/Al_2O_3/PTFE composite coatings. Dissertation, Warsaw University of Technology
3. Xia D, Gong F, Pei X, Wang W et al (2018) Molybdenum and tungsten disulfides-based nanocomposite films for energy storage and conversion: a review. Chem Eng J 348:908–928. https://doi.org/10.1016/j.cej.2018.04.207
4. Chang Y, Changb Y, Linb Ch (1997) Process aspects of the electrolytic deposition of molybdenum disulfide with nickel. Electrochem Acta 43:315–323. https://doi.org/10.1016/S0013-4686(97)00072-8
5. Khan E, Oduoza C, Pearson T (2007) Surface characterization of zincated aluminum and selected alloys at the early stage of the autocatalytic electroless nickel immersion process. J Appl Electrochem 37:1375–1381. https://doi.org/10.1007/s10800-007-9397-y
6. Kucharska B, Trzaska M (2010) The Ni-P/Si_3N_4 nanocomposite layers produced by chemical reduction method on AA 7075 aluminum alloy. Composites 2:100–104

The Effect of Organic Additives on the Microstructure, Microhardness and Friction Coefficient of Ni/WS_2 Composite Coatings

B. Kucharska, M. Ptaszek, J. R. Sobiecki, A. Zagórski and Ja. Mizera

Abstract Composite electroplating enables the production of a wide range of coatings with improved tribological properties. One of the most perspective composite coatings due to its lubricating properties are the Ni/WS_2 coatings. The aim of this study was to investigate the effect of organic additives to Watts bath on the microstructure and mechanical properties of Ni/WS_2 composite coatings produced by the electrochemical method. The study included the composite coatings deposited in a Watts bath modified by saccharine, polyethylene glycol (PEG), sodium dodecyl sulfate (SDS) and 2-butyne-1,4-diol with disperse phase of WS_2 in the form of triturated powder. The particular component was used to obtain a good-quality coating and to enable the proper wettability of tungsten disulfide. The coatings were deposited with direct current. The proper stability of the suspension was ensured by application of the mechanical stirring (400 rpm). The use of the properly selected organic additive in the electrodeposition process results in a significant improvement of the coatings homogeneity, smoothness and the higher microhardness in comparison with composite coatings deposited without additives. The coatings are characterized by increased adhesion to the substrate (low acoustic emission) as well as the lower value of the friction coefficient.

Keywords Electrochemical method · Composite coating Ni/WS_2 · WS_2 · Microhardness · Friction coefficient

1 Introduction

To improve the mechanical properties of light aluminum alloys, nickel composite layers with embedded solid lubricants (MoS_2, WS_2, graphite) can be produced on their surface [1, 2]. It allows widening the spectrum of application of these materi-

B. Kucharska (✉) · M. Ptaszek · J. R. Sobiecki · A. Zagórski · Ja. Mizera
The Faculty of Materials Science and Engineering, Warsaw University of Technology, Warsaw, Poland
e-mail: bhkucharska@gmail.com

© Springer Nature Singapore Pte Ltd. 2019

A. D. Pogrebnjak and V. Novosad (eds.), *Advances in Thin Films, Nanostructured Materials, and Coatings*, Lecture Notes in Mechanical Engineering,
https://doi.org/10.1007/978-981-13-6133-3_15

als, in particular in places where the maximum weight reduction is important with good corrosion resistance and abrasive wear. These composites can be produced by various methods, e.g., chemically and electrochemically [3–6], by thermal spraying [7] and by PVD methods (physical vapor deposition) or by sintering [8, 9]. However, these latter techniques often require high temperatures, high pressures, and expensive equipment as opposed to a relatively cheaper and simple electrochemical method. Transition metal disulfides are among the most commonly used solid lubricants. Tungsten disulfide (WS_2) has a very low friction coefficient of 0.03, which makes it possible to obtain very good lubricity even in dry conditions. Correctly formed electrochemical nickel coatings with embedded WS_2 are characterized by a low coefficient of friction and good abrasion resistance [5]. They can be used in many industries, e.g. in the aerospace, automotive and space industries, especially due to the high stability of WS_2 in a wide temperature range and in a vacuum [10]. The incorporation of heavy metal disulfide particles (WS_2, MoS_2) into a nickel matrix in an electrochemical process depends on many interrelated factors such as hydrodynamic conditions prevailing in the galvanic bath, current density, content, and type of powder in the bath or bath composition [11]. In this work, the influence of various organic additives on the quality, microstructure and mechanical properties of coatings was investigated. The main reason for the use of organic additives is to reduce the internal stresses, increase the gloss and adhesion of the coatings to the substrate. In the case of surface-active compounds (SACs) their key role is to give the particles of dispersive phase of the appropriate charge, which enables their incorporation into the nickel matrix. In addition, the task of SACs is to improve the even dispersion of particles in the entire volume of the bath. It is especially important for tungsten disulfide particles due to their poor wettability and difficulty in obtaining homogeneous composites [12]. The main goal of the research was to control the composition of the bath to obtain homogeneous and compact Ni/WS_2 coatings characterized by a relatively low surface development. It was found that using SACs (PEG-polyethylene glycol, SDS—sodium dodecyl sulphate) one can control the content and the way of incorporation of WS_2 in the coating, which greatly influences the development of the surface of the layers. It was observed that only baths containing an appropriate set of organic compounds allow to obtain functional coatings. Composite coatings produced with this additive mixture were characterized by a much higher microhardness compared to nickel coatings produced in the same bath without WS_2 particles.

2 Experimental Details

2.1 Production of Coatings

Ni/WS_2 layers were produced by electrochemical reduction on aluminum and aluminum alloy AA 2017. Ensuring good adhesion of the deposited layers to the aluminum substrate required proper preparation of its surface, which included pre-

Fig. 1 Morphology of WS_2 powder

Table 1 Bath composition and process parameters for coatings deposition

Component	Concentration (g/dm^{-3})
Bath composition	
$NiSO_4 \cdot 6\ H_2O$	300
$NiCl_2 \cdot 6\ H_2O$	40
H_3BO_3	35
WS_2	1
PEG	0.2
SDS	0.04
Saccharin	2.5
2-butyn-1,4-diol	0.1
Process parameters	
pH	4.0 ± 0.2
Temperature	45 °C
Current density	3 A/dm^2

degreasing, alkaline degreasing, etching and galvanizing. Ni/WS_2 composite coatings were produced in a Watts bath (KW) modified with organic additives and WS_2 powder (Fig. 1) with the composition and manufacturing parameters shown in Table 1. To facilitate the transport of particles of the dispersive phase to areas close to the cathode, the bath was subjected to mechanical stirring at 400 rpm. Pure nickel coatings were also produced and tested for comparison.

2.2 Examination of Coatings

The examination of the microstructure of WS_2 powder, as well as nickel and composite Ni/WS_2 layers, were carried out with the aid of the HITACHI SU-70 scanning electron microscope, equipped with EDS microanalysis X-ray system by Noran and optical microscope NIKON ECLIPSE LV150 N. Based on the analysis of images of cross-sections of Ni/WS_2 layers, the WS_2 particle distribution was determined in the nickel matrix and the thickness of the coatings. The microhardness of the produced layers was examined using microhardness tester HMV-G from Shimadzu with the Vickers method with a load of 98.07 mN (HV0.01). Adhesion tests of layers to an aluminum substrate were carried out using a CSM Instruments Scratch tester, equipped with a Rockwell diamond indenter. A pressure of 1–20 N was applied over a distance of 5 mm. The adhesion of the layers was evaluated on the basis of measurements of acoustic emission and images created during scratch tests. Tribological tests were carried out using the ball-on-disc method using the T-21 tribotester produced by ITE (Institute for Sustainable Technologies) in Radom. It was used to test the wear resistance in the ball-disc combination (point contact of the friction pair) working in sliding motion. The measurements were carried out under dry friction conditions without using an external lubricant. The steel ball (100Cr6), after a thermal improvement of ø 10 mm in diameter and a hardness of 62 HRC, was pressed with a force of 1 N to rotate at 143 rpm. to samples with nickel and composite Ni/WS_2, respectively.

3 Results and Discussion

3.1 Visual Assessment of Coatings

The Ni/WS_2 coating deposited in a pure Watts bath is characterized by high roughness and is devoid of gloss. The addition of PEG to the bath causes that the layers produced in such an electrolyte are shiny and smoothed, however, local cracks are observed on their surface, which is caused by high internall stress of the coatings. To counteract these effects, a set of organic bath additives (saccharin, SDS, PEG, 2-butyn-1,4-diol) was used. Coatings produced in such a bath were free from cracks, mostly shiny. However, there was a tendency to the heterogeneity of the microstructure between the edges and the rest of the sample. For all composite coatings, the appearance of the coatings varied on each side depending on how the sample was placed in the electrolytic cell.

3.2 Microstructure of Coatings

The structure of nickel coatings (Fig. 2a) and Ni/WS_2 (Fig. 2b–d) in cross-section is shown in Fig. 2. The Ni/WS_2 coating produced without surfactants (Fig. 2b) was characterized by a much smaller thickness compared to all other coatings due to the occurrence of edge effects (the coating was grown excessively on the edges). The addition of PEG caused that the coating thickness was evened out over the entire analyzed surface (Fig. 2c), no edge effects were observed in this case. The use of a set of additives (PEG + saccharin + SDS + 2-butyn-1,4-diol) smoothed the coating and, as in the case of PEG alone, offset the edge effects. Based on the analysis of the morphology and topography images of the nickel and composite layers, it was found that all composite layers (Fig. 3b–d) are characterized by a much larger surface development in relation to the nanocrystalline nickel (Fig. 3a). In the case of the coating produced in Watts's bath, local surfaces of cracking are observed. The use of a set of organic additives smoothed the nickel matrix. However, the introduced WS_2 particles cause a high surface roughness.

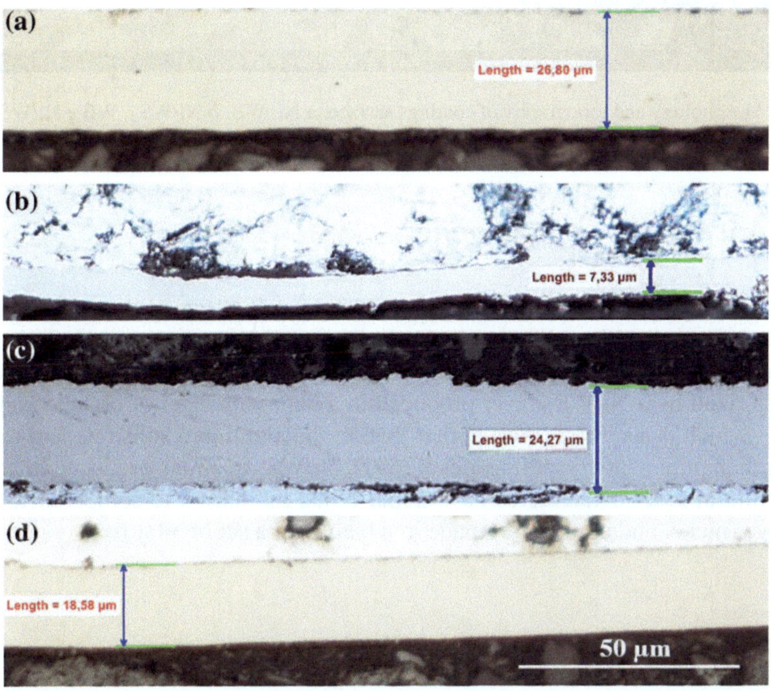

Fig. 2 Coating microstructure in cross-section: **a** Ni, WB, **b** Ni/WS_2, WB **c** Ni/WS_2, WB + PEG, **d** Ni/WS_2, WB + PEG + saccharin + SDS + 2-butyne-1,4-diol

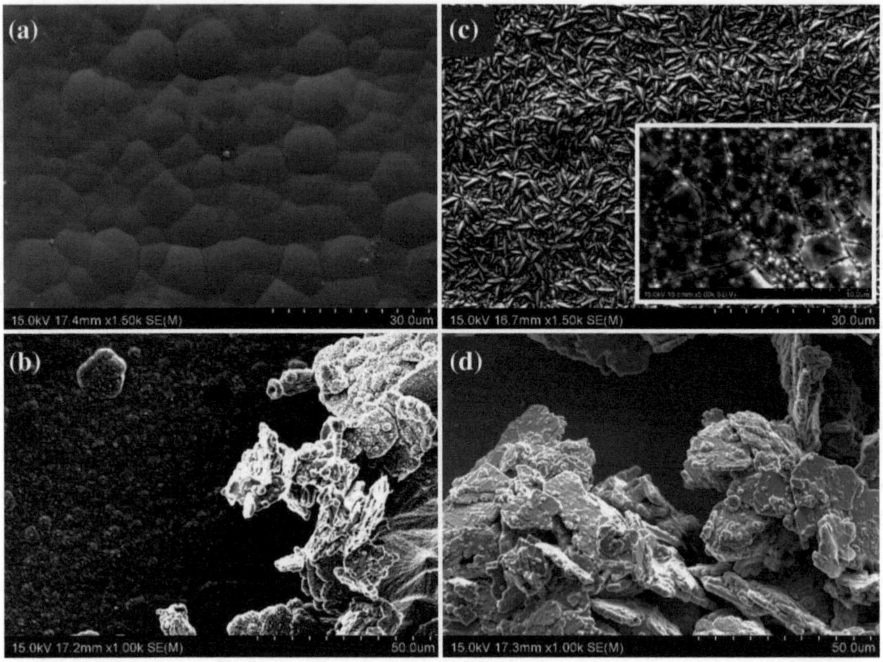

Fig. 3 Morphology and topography of coatings surface: **a** Ni, WB, **b** Ni/WS_2, WB **c** Ni/WS_2, WB + PEG, **d** Ni/WS_2, WB + PEG + saccharin + SDS + 2-butyne-1,4-diol

3.3 Microhardness of Coatings

Both the composition of the bath (affecting the crystallite size of the matrix) as well as the tungsten disulfide phase embedded in the nickel matrix have an effect on the hardness of the layer material, as shown in Fig. 4. The highest microhardness is characterized by a Ni/WS_2 layer produced in a bath with a set of organic additives (751)—which is a several times higher than in the aluminum substrate and slightly higher than pure nanocrystalline nickel (697). Ni/WS_2 coatings produced in a Watts bath without a set of additives have about twice the hardness compared to those coatings (nickel and composite) made in a bath with a set of additives.

3.4 Coefficients of Friction of Coatings

(1) **Scratch test**

The friction coefficients of the Ni/WS_2 coatings produced in the PEG-modified bath (Fig. 5a) and the organic additive set (Fig. 5b) in the first test stage of the scratches are characterized by low values (less than 0.2), with the first coating already at approx.

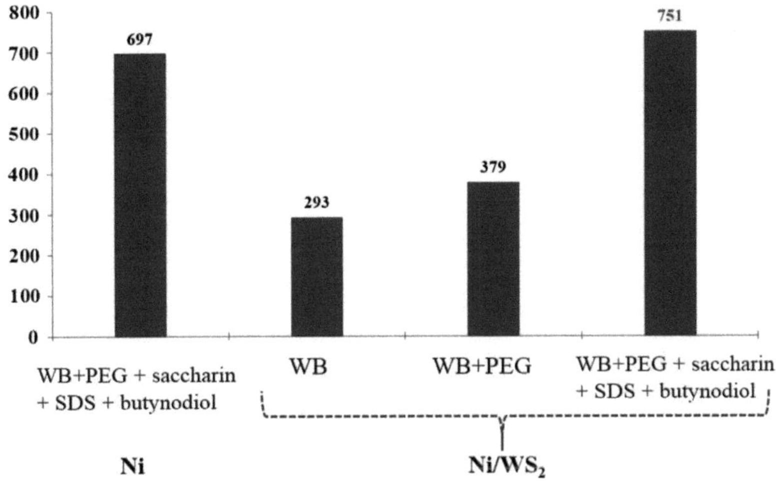

Fig. 4 Microhardness of HV0.01 coatings tested on cross-section

3 N comes to its rapid growth. In the case of a coating produced in the presence of a set of organic additives, a rapid increase in the coefficient of friction is observed at a force of approx. 7 N. In this case it is caused by delamination of the layer (as indicated by the damage of scratches and acoustic emission peaks).

(2) **Ball-on-disc test**

Based on the results of the ball-on-disc test (Fig. 6), the addition of WS_2 improved the lubricating properties of composite coatings against pure nickel (in both cases the coating was produced in the presence of a set of organic additives). The friction coefficient of Ni/WS_2 layers is very low and does not exceed 0.2 throughout the test period. In the case of nickel coating, the coefficient of friction does not stabilize during the test, and its value is high and ranges from 0.6 to 1.4.

4 Conclusions

Obtaining a good quality Ni/WS_2 composite coating requires the use of appropriate electrolytes and process parameters. Coatings deposited only in Watts bath or a bath modified only with SACs (PEG) are characterized by too poor mechanical properties in relation to the expected. Selection of appropriate organic additives, including surface-active compounds or compounds that reduce the internal stresses allow to obtain compact, non-porous and relatively homogeneous coatings. These coatings are characterized by twice higher microhardness compared to the corresponding composites made in Watts bath and modified by PEG.

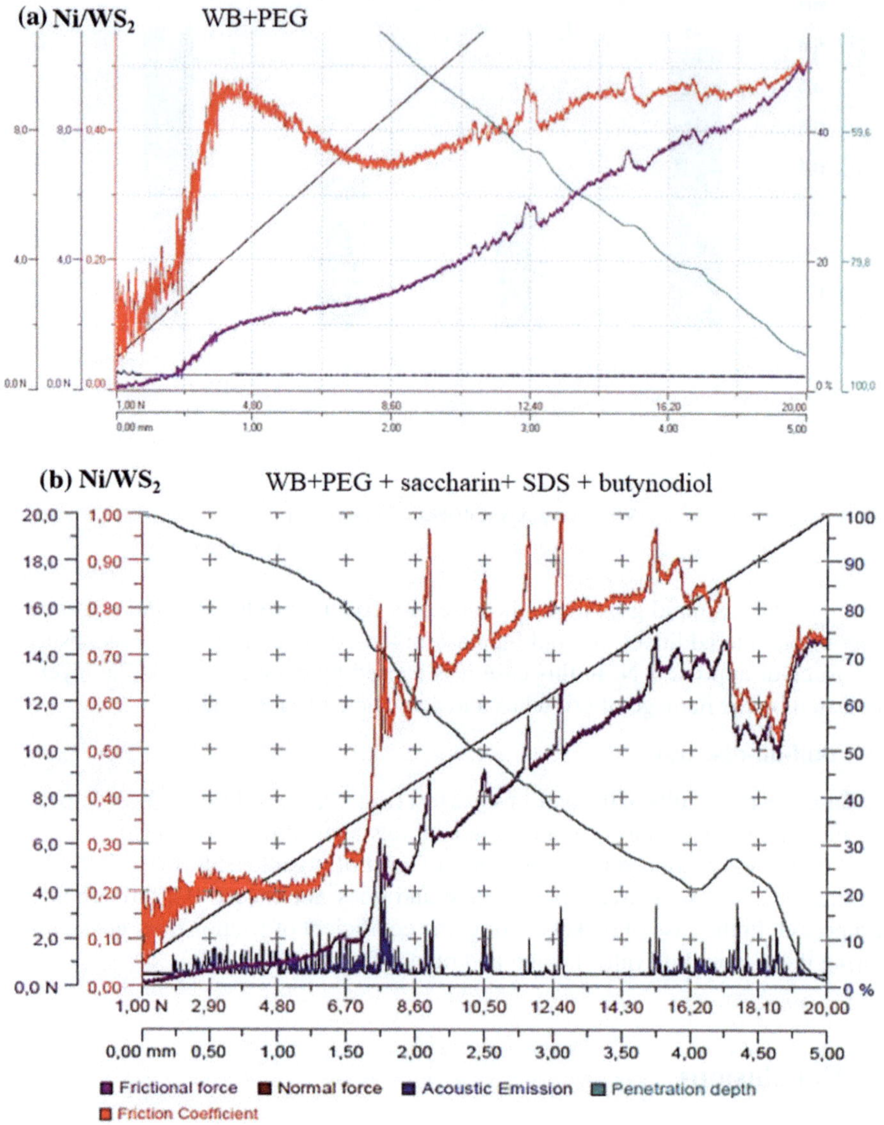

Fig. 5 Scratch test of nickel and Ni/WS_2 composite coatings deposited in Watts bath modified by series of organic compounds

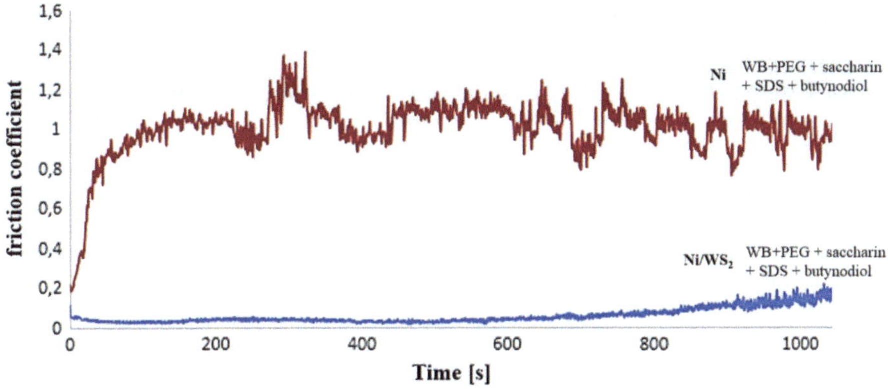

Fig. 6 Graph of the friction coefficient of Ni/WS$_2$ composite coatings deposited in Watts bath modified by **a** PEG and **b** series of organic compounds

If an appropriate set of organic additives is used, the incorporation of the WS$_2$ dispersive phase also allows to obtain much lower coefficients of friction recorded during the test of the ball-on-disc in relation to pure nickel coatings.

References

1. Brooman E (2004) Wear behavior of environmentally acceptable alternatives to chromium coatings: nickel-based candidates. Met Finish 102:75–82. https://doi.org/10.1016/S0026-0576(04)84678-6
2. Pagnan Furlan K, Biasoli de Mello JD, Klein AN (2018) Self-lubricating composites containing MoS$_2$: a review. Tribol Int 120:280–298. https://doi.org/10.1016/j.triboint.2017.12.033
3. Chang Y, Chang Y, Lin C (1998) Process aspects of the electrolytic codeposition of molybdenum disulfide with nickel. Electroch Acta 43:315–324. https://doi.org/10.1016/S0013-4686(97)00072-8
4. Cardinal MF, Castro PA, Baxi J et al (2009) Characterization and frictional behavior of nanostructured Ni–W–MoS$_2$ composite coatings. Surf Coat Technol 204:85–90. https://doi.org/10.1016/j.surfcoat.2009.06.037
5. García-Lecina E, García-Urrutia I, Díez JA et al (2013) Codeposition of inorganic fullerene-like WS$_2$ nanoparticles in an electrodeposited nickel matrix under the influence of ultrasonic agitation. Electroch Acta 114:859–867. https://doi.org/10.1016/j.electacta.2013.04.088
6. André B, Gustavsson F, Svahn F et al (2011) Performance and tribofilm formation of a low-friction coating incorporating inorganic fullerene like nano-particles. Surf Coat Technol 206:2325–2329. https://doi.org/10.1016/j.surfcoat.2011.10.012
7. Zhang X, Zhang X, Wagn A et al (2009) Microstructure and properties of HVOF sprayed Ni-based submicron WS$_2$/CaF2 self-lubricating composite coating. Trans Nonferrous Met Soc China 19:85–92. https://doi.org/10.1016/S1003-6326(08)60233-2
8. Parucker M, Klein AN, Binder C et al (2014) Development of self-lubricating composite materials of nickel with molybdenum disulfide, graphite and hexagonal boron nitride processed by powder metallurgy: preliminary study. Mater Res 17:180–185. https://doi.org/10.1590/S1516-14392013005000185

9. Wu Y, Wang F, Cheng Y et al (1997) A study of the optimization mechanism of solid lubricant concentration in $NiMoS_2$ self-lubricating composite. Wear 205:64–70. https://doi.org/10.1016/S0043-1648(96)07299-7
10. Xu S, Weng L, Liu Y et al (2017) Microstructure evolution and enhanced vacuum tribological performance of Ni-doped WS_2 composite coating. Surf and Coat Techn 325:81–88. https://doi.org/10.1016/j.surfcoat.2017.06.036
11. Güler ES (2013) Electrocodeposition of molybdenum disulfide particles in nickel matrix. Dissertation, Middle East Technical University
12. Alberdi A, Hatto P, Díaz B, Csillag S (2011) Tribological behavior of nanocomposite coatings based on fullerene-like structures. Vacuum 85:1087–1092. https://doi.org/10.1016/j.vacuum.2010.11.019

Effect of Temperature on the Growth of Pores in Binary Bi/Sn Films

S. I. Petrushenko, S. V. Dukarov, Z. V. Bloshenko, I. G. Churilov and V. N. Sukhov

Abstract The work is devoted to the study of through pores that arise in binary Bi/Sn films when they are annealed near the melting point. It is shown, that the study of the temperature dependence of the average pores size can be used to measure the activation energy of diffusion processes that ensure the de-wetting of the samples under study. Concentration dependence of the activation energy of diffusion in bilayer Bi/Sn films is obtained. It is shown, that the diffusion activation energy has a minimum in the homogeneous region and is practically constant at a component concentration corresponding to the two-phase section of the phase diagram.

Keywords Thin films · Binary systems · Diffusion · Through pores

1 Introduction

Thermal stability of thin-film systems largely determines the scope and the possibility of their application. In particular, the limited thermal stability of films used as conductive, buffer or protective layers limits further miniaturization of various electronic devices. Thus, the processes of pore formation, which are caused by diffusion and actively occur in films near their melting point [1], can lead to disruption of the continuity of the functional layers and the failure of the entire device. Also, processes of mutual solubility of components and diffusion homogenization of the sample can contribute to degradation [2, 3]. These processes are of a diffusive nature. With a reduction in the characteristic size of the element, the distance that a substance must overcome to violate its operational properties decreases. This leads to a significant acceleration of degradation. An even more important factor in the intensification of the thermal de-wetting of films with a decrease in their thickness is the size effect of diffusion [4, 5]. Thus, the study of the through porosity that occurs in continuous films during their annealing has both applied and general scientific significance. Such

S. I. Petrushenko (✉) · S. V. Dukarov · Z. V. Bloshenko · I. G. Churilov · V. N. Sukhov
V. N. Karazin Kharkiv National University, Kharkiv, Ukraine
e-mail: petrushenko@univer.kharkov.ua

© Springer Nature Singapore Pte Ltd. 2019
A. D. Pogrebnjak and V. Novosad (eds.), *Advances in Thin Films, Nanostructured Materials, and Coatings*, Lecture Notes in Mechanical Engineering,
https://doi.org/10.1007/978-981-13-6133-3_16

studies make it possible to obtain important information about the characteristics of the diffusion processes taking place in these systems. Note that the use and study of low-dimensional systems are greatly complicated by their high sensitivity to impurities and the conditions of obtaining. It is possible to significantly reduce the impact of uncontrolled factors if you use vacuum methods of preparation. Such methods make it possible to obtain samples under conditions where the impurity concentration can be made negligible. It is also very advantageous to combine vacuum methods with techniques that allow one to create a series of samples in one experiment that differ by parameter being studied. One method of practical implementation of this approach is the method of samples of variable composition and variable state. In this method, due to a special arrangement of the evaporators, the substrate, moving and fixed shields, it is possible to obtain a series of samples with smoothly or discretely varying thickness or concentration of components. Samples corresponding to different states within the framework of this technique are obtained by creating a certain temperature gradient along the substrate. As a result, different sections of the film, which are deposited under completely identical conditions, undergo heating to different temperatures.

This approach has particular importance for studying the effect of concentration on the behavior of multicomponent systems. To carry out such studies the preparation of a series of samples is required. Such samples are usually obtained in different experimental cycles. This is inevitably accompanied by some change in the uncontrolled parameters of their preparation. At the same time, using the method of samples of variable composition and variable state (at least in the case of binary systems), it is possible to obtain a set of samples corresponding to any given concentration interval within a single vacuum cycle.

The use of various modifications of the method of samples of variable composition and variable state made it possible to obtain a number of results that are interesting both from the point of view of fundamental science and applied developments. Thus, in [6], due to obtaining in the same experiment films of different thicknesses and experiencing different thermal effects, results concerning the initial stage of the growth of film systems were obtained. Also, an approach based on the method of samples of variable composition and the variable state can be used to determine supercooling values in disperse systems. In this case, the crystallization temperature of the supercooling liquid phase is assumed to be equal to the temperature of the change from the condensation mechanism of vapor-crystal to the condensation mechanism of the vapor-liquid. It is registered ex situ [7]. In contrast to in situ methods [8–10], using this method, it is easy to obtain a whole series of samples differing in the desired parameter. This is convenient for studying the effect of wetting, composition or size on the supercooling.

The use of the method of samples of variable composition and variable state to study the thermal stability of vacuum condensates and the diffusion processes occurring in them is performed in [1, 11, 12]. In [11], for copper films, a size dependence of the dispersion temperature was obtained, and it was shown in [1] that the study of the temperature dependence of the mean pore size that appears in films near their melting point could be used to determine the activation energy of diffusion in these systems.

Note that the study of the processes of pore formation occurring in film systems not only makes it possible to establish the features of the diffusion processes taking place in them but also has an important application value. However, most of the works aimed at covering such issues relate to the study of one-component films. At the same time, the solution of many applied problems requires the study of more complex systems, for example, binary ones. This paper is devoted to the study of pore formation and diffusion processes occurring in Bi/Sn films. The choice of the object of investigation is determined both by methodological considerations, namely, by low melting points and a simple phase diagram, and by the applied value of this contact pair, which along with other fusible eutectics [13] is actively study [14] and is considered a promising replacement for toxic solder based on lead.

2 Experimental

Two-layer Bi/Sn films with a total thickness of approximately 300 nm, obtained by using a method of samples of variable composition and variable state, were chosen for this study. Samples were formed by layer-by-layer condensation of components from separate sources in a vacuum of 10^{-7} Torr at the extended steel substrate measuring $170 \times 60 \times 4$ mm. Molybdenum (Bi) and tantalum (Sn) boats heated by electric current were used as evaporators. Immediately before the deposition of the components, an amorphous carbon layer was deposited on the substrate, which isolated the system under study from the substrate material. Practically complete non-wetting of the carbon by melts of selected metals somewhat allows us to treat the films under study as free films. The mutual arrangement of evaporators, shields and a substrate ensured the preparation of films, which made the concentration of the components in the given limits continuously vary along the width of the substrate. The thicknesses of the films and, consequently, the concentration of the components were measured on the basis of geometric considerations from the data of two quartz resonators located on opposite sides of the substrate. Due to the appropriate arrangement of the equipment elements, the quartz sensors measured the mass thickness of each of the deposited metals separately.

After completion of the condensation, a temperature gradient was created along a steel substrate by means of a heating block mounted on one end of the substrate and forced cooling of the other. The temperature gradient was perpendicular to the concentration gradient and was monitored from the data of several K-type thermocouples welded to the rear of the substrate. Samples, after being set a steady temperature distribution, were held in this state for an hour.

After the heating and the appropriate thermal annealing, the substrate with the film was cooled in vacuum to room temperature, removed from the vacuum chamber and examined by scanning electron microscopy in the JEOL JSM-840 electron microscope.

3 Results and Discussion

Electron-microscopic studies of Bi/Sn films showed that in the region of the substrate, the temperature of which ensures complete melting of the sample, the film consists of spherical particles (Fig. 1a). Such morphology indicates the melting of this part of the sample [15]. At temperatures below the liquids and above solidus, the films continue to be islands. However, in this temperature region, irregularly shaped formations are observed along with spherical islands. This indicates the coexistence of liquid and crystalline phases in this region (Fig. 1b). The relative amount of the liquid phase decreases with decreasing temperature, and it disappears after reaching the solidus line.

Below the solidus line, the films remain solid polycrystalline. At the same time, in the immediate vicinity of the melting start temperature, through pores are detected in the films. Their size for the same duration of thermal exposure is determined by the temperature to which this region was heated and the concentration of the components of the binary film (Fig. 2). It should be noted that the morphology of the films somewhat differs in parts corresponding to bismuth concentrations above and below its maximum solubility in tin (i.e., in the two-phase and homogeneous regions of the phase diagram of this contact pair, respectively). In a homogeneous region, large inclusions of bismuth are observed on the surface of the film after cooling. In a two-phase region, these inclusions have a substantially smaller size and form a generally homogeneous structure (Fig. 2).

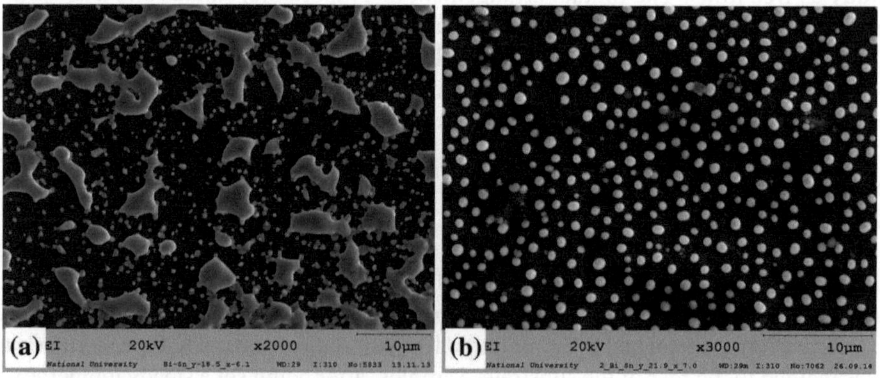

Fig. 1 SEM images of samples corresponding to parts of the film whose temperature exceeds the liquidus temperature (**a**) and is between the solidus and liquidus temperatures (**b**)

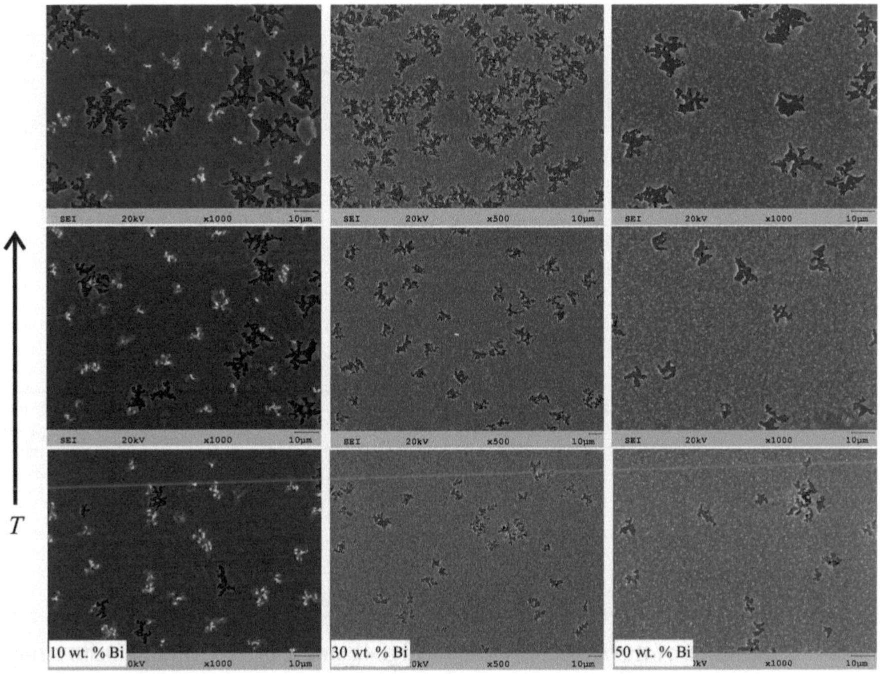

Fig. 2 SEM images of Bi/Sn films corresponding to different component concentrations and annealing temperatures

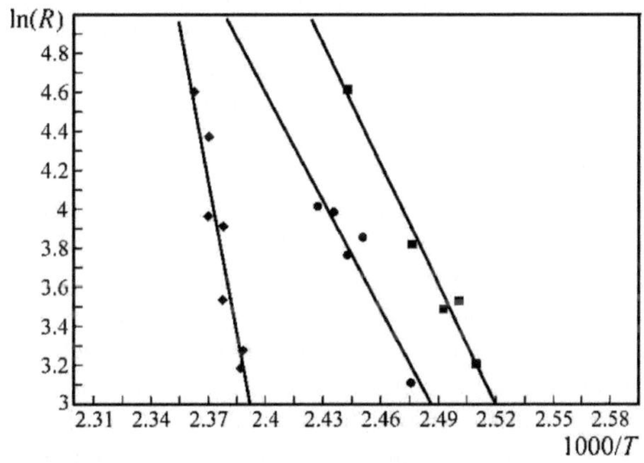

Fig. 3 Arrhenius graphs corresponding to different concentrations of the Bi-Sn contact pair components

The temperature dependence of the average pore size (r), for films with different component concentrations, in $\ln R - 1/T$ coordinates (Fig. 3) is well described by a linear dependence. This can be used to estimate the activation energy of diffusion, which provides porosity in the binary films under study [1, 11].

Then the activation energy of surface diffusion can be determined from the expression [1]:

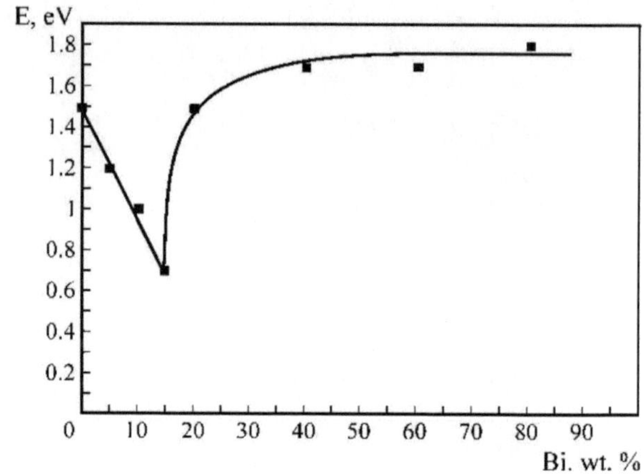

Fig. 4 Dependence of the diffusion activation energy on the concentration of bismuth in Bi/Sn films

$$E_a = \alpha k$$

where α is the slope of the Arrhenius graph (Fig. 4), and k is the Boltzmann constant.

According to the studies carried out, the activation energy of diffusion in Bi/Sn films depends significantly on the concentration of the components of the contact pair (Fig. 4). Thus, initially, with an increase of the bismuth concentration, the activation energy decreases rapidly to a minimum of 0.6–0.7 eV at a concentration approximately corresponding to the boundary solubility of bismuth in tin. After that, it rapidly increases, reaching a value of 1.6 eV at a concentration of 25–30 wt%, and then practically does not change. This means that the activation energy of surface diffusion in a two-phase region is independent of the concentration of the components.

The presence of a minimum on the concentration dependence of the diffusion activation energy was observed by authors of [16] for Mn–Bi alloys. The minimum value of the activation energy is observed at 55 at.% Mn. However, a small number of concentrations studied by the authors make it difficult to compare their results with the phase diagram of the contact pair, which has a rather complex form [17]. Nevertheless, the concentration corresponding to the minimum value of activation energy of diffusion which was found in [16] is close to the concentration boundary for the existence of an alpha solid solution based on manganese [17].

The intensification of diffusion processes in a homogeneous concentration region is also evidenced by results of [18, 19], in which the kinetics of the formation of intermetallides was studied upon contact of the eutectic [18] and homogeneous [19] Bi–Sn alloys with copper. The nonmonotonic character of the dependence of activation energy on concentration for the processes providing mass transfer was also found in [20], in which the processes of liquid-phase mass transfer are theoretically researched. The authors showed that the mixing activation energy in binary Pb–Li melts has a minimum observed at the concentration of the components corresponding to the beginning of the eutectic part of the phase diagram.

The concentration dependence of the parameters corresponding to the transport phenomena was also studied in [21]. The authors of [21] showed that the temperature coefficient of electric resistivity of Cu–Sn alloys has a minimum at a concentration of components corresponding to the boundary solubility of tin in copper. Moreover, when studies the viscosity of melts, it was established in [21] that it has a maximum value at the same concentration of components. The authors associated this with the formation of clusters in the melt, the ratio of the elements in which corresponds to the concentration providing the minimum value of the temperature coefficient of resistance.

Thus, the results of many studies [16–22] indicate that the concentration corresponding to the boundary of the existence of a solid solution on the basis of one of the components corresponds to the region where the concentration dependence of the transfer parameters is monotonic. At the same time, the considerations in the literature regarding the nature of this phenomenon are insufficient.

So it is worth paying attention to the phenomenon studied in [12] in which it is shown that the activation energy of diffusion in copper films can sharply decrease in the presence of liquid lead. In our binary film systems, with a concentration of bismuth below its limiting solubility in solid tin, due to contact melting [23–26], one can expect not only the formation of a solid solution but also a liquid phase of a eutectic composition. Based on the results of [12], it can be expected that the liquid phase can simplify the diffusion processes in the samples. In this regard, the decrease in the activation energy with increasing bismuth concentration seems to be natural, since the concentration of the liquid phase should increase with increasing bismuth content. The decrease in the intensity of the diffusion processes in the two-phase region and the corresponding increase in the activation energy can be due to the presence of two dissimilar phases. It is reasonable to assume that the interphase boundaries, unlike, for example, grain boundaries, are diffusion stoppers, which naturally complicate the diffusion decay of the films.

The intensification of diffusion processes, which according to our results and works [18, 19] takes place in homogeneous samples, explains the fact that larger pores are observed in this concentration region than in the two-phase part (Fig. 2).

4 Conclusions

When studying the through porosity formed in binary Bi/Sn films, it is shown that the kinetics of diffusion processes is determined by the concentration of the components. At first, with an increase in the bismuth content, a decrease in the activation energy of mutual diffusion, which reaches a minimum at the concentration boundary of the existence of a homogeneous region, is observed. After that, the activation energy rapidly increases and remains almost constant starting from a bismuth concentration of 25–30 wt%.

Acknowledgements This work was supported by the Ministry of Education and Science of Ukraine.

References

1. Dukarov SV, Petrushenko SI, Sukhov VN et al (2014) Effect of temperature on the pores growth in the polycrystalline films of fusible metals. Probl At Sci Technol 89(1):110–114
2. Dukarov SV, Petrushenko SI, Sukhov VN et al (2016) In situ research on temperature dependence of the lattice parameters of fusible metals in thin Cu-Pb and Cu-Bi films. Funct Mater 23(2):218–223. http://dx.doi.org/10.15407/fm23.02.218
3. Bi J, Gao Q, Ao J et al (2018) Influence of Cu on Ga diffusion during post-selenizing the electrodeposited Cu/In/Ga metallic precursor process. Sol Energy Mater Sol Cells 182:92–97. https://doi.org/10.1016/j.solmat.2018.03.007
4. Unutulmazsoy Y, Merkle R, Fischer D et al (2017) The oxidation kinetics of thin nickel films between 250 and 500 °C. Phys Chem Chem Phys 19:9045–9052. https://doi.org/10.1039/C7CP00476A
5. Kryshtal AP, Bogatyrenko SI, Sukhov RV et al (2014) The kinetics of the formation of a solid solution in an Ag–Pd polycrystalline film system. Appl Phys A Mater Sci Process 116:1891–1896. https://doi.org/10.1007/s00339-014-8349-8
6. Dukarov SV, Petrushenko SI, Sukhov VN (2018) Growth of island films during vapor-liquid condensation. J Nano Electron Phys 10(1):01023. https://doi.org/10.21272/jnep.10(1).01023
7. Gladkikh NT, Dukarov SV, Sukhov VN et al (2011) Condensation mechanism of AgCL and NaCl island films on a nickel substrate. Funct Mater 18(4):529–533
8. Petrushenko SI, Dukarov SV, Sukhov VN (2016) Stability limits of the liquid phase in the layered Mo/Pb/Mo, Mo/Bi/Mo and Mo/In/Mo film systems. J Nano Electron Phys 8(4(2)):04073. http://dx.doi.org/10.21272/jnep.8(4(2)).04073
9. Bazlov AI, Tabachkova NYu, Zolotorevsky VS et al (2018) Unusual crystallization of Al85Y8Ni5Co2 metallic glass observed in situ in TEM at different heating rates. Intermetallics 94:192–199. https://doi.org/10.1016/j.intermet.2017.12.024
10. Kolendovskii MM, Bogatyrenko SI, Kryshtal AP et al (2012) Piezoquartz resonator as an in situ method for studying the phase transitions in thin metal and alloy films. Tech Phys 57(6):849–855. https://doi.org/10.1134/S1063784212060175
11. Petrushenko SI, Dukarov SV, Sukhov VN (2016) Growth of through pores and thermal dispersion of continuous polycrystalline films of copper. Metallofizika i Noveishie Tekhnologii 38(10):1351–1366. https://doi.org/10.15407/mfint.38.10.1351
12. Petrushenko SI, Dukarov SV, Sukhov VN (2017) Effect of lead on the thermal dispersion of continuous polycrystalline copper films. Vacuum 142:29–36. https://doi.org/10.1016/j.vacuum.2017.04.037
13. Kashezhev AZ, Kumykov VK, Kutuev RA et al (2016) Density and surface tension of diluted Sn–In alloys. Bull Russ Acad Sci Phys 80(6):746–749. https://doi.org/10.3103/S1062873816060150
14. Bogatyrenko SI, Minenkov AA, Kryshtal AP (2018) Supercooling under crystallization of Bi-Sn eutectic alloy in contact with Bi, Sn and amorphous C. Vacuum 152:1–7. https://doi.org/10.1016/j.vacuum.2018.02.039
15. Petrushenko SI, Dukarov SV, Sukhov VN et al (2015) Inner size effect in the polycrystalline metal films of fusible metals. J Nano Electron Phys 7(2):02033

16. Xiang Z, Wang T, Ma S et al (2018) Microstructural evolution and phase transformation kinetics of MnBi alloys. J Alloys Compd 741:951–956. https://doi.org/10.1016/j.jallcom.2018.01.147
17. Wang CP, Yu WJ, Li ZS et al (2011) Thermodynamic assessments of the Bi–U and Bi–Mn systems. J Nucl Mater 412(1):66–71. https://doi.org/10.1016/j.jnucmat.2011.02.021
18. Yoon JW, Lee CB, Jung SB (2002) Interfacial reactions between Sn-58 mass%Bi eutectic solder and (Cu, Electroless Ni-P/Cu) substrate. Mater Trans 43(8):1821–1826. https://doi.org/10.2320/matertrans.43.1821
19. Yoon JW, Jung SB (2003) Investigation of interfacial reactions between Sn–5Bi solder and Cu substrate. J Alloys Comp 359:202–208. https://doi.org/10.1016/S0925-8388(03)00291-3
20. Wang B, Xiao S, Gan X et al (2014) Diffusion properties of liquid lithium–lead alloys from atomistic simulation. Comput Mater Sci 93:74–80. https://doi.org/10.1016/j.commatsci.2014.06.020
21. Jia P, Zhang J, Hu X et al (2018) Correlation between the resistivity and the atomic clusters in liquid Cu-Sn alloys. Physica B Condens Matter 537:58–62. https://doi.org/10.1016/j.physb.2018.02.003
22. Porth CB, Cahoon JR (2010) Interdiffusion of Bi in liquid Sn. J Phase Equilib Diffus 31(2):149–156. https://doi.org/10.1007/s11669-010-9681-0
23. Kryshtal AP, Sukhov RV, Minenkov AA (2012) Critical thickness of contact melting in the Au/Ge layered film system. J Alloys Comp 512(1):311–315. https://doi.org/10.1016/j.jallcom.2011.09.086
24. Akhkubekov AA, Akhkubekova SN, Bagov AM et al (2016) Effect of electrotransfer and indium additions on contact melting and phase formation in bismuth–tin systems. Bull Russ Acad Sci Phys 80(11):1322–1325. https://doi.org/10.3103/S1062873816110034
25. Gudieva OV, Kambolov DA, Korotkov PK et al (2015) Contact melting temperature of small-dimensional phases. Bull Russ Acad Sci Phys 79(6):784–785. https://doi.org/10.3103/S106287381506012X
26. Gladkikh NT, Kryshtal'AP, Sukhov RV (2010) Contact melting in layered film systems of the eutectic type. Phys Solid State 52(3):633–640. https://doi.org/10.1134/S1063783410030273

Method for Identification of Optical Resonances of Metal Films

M. Yu. Barabash, G. G. Vlaykov, A. A. Kolesnichenko and L. V. Ryabov

Abstract In this paper, the properties of thin (10–100 nm, R > 7 Ω) copper films deposited on glass substrates at T = 300 K are considered. The thin films were produced by thermal evaporation in a vacuum with residual gas pressure (10^{-2}–10^{-3} Pa). The substrates had luminescence under exciting by Raman light (Raman scattering) of 30 mW power, with a wavelength of 785 nm. The resonances were identified by comparing the absorption and Raman spectra at excitation wavelengths of 633 and 785 nm. Among films with a close absorption level, films with the largest amplitude of the Raman signal had a structure with the maximum value of optical resonances. Under the influence of exciting light of 30 mW, the Raman spectrum and the color of these films changed. Copper films deposited on glass substrates at a temperature of 300 °C with an electrical resistivity R < 0.5 Ω/square did not have pronounced Raman peaks, their Raman spectrum and color did not change under 300 mW excitation light with a wavelength λ of 785 nm.

Keywords Optical resonance · Thin films · Raman scattering · Clusters

1 Introduction

Investigation of the plasmon nanostructures optics has led to the development of studies in photonics [1], plasmonics [2], sensors [3–7], biomedicine [8], in research of Raman scattering amplification (SERS) [9]. In the overwhelming majority of plasmon resonant devices, metal structures in the form of nanoparticles or layers are used. These films, consisting of small particles, have anomalously high nonlinear optical characteristics, the unifying features of which are the effects of π-delocalization of electrons. As a rule, their characteristics are investigated by optical spectroscopy [10, 11]. The spectra give information about the size, aggregation degree, the thickness

M. Yu. Barabash (✉) · G. G. Vlaykov · A. A. Kolesnichenko · L. V. Ryabov
Technical Centre of National Academy of Science of Ukraine, 13, Pokrovskaya Str., Kiev 04070, Ukraine
e-mail: mbarabash@nasu.kiev.ua

© Springer Nature Singapore Pte Ltd. 2019

A. D. Pogrebnjak and V. Novosad (eds.), *Advances in Thin Films, Nanostructured Materials, and Coatings*, Lecture Notes in Mechanical Engineering,
https://doi.org/10.1007/978-981-13-6133-3_17

of nanoparticles, and the change of electronic structure due to the transition from macroobjects to nanoobjects. The absorption spectra of these films are characterized by the presence of an intense absorption band that is absent in macro-samples and lies in the UV or visible spectral range [11, 12]. The appearance of this band is associated with the manifestation of local surface plasmon resonance as a result of the collective motion of electrons in the field of the electromagnetic wave, under the influence of the particle surface. However, in many cases, the absorption spectra are formed by very wide unstructured bands. This significantly reduces the amount of information obtained about the structure of the films. For this reason, for probing specific objects spontaneous Raman scattering is used. The Raman process involves the exchange of energy between a scattered photon and a scattering molecule, accompanied by a background luminescence. The spectral components of the nonresonant Raman scattering are shifted relative to the frequency of incident radiation by an amount equal to the intervals between the energy levels of the scattering atoms or molecules, which does not depend on the wavelength the incident radiation. The intensity of the Raman band is proportional to the number of molecules in the initial state, the transitions from which generate the given band. The luminescence appears as background; it is a spontaneous emission of a photon by atoms or molecules after their transition to an excited state upon absorption of incident radiation with a frequency lying within the absorption band. If excited atoms and molecules experience collisions leading to redistribution over other excited levels as a result of nonradiative transitions, then broadband luminescence is observed that has an almost continuous spectrum. When investigating the luminescence emitted by an ensemble of particles, it is necessary to distinguish between Raman scattering and luminescence. For this purpose, various closely interrelated properties of both secondary luminescence and the system itself are used, which allow to determine whether the acts of annihilation and birth of photons occur immediately one by one, or other processes intervene between them. This is not a trivial task. In the case of stimulated Raman scattering, a frequency criterion is used, according to which if the lines in the secondary emission spectrum follow the exciting line, then the luminescence is referred to the Raman process. Otherwise, it is believed that the incident radiation is converted into luminescence [13]. In silicate glasses, the Raman signal is formed by impurities, and it can be forced and accompanied by luminescence [14]. As a rule, it is caused by the intrinsic luminescence of the glass matrix as a result of the recombination of charged particles that were formed under the action of exciting Raman light [15]. Usually, the luminescence from glass substrates plays the role of a parasitic background. In the present work, it performed the role of test radiation for analyzing the properties of copper films.

Thin copper films with a thickness of the order of several and hundreds nanometers were obtained by the method of vacuum deposition. This method ensures the production of films of a specified composition with a controlled structure with minimal impurities. The structure of a thin metallic film on the surface of a dielectric substrate starts to form with the appearance of individual clusters (nuclei). Then these clusters grow and aggregate, and Ostwald ripening, and the formation of a continuous film occur. In many cases, the clusters in the process of growth form a system of interacting nanoparticles with fractal properties. These ensembles of particles possess the property of scale self-similarity in a certain region of their dimensions and location, and special nonlinear optical properties [16]. The formation of the film structure occurs in layered, island or intermediate growth regimes, depending on the deposition parameters, on the substrate material and on the type of its surface [17, 18]. Given the complexity of these phenomena, for obtaining films of a given composition and structure, it is necessary to select empirically the parameters of their deposition. At a certain thickness, the film becomes electrically continuous with ohmic conductivity. The electrical conductivity at which the transition from a nonconducting state of condensate on a dielectric substrate to an electrically continuous layer is observed is the percolation threshold. This value also depends on many parameters: deposition mode, the substrate temperature during condensation, film deposition rate, etc. It is related to the kinetics of film formation and serves as a parameter of the film structure.

The purpose of this work is to study the effects of optical resonances in thin copper films on a dielectric substrate using absorption and Raman spectra.

2 Methods and Materials

The substrates were silicate glass (cat. no. 7102, CHINA) plates of 10^{-3} m thickness and dimensions of (2.544) $\times$ 10^{-2} m, with a refractive index $n_0 = 1.51$, transparent in the spectral range of 300–100 nm. During cleaning they were treated in boiling soapy water, followed by washing in distilled water, 30% hydrogen peroxide and drying. Copper contacts (R = 1 Ω) with a width of 5 $\times$ 10^{-3} m were preliminarily deposited by thermal spraying on the substrates preheated to 300 °C. They were used to measure the electrical conductivity of layers during their deposition. The deposition was carried out at a modernized vacuum universal station VUP-5M. The pressure of residual gases in the chamber did not exceed 10^{-2}–10^{-3} Pa. The samples in the form of thin layers (10–100 nm) of copper on a glass substrate were obtained at room temperature by electron evaporation in a vacuum in electric field E = 60–80 V/cm^2. This method has peculiar efficiency and good controllability. The deposited films had a resistance of 7–3.5 $\times$ 10^6 Ω. The distance from the evaporator anode to the substrate was 7 $\times$ 10^{-2} m, that provided a uniform deposition of thin films. The process of copper film deposition was controlled by a developed software with data output (thickness, temperature, absorption, electrical conductivity) to the frequency meter and to the monitor of a personal computer (PC) using a 4-channel ADC with interface board. The thickness of the films was estimated by the shift of resonance frequency (f_{res} = 8 MHz) of a quartz sensor. Particular attention was paid during the deposition process to the electrical conductivity of the film. After deposition, the copper films were held in a vacuum for 30 min. The samples were analyzed by means of optical spectra. The absorption spectra of the glass substrate and copper films on the substrate were obtained on a Shimadzu UV-2401PC spectrometer. Raman spectra of the samples were taken with a Renishaw InVia spectrometer, with two excitation

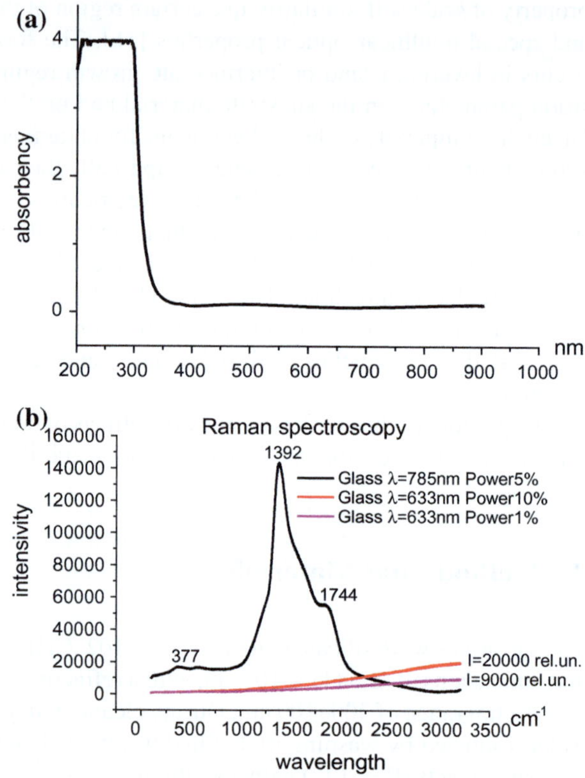

Fig. 1 Glass (cat. no. 7102, China); **a** absorption spectrum; **b** Raman spectrum. Laser radiation powers (633, 785 nm) are 100 and 300 mW

wavelengths of 633 and 785 nm with an analysis of the frequency scattering criterion. The absorption spectra of copper films on glass substrates are shown in Fig. 1.

3 Results and Discussion

Figure 1 shows the absorption spectrum in the visible range and the Raman spectrum of the glass substrate at two wavelengths of exciting light (633 and 785 nm).

The absorption spectrum (Fig. 1a) showed that the glass cat. no. 7102 has a band with an absorption maximum of the order of 5 eV. The spectral dependence of the absorption coefficient in the low-energy section of this band is described by the exponential dependence on the energy of quanta (Urbach's rule). This feature of the fundamental absorption edge is valid for amorphous substances and is due to static variations of the lengths and angles of the bonds, and fluctuations in the potential of the field of the glass matrix. The edge of the absorption band lies near 350 nm, which corresponds to a forbidden band of 3.5 eV. The registration of Raman spectra (Fig. 1b) occurs in two modes: with exciting light ($I_{exc} = 5.0$ kW/cm^2) with a wavelength of

633 nm (1.96 eV), and exciting light (I_{exc} = 7.5 kW/cm^2) with a wavelength of 785 nm (1.6 eV). The spectra have several features, different shapes, and depend on the wavelength of exciting light. The spectrum (I_{exc}, λ = 785 nm) is a set of wide contours with implicitly expressed maxima. It does not have narrow peaks, which, as a rule, correspond to microimpurities in the glass matrix. The amplitude of the scattering signal with excitation light with a wavelength of 633 nm is lower than with λ = 785 nm. The spectrum reveals a monotonic increase in the signal in the low-energy region, which is confirmed by the presence of deep defect levels. If the intensity of exciting light is increased by a factor of 10, the magnitude of the signal increases by a factor of 2. This dependence of the scattering signal is not typical for spontaneous and stimulated Raman scattering, and can indicate to the recombination luminescence of the glass under exciting Raman light. In this case, certain centers are ionized in the excited glass, and during this process two independent particles with different charges appear. The radiation occurs when these particles recombine, and it is characterized by a significant Stokes shift [15].

Figure 2a shows the absorption spectra of copper films with a resistance of 7–3.5 × 10^6 Ω, which were deposited on the glass substrate at T = 300 K. Films with higher resistance have less absorption. In the spectra, one can distinguish the absorption bands characteristic for copper films (the minimum value at λ = 560 nm) [18], which disappear with increasing film resistance (with a decrease in the thickness of the films). Films with a resistance of 7–10 Ω have a maximum of absorption curve near 700 nm, which can be related to the plasmon absorption. Thinner films do not have this maximum. An estimation of the long-wavelength absorption threshold of the spectral curves (λ = 560–570 nm) shows that the optical band gap of the film material is Eg = 2.17–2.221 eV. This corresponds to the data for copper oxide (E_g = 2.18 eV) [19]. The smoothing of the spectral absorption curves of films with a decrease in their thickness can be explained by an increase in the proportion of copper oxide in thin films under atmospheric conditions since copper oxide in the long-wavelength region (λ = 900 nm) does not absorb light [18], whereas copper absorbs. This is complemented by observing a 20% increase in resistance of copper films when air is blown into the evaporation chamber. Figure 2b shows the Raman spectra of a copper film with a resistance of 1 Ω deposited on a glass substrate with a temperature of 300 °C at various intensities of the exciting light. At this deposition mode, copper atoms are introduced into the glass matrix, attach to bridging oxygen, and are resistant to mechanical influences. The Raman spectra contain copper band (615 nm) [20] and a band (1340 nm) typical for the glass substrate. Figure 3a, b shows the Raman spectra of thin films of copper with a resistance of 7, 8, and 10 Ω deposited on glass substrates at different intensities (E = 80, 74, and 70 V/cm) of the field of the electronic evaporator of vacuum unit.

The Raman spectra of films have different forms when excited by light with a wavelength of 630 and 785 nm, because of the different structure of the films and the spectral composition of the glass radiation. The film with the resistance of 8 Ω absorbs more strongly than a film with the resistance of 7 Ω, and weaker than the film with the resistance of 10 Ω (Fig. 2a). Nevertheless, it has the highest intensity of Raman scattering (1600 and 1850 cm^{-1} bands), which indicates the presence

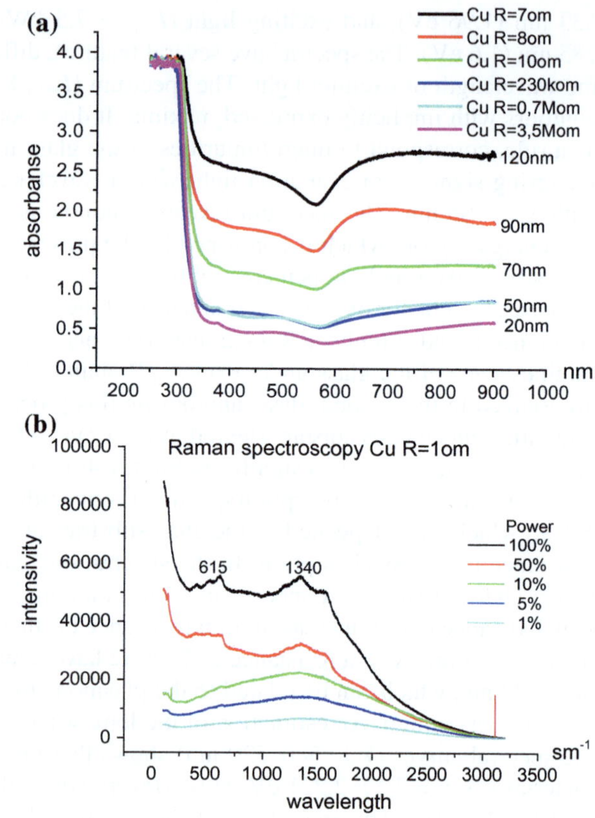

Fig. 2 Copper films on glass (cat. no. 7102) substrate: **a** absorption spectrum, **b** Raman scattering of a copper film deposited on a substrate heated to 300 °C

of nonlinear optical properties that are associated with the fractal structure of the film aggregates. It is known that the fractal structure of aggregates leads to the appearance of significant optical nonlinearities at the surface plasmon frequency. Due to the local anisotropy of the neighborhood of each particle, inherent to the fractal aggregates, no mutual compensation of the local fields occurs. Therefore, optical radiation induces electrodynamic interactions of fractal particles, and near the particles composing of local fractal fields appear tens of times higher than the average field that excites plasmon oscillations [16]. This is accompanied by the appearance of electromagnetic field resonances inside the metal particles of the film, and the transfer of electrons from the metal particles to the glass matrix, which lead to an increase in the radiation intensity. Similar effects were observed in substrates with copper layers with a resistance of a few $k\Omega/cm^2$, although the intensity of the Raman spectrum was lower.

Figure 4 shows the result of a single action of the exciting Raman light on the surface of a copper film, which was deposited in different modes onto a glass substrate. In the first case (Fig. 4a), the film changed its size, color, and structure to a greater extent than the size of the aperture of the Raman light spot. Under the influ-

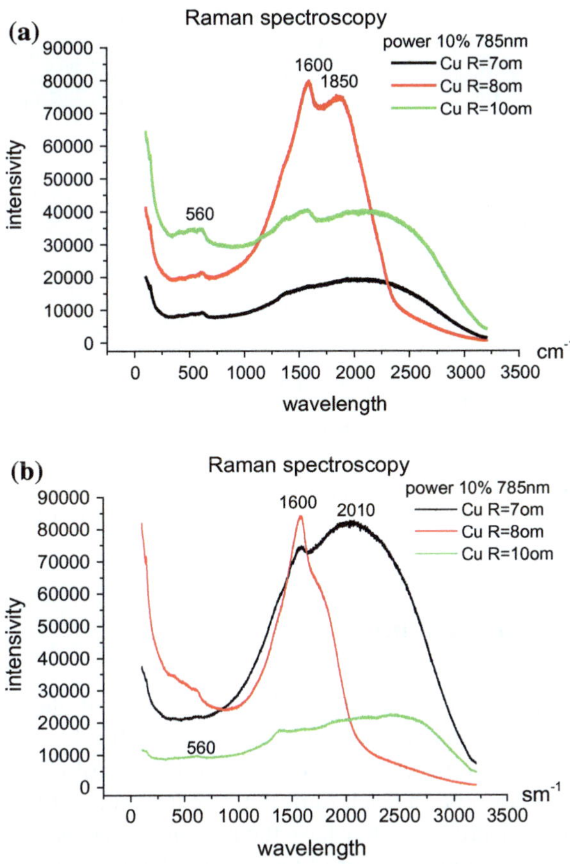

Fig. 3 Raman spectra of copper films on glass substrate: **a** single exposure, **b** double exposure

ence of external factors, various processes can occur inside the films: for example, the reduction of copper film as a result of exposure to light. This is because the substrate-film system is in a nonequilibrium state, as a consequence of the stresses in the film and substrate. Under external influences, these stresses relax. This leads to loss of film continuity and to activation of recrystallization processes [21]. In the second case (Fig. 4b), the surface of copper film under the influence of exciting light does not change the size, color, and structure. The film has a lower resistance (greater thickness) after deposition on heated glass above the glass-transition temperature. The thin film is in equilibrium with the substrate, has good adhesion and is resistant to external impacts.

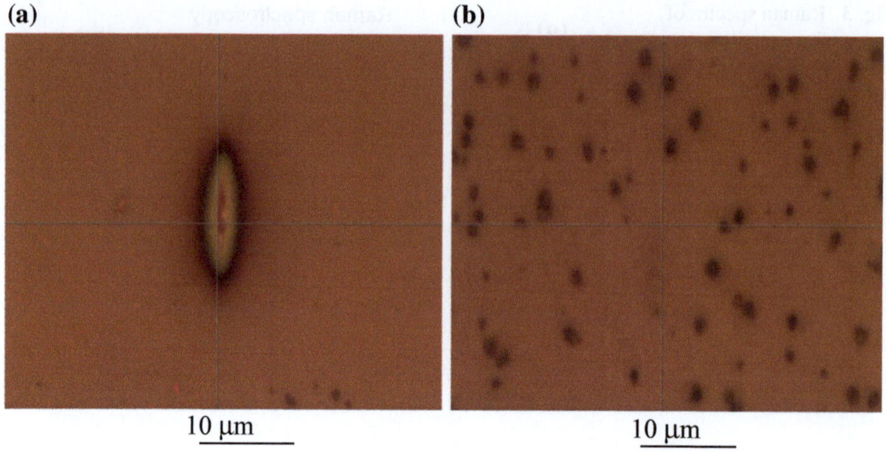

Fig. 4 The surface of copper film on glass substrate after exposure to the exciting light: a film with resistance of 8 Ω, (Pexp-30 mW), б—film with resistance of 0.5 Ω/square (Pexp-300 mW) deposited on glass substrate heated to 300 °C

4 Conclusions

A thin copper film deposited on a luminescent glass substrate at room temperature has optical resonances. The resonances can be identified by comparing the absorption and Raman spectra at different wavelengths of exciting light. Among films with a close absorption level, films with the largest amplitude of the Raman signal have an optimal structure; they have the maximum values of optical resonances. The Raman spectrum and the color of these films change under the impact of exciting light.

Acknowledgements The work was carried out within the framework of Fundamental problems of new nanomaterials and nanotechnologies (2015–2019) of The National Academy of Science of Ukraine.

References

1. Joannopoulos JD, Johnson SG, Winn JN et al (2008) Photonic crystals. Molding the flow of light. Princeton University Press
2. Klimov VV (2009) Nanoplasmonica. Fizmatlit, Moscow
3. Toropov A, Shubina T (2015) Plasmonic effects. In: Metal-semiconductor nanostructures. Oxford Science Publications, Press Oxford University. https://doi.org/10.1093/acprof:oso/9780199699315.001.0001
4. Komisarenko FE, Zhukov MV, Muhin IS et al. (2017) Formirovanie metallicheskih nanoostovkov pri elektronnom obluchenii tonkoy plenki zolota na stekle. J Tech Phys 87(2):306–309 (in Russian). https://doi.org/10.21883/JTF.2017.02.44143.1784
5. Samsonov VM, Kuznezova YV, D'yakova EV (2016) O fractal'nih svoystvah agregatov metallicheskih nanoclasterov na tverdoy poverhnosti. J Tech Phys 86(2):306–309 (in Russian)

6. Korolenko PV, Rizhikova YV (2015) Konstructivnie fractali v modelyah nanoclusterov. In: Proceedings of the conference "M.V. Lomonosov readings, physics section". Moscow State University, Moscow, p 5 (in Russian)
7. Gryn'ko DO, Barabash MY, Borshagivskiy EG et al (2008) Template as a tool of nanotechnology group. Nanosyst Nanomater Nanotech 6(N1):91–103
8. Hlebzov NG (2008) Optica i biofotonica nanochastiz s plasmonnim rezonansom. Quantum Electron 38(N6):504–529 (in Russian)
9. Kim YK, Ok G, Choi SW et al (2017) The interfacing structural effect of Ag/grapheme oxide nanohybrid films on surface enhanced Raman scattering. Nanoscale 9:5872. https://doi.org/10.1039/c7nr00308k
10. Petrov YuI (1986) Klasteri i malie chastizi. Nauka, Moscow (in Russian)
11. Kreibig U, Vollmer M (1995) Optical properties of metal clusters. Springer, Berlin
12. Karpov SV, Slabko VV (2003) Optichiskie I fotofizicheskie svoystva fractal'no strukturirovanih zoley metallov. SO RAN, Novosibirsk (in Russian)
13. Bobovich YS (1972) Lazernaya spektroskopiya spontannogo kombinatsionnogo rasseyaniya slabo vzaimodeystvuyuschih molekul i ee prilozheniya. UFN 108:401 (in Russian). https://doi.org/10.3367/UFNr.0108.197211a.0401
14. Arbuzov VI (2008) Osnovi radiazionnogo opticheskogo materialovedeniya. SPbGUITMO, SPb (in Russian)
15. Trofimov VI, Osadisenko VL (1993) Rost i morfologiya tonkih plenok. Energoatomizdat, Moscow (in Russian)
16. Lewis B, Anderson JC (1978) Nuclation and growth of thin films. Academic Press, New York
17. Lazarev VB, Sobolev VV, Shapligin IS (1983) Himicheskiev i fizicheskie svoystva prostih oksidov metallov. Nauka, Moscow (in Russian)
18. Byub P (1962) Fotoprovodimost' tverdih tel. Foreign. liter-ra, Moscow (in Russian)
19. Colomban P, Henry D (2005) Raman signature modification induced by copper nanoparticles in silicate glass. J Raman Spectrosc 36(9):884–890
20. Kukushkin SA, Osipov AV (1998) Prozessi kondensazii tonkih plenok. UFN 168:1083–1116 (in Russian). https://doi.org/10.3367/UFNr.0168.199810b.1083
21. Morozov NF, Paukshto MV, Tovstik PE (1997) Proceedings of international conference and exhibition: "Micro-Mat 97", Berlin, p 218

Complex Method of the Composite Nanocoatings Formation

A. E. Stetsko and Ya. T. Stetsko

Abstract The Complex method of chemical treatment and diffusive chrome-plating is offered for the strengthening details of machines, diffusive layers restored method the given are investigated.

Keywords Complex method · Chemical heat treatment · Hardened diffusion layer · Microhardness

1 Introduction

Technological methods are the most effective for increasing the resource details. They allow to improve the accuracy of parts manufacturing and assemblies of machine nodes, as well as to provide the optimal (for the given operating conditions) state of the surface layer.

The most efficient method of influencing the conjugated surfaces of machine parts is chemical heat treatment. Chemical heat treatment is one of the most effective methods of influencing the conjugated surfaces of machine parts [1–4]. There are diffusion layers of high quality. After implementation of this layers at operational surfaces they at pass to the core material, which is positive in terms of strength and stability. Technology, engineering and service company with trained professionals and experience of existing equipment are also methods of chemical and thermal processing.

A. E. Stetsko (✉) · Ya. T. Stetsko
Faculty of Computer Engineering Printing, Ukrainian Academy of Printing, UAP, Lviv, Ukraine
e-mail: andrew73@ukr.net

© Springer Nature Singapore Pte Ltd. 2019

A. D. Pogrebnjak and V. Novosad (eds.), *Advances in Thin Films, Nanostructured Materials, and Coatings*, Lecture Notes in Mechanical Engineering,
https://doi.org/10.1007/978-981-13-6133-3_18

2 Materials and Methods

Diffusion carbide coating is an extremely effective method at presents protecting metals and alloys corrosive environments from corrosion. Promising is the use of Complex carbide coatings when combining the positive properties of one-component coatings. Chemical processing must have higher protective properties.

Nevertheless, chemical and thermal processing restricts the use of physical and mechanical properties of the material.

Consequently, there is a problem of worked surfaces restoration of machine parts by methods that are technologically simple, using universal equipment and existing qualifications of employers, studying the types of recovered layers to establish their application specifically to the working conditions of machine parts.

The purpose of this work is the formation of surface diffusion layers on the parts of machines made by a Complex method of chemical processing and diffusive chromium.

For the restoration of machine parts, an offered Complex method of chemical processing and diffusion chromium is proposed [5–13]. It consists in depositing on the prepared surface of a part of a Ni–Co–P chemical coating in aqueous solution of certain formulations and diffusion chromium. As a result of strengthening, a diffusion hardened layer on the surface of the parts is formed. Its structure, depending on the applied modes of the method, consists of several zones, the working of which is an outer composite zone, which reaches 250 μm. In the process of recover, by the Complex method universal equipment is used.

3 Discussion

Chemical treatment consists in applying to the surface of the part that has undergone a preliminary mechanical treatment, refined, degreased, decaffeinated and washed, of chemical precipitation in aqueous solution of a certain formulation. Ni–Co–P chemical coating of amorphous type with thickness 8–12 μm was obtained.

Chemical–thermal treatment (chromium diffusion) was carried out at a temperature of 1050 °C. The part is placed in a retort with a powder mixture of ferrochromium, aluminum oxide and ammonium chloride and sealed with a fused shutter. To form the diffusion layer of the desired structure, an isothermal exposure of 700 °C or 800 °C for a duration of 1 or 1.5 h is used.

Traditional diffusion chromium, which produces thickened layers of 15–30 μm thick, which consist of $Cr_{23}C_6$ and Cr_7C_3 chromium carbides. The Complex method of chemical treatment and diffusion chromium forming a diffusion layer, which has four diffusion zones on carbon steels: the outer composite zone 1, that consists of columnar chromium carbides in the matrix of solid solution Cr in α–Fe thickness of 100–250 μm and the integral microhardness of 11–15 GPa; Zone 2 of a solid solution of chromium in α-iron in the thickness of 10–50 μm and microhardness of 4.5 GPa;

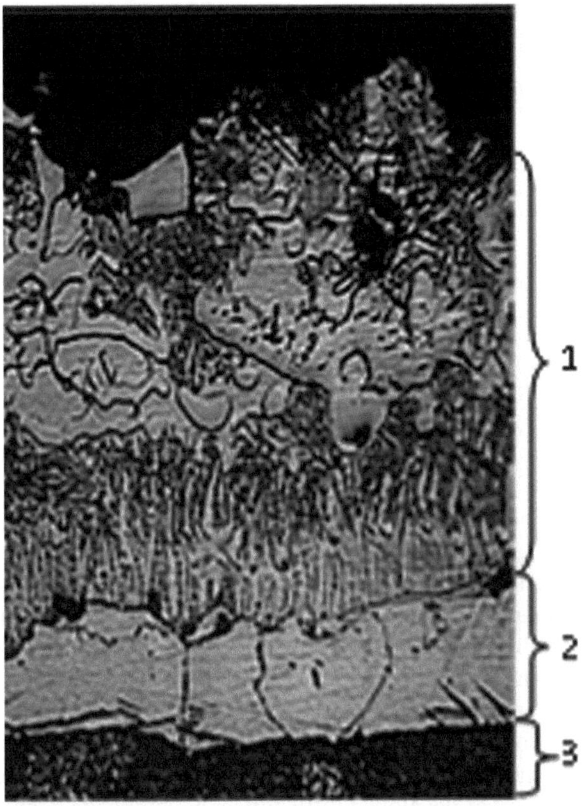

Fig. 1 The microstructure of hardened diffusion layer of chemically Ni–Co–P coating and for the diffusion Cr modes: isothermal holding 1 h at 700 °C, diffusive saturation Cr 5 h at a temperature of 1050 °C; magnification 600×

eutectoid zone 10–30 μm and microhardness of 4 GPa; a degraded zone with a thickness of 100–180 μm and a microhardness of 1.4–1.6 GPa (ferrite component) and a core of the part. The composite structure of the zone allows to significantly increase the work life due to the relaxation of internal micro-stresses in the soft phase—a solid solution of chromium in α-iron during the course of work, while the main load will be the solid phase—columnar grains of high hardness chromium carbide (up to 18 GPa).

After realization of the modes of the Complex chemical treatment and diffusion chromium saturation with 5 h diffusion chromium at 1050 °C and the hourly isothermal shutter at 700 °C, we obtained a steel (0.45%C) reinforced layer consisting of 4 main zones is (Fig. 1). The outer composite zone 1 (thickness up to 100 μm) consists of groups of transcrystale micrograins of chromium carbides. At the same time, there is an incident of carbide grains, which are mainly located near the physical surface (closer to the source of chromium).

The integral microhardness of composite zone 1 of these samples reaches 11 GPa. Here you can observe solid colonies of carbide micrograins. The formation of colonies of carbide micrograin in grains of the source material is observed.

Table 1 Phase analysis

#	Parameters of phase		
	2Θ	d	Phase
1	56.75	2.41	Unidentified
2	65.3	2.12	Cr_7C_3
3	66.8	2.08	Unidentified
4	68.8	2.027	α-Fe (110)
5	88.9	1.625	Unidentified
6	105.8	1.43	α-Fe (002)
7	110.9	1.391	Unidentified
8	151.9	1.181	Cr_7C_3
9	155.8	1.171	α-Fe (112)
10	159.8	1.163	Cr_7C_3

The phase analysis (Table 1) performed on these samples showed the presence of Cr_7C_3 chromium carbides and α-iron.

The distribution pattern of diffusion elements indicates the presence of a large amount of Ni and Co in zone 2 of a solid chromium solution in α-iron (Fig. 2), which indicates active diffusion processes. The peak of chromium and, consequently, the sharp drop in the concentration of other elements in these places indicates the presence of formed carbide colonies in the composite zone.

After comprehensive restoration of steel parts (0.45%C) with chemical coating got morphology developed a diffusion layer composite zone structure (Fig. 3) morphology of developed diffusion layer composite zone's structure has been obtained. The composite layer 1, a thickness of 250 μm, typical developed network stretched to the physical surface of carbide grains, which are placed in a matrix of a solid solution of chromium in α-Fe. On the border of zone 2, these grains fused into a continuous strand of carbides.

The carbides grain does not get the physical surface to 30–50 μm. The integrated composite microhardness zone has 12 GPa.

Schedule distribution diffusion elements (Fig. 4) confirm that the surface area of a solid solution of chromium in α-Fe, in which the diffusion elements of concentration are stable. The concentration of chromium peak (above 50%) of the concentrations of chromium is reached (under falling concentration in these areas of other elements) by depth location carbide grains. Zone 2 of homogeneous solid solution of chromium in α-Fe content is characterized by high values of Ni (10%) and Co (up 3%). Thickness of Zone 2 consist of average 25–40 μm. Definitely, the nickel and other elements pushes carbon from the subsurface zone, and observes the formation of elongated carbide grains, which are located on the border zones 1 and 2 toward physical surface. The use of the Complex method of chemical treatment and isothermal holding allows reinforcing development at a fairly great depth. Composite Zone 1 with this layer, in parts friction pairs reaches 250 μm. It is providing an increased service life. Phase analysis (Table 2) was performed on this sample twice through a large unidentified

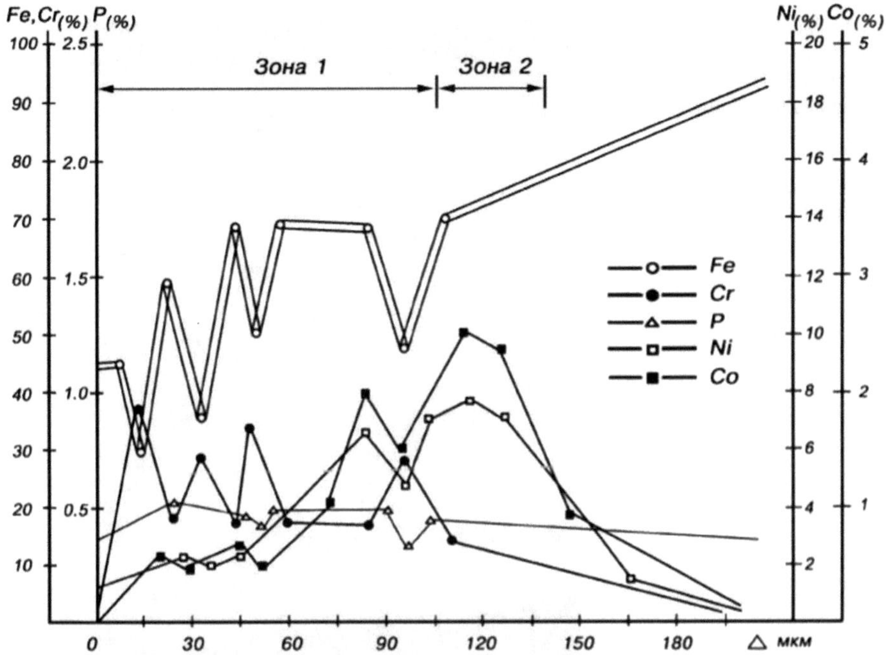

Fig. 2 Diffusion concentration distribution of elements diffusion layer of hardened diffusion layer of chemically Ni–Co–P coating and for the diffusion Cr modes, steel (0.45%C)

Table 2 Phase analysis

#	Parameters of phase		
	2Θ	d	Phase
1	44.4	2.56	Unidentified
2	55.1	2.09	Unidentified
3	74.2	1.606	Unidentified
4	89.1	1.381	Unidentified
5	145.3	1.051	α-Fe (002)

peaks, indicated Complex state of stress hardened layer. But it is possible to say about the availability of a large number of Cr_7C_3 and α-Fe.

Diffusion layer on steel (1.0%C) (Fig. 5), obtained after implementation of regimes of the Complex recovery method: 7 h. diffusion chromium (at 1050 °C) with a pre-hourly isothermal endurance (at 800 °C) with chemical treatment, is characterized by high volumetric content of chromium carbide grains in composition zone 1. These grains have an elongated shape, and on the boundary of Zone 2 form a solid carbide colony with carbide chromium seeds, which have merged among themselves. This visually observes the difference between these types of grains. The extracted

Fig. 3 The microstructure of hardened diffusion layer of chemically Ni–Co–P coating and for the diffusion Cr modes: isothermal holding 1 h at 800 °C, diffusive saturation Cr 7 h at a temperature of 1050 °C; magnification 600×

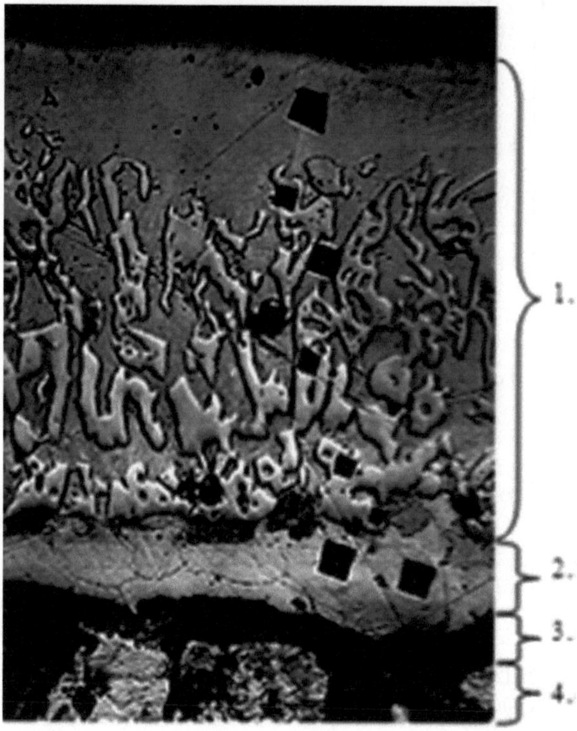

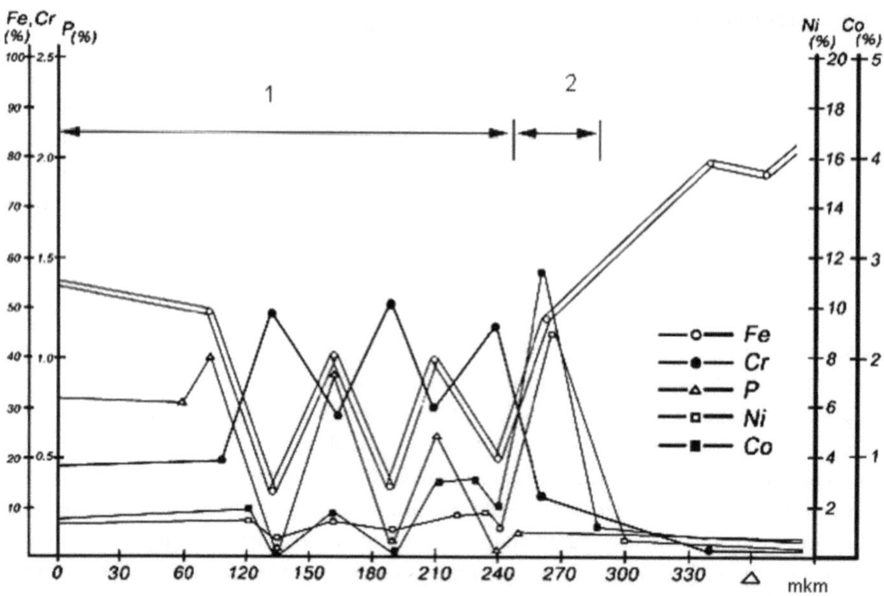

Fig. 4 Diffusion concentration distribution of elements diffusion layer of hardened diffusion layer of chemically Ni–Co–P coating and for the diffusion Cr modes steel (0.45%C)

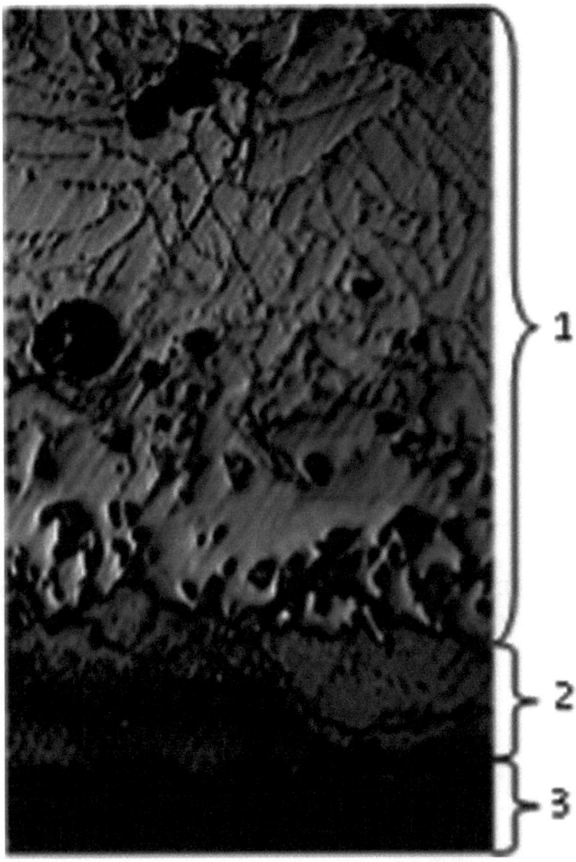

Fig. 5 The microstructure of steel (1.0%C) of hardened diffusion layer of chemically Ni–Co–P coating and for the diffusion Cr modes: isothermal holding 1 h at 800 °C, diffusive saturation Cr 7 h at a temperature of 1050 °C; magnification 600×

carbide grains are $Cr_{23}C_6$, and the carbide colony is Cr_7C_3. Integral microhardness of composite zone 1 reaches 15 GPa.

Part of the distribution diagram of the diffusion elements (Fig. 6), which reflects the zone 1 in depth, shows a fairly even content of diffusion elements with small fluctuations, and only at the end of the zone, at the location of the solid carbide colony, growth of chromium content (up to 50%) is observed, and, correspondingly, a drop in the concentration of other elements. The content of diffusion elements in zone 1 is sufficiently high and stable in the thickness of the zone (except for the location of solid carbides near the boundary with zone 2). The homogeneous zone 2 of the Cr solid solution in α-Fe is characterized by an unstable thickness, and in places the zone 1 comes close to zone 3. This sample has a maximum Ni and Co content in zone 2 (about 18% Ni and over 3% Co).

After strengthening with a Complex method of chemical processing and diffusion chromium on steel (1.0% C) at us, similarly as on steel (0.45% C), all characteristic zones, including eutectoid zone 3, are observed. Phase analysis (Table 3), conducted

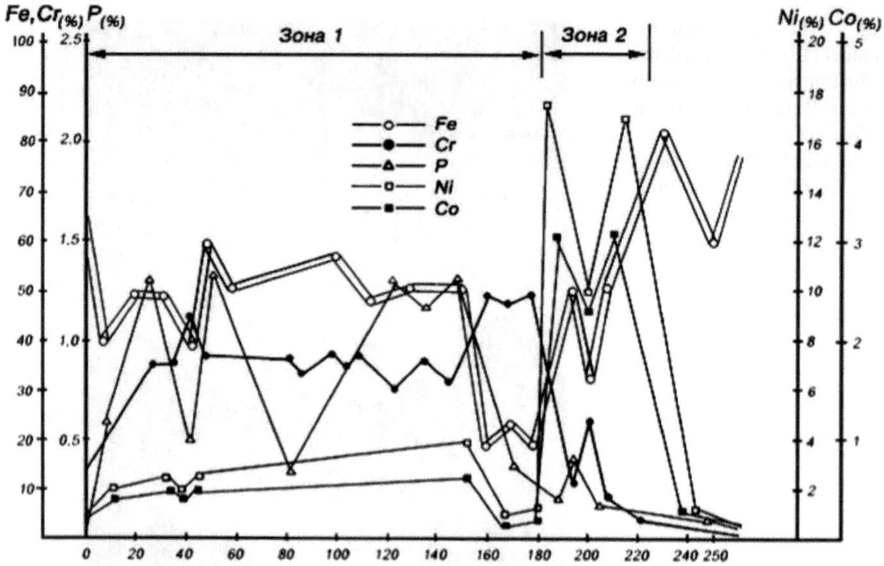

Fig. 6 Diffusion concentration distribution of elements diffusion layer of hardened diffusion layer of chemically Ni–Co–P coating and for the diffusion Cr modes steel (0.45%C)

Table 3 Phase analysis

#	Parameters of phase		
	2Θ	d	Phase
1	65.5	2.12	Cr_7C_3
2	67	2.075	γ-Fe
3	68.8	2.027	α-Fe(110)
4	79.1	1.8	γ-Fe
5	85.1	1.69	Unidentified
6	91.2	1.603	Unidentified
7	105	1.44	α-Fe
8	125.7	1.29	Unidentified
9	128.8	1.27	γ-Fe
10	148	1.19	Unidentified
11	154	1.17	α-Fe(112)

twice showing that in this layer there are chromium carbides of Cr_7C_3, $Cr_{23}C_6$, as well as α- and γ-iron.

Distribution graphs of diffusion elements (Ni, Co, P) show (Figs. 2, 4 and 6) that their higher concentration is placed between the chromium carbide grains. For example, for Ni, which, despite the increase in strength, simultaneously increases and the plasticity of the material is observed its increase in 10%. Thus, the carbide grains seem to be covered with plastic material, which can relieve the internal stresses that arise during the work of the part, which increases the lifetime of the workpiece, restored by a Complex chemical treatment and diffusion chromium.

4 Conclusions

The implementation of a Complex method of chemical processing and diffusion chromium plating for the restoration of machine parts allows to obtain diffusion layers of different structure, thickness and hardness.

The optimums for chemical treatment are the parts of machines made of structural medium carbon steel. It is obtained on the reconstructed parts that are made of steel (0.45%C), diffusion layers, the thickness of the outer composite zone reaches 250 μm, and the integral microhardness of 12 GPa. The phase composition of the composite zone consists of chromium carbides $Cr_{23}C_6$, Cr_7C_3 and α-Fe (002).

The Complex method of chemical processing and diffusion chromium can strengthen the parts made of high-carbon steels. The diffusion layer obtained on them has a composite zone of high integral microhardness (about 15 GPa) and a thickness up to 200 μm with densely located carbide grains. The phase composition of the composite zone consists of chromium carbides $Cr_{23}C_6$, Cr_7C_3 and α-Fe (002).

The parts made of medium carbon steel and working under dynamic loads should be strengthened under the regimes that provide 5 h diffusion chromium with an hourly isothermal shutter at 700 °C. This allows obtaining diffused restored layers, the outer composite zone of which consists of fine-dispersed, filamentous chromium carbides in a Cr crystalline matrix in α-iron with a thickness of about 100 μm and an integral microhardness of up to 11 GPa. The phase composition of these layers is chromium carbide Cr_7C_3, α-Fe (110) and α-Fe (002).

References

1. Muralil M, Sambathkumar M, Senthil MS et al (2014) Micro structural and mechanical properties of AA 7075/TiO_2 in situ composites. Univ J Mater Sci 2(3):49–53
2. Kulka M, Pertek A (2003) Characterization of complex (B + C + N) diffusion layers formed on chromium and nickel-based low-carbon steel. Appl Surf Sci 218(1–4):114–123
3. Oberg E (2017) Heat-treatment of steel—a comprehensive treatise on the hardening. Read Books Ltd.
4. Campbell FC (2008) Surface hardening of steel. In: Elements of metallurgy and engineering alloys. ASM International, Materials Park
5. Stetsko AE (2017) Effect of chemical vapor deposition on the morphology of the nanocomposite layer in the complex method. In: Fesenko O, Yatsenko L (eds) Nanophysics, nanomaterials, interface studies, and applications. NANO 2016. Springer proceedings in physics, vol 195. Springer, Cham. https://doi.org/10.1007/978-3-319-56422-7_46
6. Stetsko A (2016) Composite Coatings Formed by Complex Methods of Surface Hardening. In: Fesenko O, Yatsenko L (eds) Nanophysics, Nanophotonics, Surface Studies, and Applications.

Springer Proceedings in Physics, vol 183. Springer, Cham, Switzerland. https://doi.org/10.1007/978-3-319-30737-4_12
7. UA 115722. C23C 10/18, C23C 10/58, C23C 22/60, 11 Des 2017 (patent)
8. UA 110046, C23C 10/02, C23C 10/40, C23C 10/32, C23C 22/02; 10 Nov 2015 (patent)
9. UA 109285, C23C 10/32, C23C 22/62, C23C 22/05, C23C 10/02; 10 Jul 2015 (patent)
10. UA 112605, C23C 10/02, C23C 10/32, C23C 10/40, C23C 22/05, 26 Sept 2016 (patent)
11. UA 115643, C23C 10/26, C23C 10/32, C23C 22/60, 27 Nov 2017 (patent)
12. UA 124311, C23C 22/36, 10 Apr 2018 (patent)
13. UA 116738, C23C 22/05, C23C 22/60, 25 Apr 2018 (patent)

The Role of a Thin Aluminum Film in the Reconstruction of Silicon's Near-Surface Layers

R. Lys, B. Pavlyk, D. Slobodzyan, J. Cebulski and M. Kushlyk

Abstract The effect of a thin film of aluminum on morphology and change in the characteristics of p-type silicon crystals was investigated in the work. It is established that a sprayed metallic film receives the nanosized complexes from the inner part of the crystal to the surface. These defects are effective centers of dislocation-related electroluminescence of structures based on p-Si. Additional elastic deformation (for several hours) of silicon leads to an increase in their concentration. This can be used to increase the efficiency of the luminescence of silicon monocrystals in the infrared region.

Keywords p-type silicon · Dislocation electroluminescence · Deformation · Near-surface layer

1 Introduction

Modern optoelectronics puts forward significant demands to materials used as radiation sources and receivers. Mono- and polycrystalline silicon is used effectively in integrated electronics, but the radiation recombination in it is by several orders lower than in straight-band semiconductors. Therefore, it is necessary to refuse from using such silicon as the basis in a closed loop receiver-transmitter of radiation or modify its properties.

R. Lys (✉) · B. Pavlyk · D. Slobodzyan · M. Kushlyk
Department of Sensor and Semiconductor Electronics, Ivan Franko National University of Lviv, 107 Hen. Tarnavskoho str., Lviv, Ukraine
e-mail: lys_r@ukr.net

D. Slobodzyan · J. Cebulski
Department of Experimental Physics, University of Rzeszów, 1, Pigonia str., Rzeszów, Poland

M. Kushlyk
Division of Physics and Technology of Wide-Band-Gap Semiconductor Nanostructures, Institute of Physics, Polish Academy of Sciences, Warsaw, Poland

© Springer Nature Singapore Pte Ltd. 2019

A. D. Pogrebnjak and V. Novosad (eds.), *Advances in Thin Films, Nanostructured Materials, and Coatings*, Lecture Notes in Mechanical Engineering,
https://doi.org/10.1007/978-981-13-6133-3_19

One of the promising ways of solving this problem is the production of light-emitting structures (LES) with the so-called dislocation electroluminescence (DEL), which was detected in the silicon with a high density of dislocations [1–3].

The idea of using DEL to increase the radiation recombination is promising, since the dislocation centers of luminescence have high-temperature stability [4, 5], and the radiation energy of the centers responsible for the long-wave region of luminescence (1.2–1.6 μ) coincides with the area of the greatest transparency of fiber optics and locates in the region of silicon transparency.

However, the presence of dislocations in near-surface layers of the semiconductor causes an increase in the diffusion of defects from the inner part of silicon and their interaction in the presence of additional external or internal mechanical and thermal fields. In particular, such mechanical stresses in the silicon substrate may be induced by building up of a metal or dielectric film. Consequently, the characteristic feature of dislocations in silicon crystals is the presence of regions around them with a high concentration of point defects (Cottrell's clouds).

For the study of DEL in p-Si crystals, the authors [5] proposed a method according to which LES were made by applying a thin Al film onto a silicon substrate. However, the mechanical stresses in the joints between the film and the substrate, which arise due to the irregularity of the parameters to their lattices, were not taken into account [6]. For example, the irregularity of the parameters of the crystal lattices of silicon and aluminum is about 25%.

The purpose of the work is to investigate the effect of the deformation fields, which have been formed due to the spraying of a metal film on the electro physical characteristics of a uniformly elastically deformed crystals of silicon of the p-type.

2 Methodology

The analyzed samples were cutout of a single-crystal plate of p-type silicon ($\rho = 10\ \Omega$ cm). After the standard procedure of cutting, mechanical and chemical polishing, the aluminum contacts in the form of films (thickness ~100 nm) were applied on both sides with the orientation (111) in a vacuum chamber VUP-5M at a pressure of 10^{-5} Pa and a temperature of 650 K, in such a way that the edges of the surface were under the film, and the central part was free.

Computer modulation was performed to determine the energy of the deformed layer of silicon.

The measurement of the electrical conductivity was carried out in a vacuum cryostat at a pressure of residual gases of about 10^{-3} Pa in the process of applying one-piece compression to the ends (in the direction [112]) with a force 15–40 MPa and at a deformation rate of 8 or 32 μm/min.

After the measurement of the mechanistic-stimulated change in electrical conductivity, aluminum etching, selective etching and structural investigation of the specimen surface (111) were performed using an optical microscope.

3 Results

The changes of mechanically stimulated electrical conductivity, along the direction of deformation of p-type conductivity silicon samples under the action of an elastic one-axis load (σ), were investigated in the work. The process of deformation (compression-release) was repeated several times during the experiment. During the deformation, the resistance of the experimental samples gradually increases and reaches saturation after four compression-release cycles (Fig. 1). A further load does not lead to a significant increase in the resistance of the samples. The change in the deformation rate practically did not affect the general view of the dependences ρ(σ).

A gradual and prolonged increase in the resistance of the samples (Fig. 1) is observed between the individual identical deformation cycles and in the case of compression suspension (within 1/2 h). This indicates the diffusion processes in the crystal-line lattice. The processes involve point defect or small accumulation of defects. It is known that the annealing of such defects, especially at room temperature is a long process [7, 8] (we have established [9, 10] that the resistance value returns to its original value seven days after the action of elastic deformation). These defective diffusions capture the main charge carriers, which leads to a decrease in the conductivity of the experimental samples. Also, due to the receiving of the defects by the surface from the inner part of the semiconductor (especially under the deposited metal contacts), their number in the near-surface layer of silicon increases. This, accordingly, also leads to a decrease in the mobility of the main charge carriers (due to increased scattering), and, consequently, to resistance increase.

At the initial stages of the deformation of the samples, their micro plastic deformation occurs due to the displacement of the growth dislocations (the mechanism of dislocations depletion). Above some critical tension, the micro creeping is controlled by the emergence of new dislocations from heterogeneous sources (the mechanism of propagation) [11].

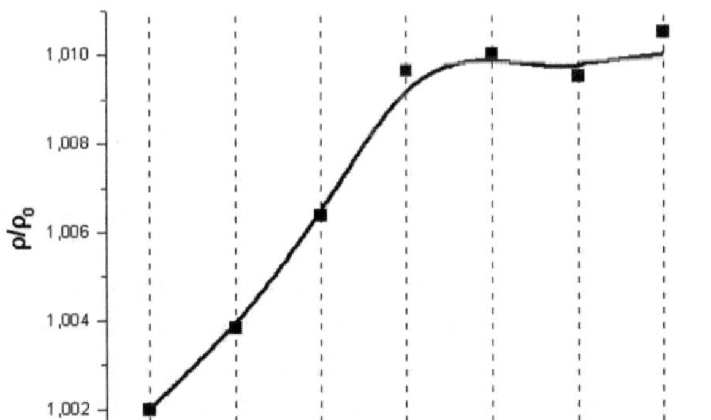

Fig. 1 The relative value of the specific resistance of p-Si samples, measured successively in compression-release deformation cycles at a constant value of the mechanical load of 25 MPa

In the paper [11], it was established that under the action of low-temperature deformation at small and medium stresses (400 MPa) the dislocation in silicon originates only in thin subsurface layers and in contrast to the growth dislocations, it appears in the form of small pits of etching. Under these conditions, the main type of structural defects that are created during a short-term (several minute) deformation are vacancies and vacancy-admixture clusters. In the long trials (several hours, days), the dislocations also occur in the near-surface layers of crystals. The deformation dislocations are of heterogeneous origin and arise on inhomogeneities that are periodically located in the "growth lanes." The growth dislocations do not manifest themselves as sources of new dislocations.

In the process of contraction, the resistance of experimental samples continues to grow [12]. This can be explained by a slight increase in the interatomic distance for Si, which causes the acceleration of diffusion of defects to the surface. In addition to the motion of dislocations, there is also a deflection of fixed at the ends dislocations in silicon under the action of a load. The return of such half loops of deflection to the initial position (release) also leads to the capture of the charge that takes part in the current transfer.

Consequently, the change of the defective background in the crystal under the action of the load affects the concentration and mobility of charge carriers with which the electrical conductivity of the semiconductor is related.

During the modeling, it was taken into account that the film applied on the surface of the silicon leads to the appearance of several transition layers, which expand towards the inner part of the crystal [13]. Thus, the Al-p-Si system is a heterogeneous structure. It consists of disordered silicon and regions of dislocation grids. In addition, the thickness of each transition layer is also affected by the presence of structural defects on the surface of the original silicon. An increase in the density of structural defects leads to an increase in the number of channels of accelerated diffusion when the metal is applied, which leads to an increase in the mechanical stress at the metal-pad boundary.

From Fig. 2 it is seen that the energy curves of the deformed layer and dislocations intersect at the point X_m—the maximum depth of capture of defects by a surface layer of edge dislocations, whose position depends on the magnitude of the external deformation and the initial parameters in the precipitation of a metallic film [6]. As the thickness of the deformed layer increases, which corresponds to the change in the conditions of the film spraying (substrate temperature, spray velocity, spray time after spraying, etc.), the displacement of the X_m point toward the depth of the crystal takes place.

The confirmation of this is the study of the surface morphology after the selective etching. Figure 3 shows microscopic photos of the surface of silicon after removing the aluminum film. It can be seen that the increased concentration of defects is formed under the deposited film after the compression and deformation processes. Based on the analysis of selectively etched surface made using AFM (Fig. 4), it is evident that these are groups of superficial structural defects, which differ from the dislocation pits of etching by the depth and shape of the edges. According to the article [6], the authors consider these defects to be an accumulation of point defects.

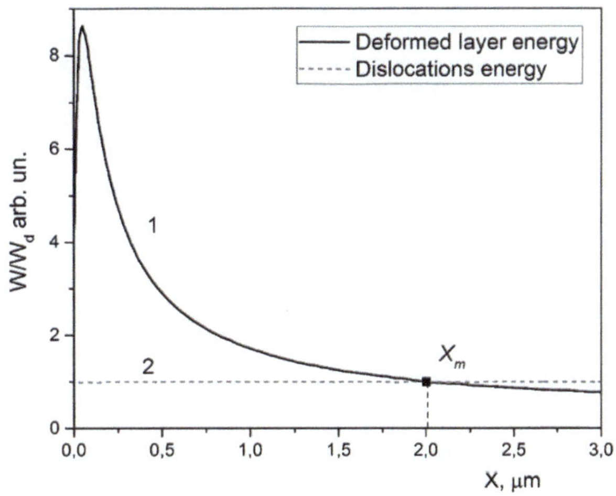

Fig. 2 Graph of the dependence of interaction energy of the defects with the silicon layer de-formed by aluminum film (curve 1) and the dislocation (curve 2) from the coordinate

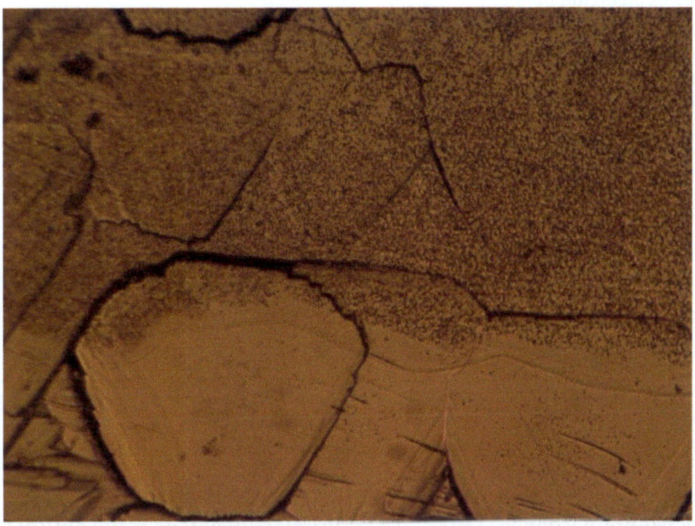

Fig. 3 Microphotography of the surface (111) of silicon coated with Al-film made with an optical microscope at X70 enhancements

In contrast to these crystals, on the samples surfaces on which the Al-film was only deposited (without any external deformation or temperature treatment), such an increase in the concentration is not observed. This allows us to assert precisely that these defects were received in the near-surface layer of the crystal from the inner part of the crystal due to the external deformation and the presence of Al-film.

So, such a technique allows us to obtain a high concentration of dislocations in Si crystals without the risk to destroy the samples. In turn, such values of dislocation concentrations on sample surface allow to obtain a high intensity of dislocation-

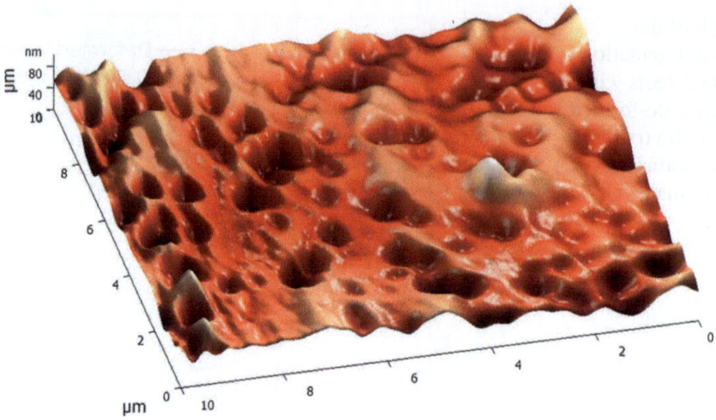

Fig. 4 Microphotography of the surface (111) of silicon coated with Al-film, made with AFM

Fig. 5 Value of changing of LES external energy efficiency, depending on values of pressure on a crystal (elastic deformation)

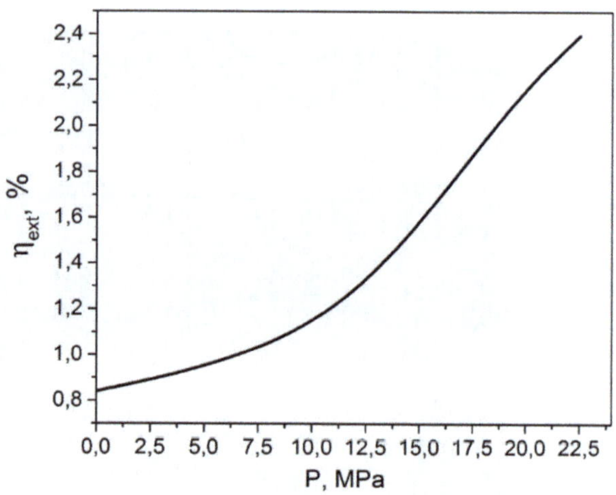

related electroluminescence (DEL). An additional advantage is the ability to observe a high-performance DEL at room temperature.

In paper [14] method to obtain external energy efficiency from the integral intensity of DEL emission are represented. These calculations show that such manufacturing method of emitting structures in combination with the elastic deformation can increase the value of η up to 2.4% (Fig. 5).

Changes of external energy efficiency dislocation-related electroluminescence can be explained by several mechanisms. Firstly, as it was shown in papers [6, 9] in subsurface layers of Si concentration of impurities, dopants and defects could increase under the influence of elastic deformation. This process takes place because of the presence of deformational potential caused by deposited Al film on silicon substrate. It needs to be admitted, that such deformation potential appears as a result of fast forming a heterogeneous structure with bordering of Al and Si layers (lattice mismatch Δa ~23%) [6]. In areas of subsurface layers with such deformation potential and near dislocation cores Cottrell atmosphere and agglomerations of point defects can generate. Also presents in such layers of semiconductor additional deformation potential accompany with the restructuring of corresponding structural defects. As a result, concentration of radiation centers connected with dislocation increase in crystals subsurface layer. Also, these experimental results are in well correspond with simulation, that was represented in the paper [6].

The second mechanism can be explained by the restructuring of band structure under the influence of elastic deformation. Energy levels of recombination centers, in the band gap, split up to several energy sublevels. This process leads to an increase of the FWHM value of some DEL bands in emission spectra. However, in the band gap of Si relative shift of the energy levels maximum and minimum are presented. Such shift affects to the possibility of radiative recombination that, in turn, leads to increasing or decreasing of DEL intensity.

4 Conclusions

It is established that the surface of silicon, which is located under the aluminum contacts has the property of receiving defects from the inner part of the crystal.

An additional elastic monovalent deformation of p-Si crystals causes an increase in the surface resistance of the samples. This indicates additional diffusion processes in the crystalline lattice, which involve point defects or small accumulation of defects.

It has been experimentally established that the long-term elastic deformation of p-Si crystals leads to the dislocations grows in near-surface layers of crystals. This fact was successfully used to increase the efficiency of the dislocation-related electroluminescence of silicon-based light emitting structures.

Acknowledgement Slobodzyan D. acknowledge the support of this work provided by the International Visegrad Fund (scholarship no. 51810525).

References

1. Drozdov NA, Patrin AA, Tkachev VD (1976) Recombination radiation at dislocations in silicon. JETP Lett 23(11):651–653
2. Sauer R, Kisielowski-Kemmerich C, Alexander H (1986) Dissociation-width-dependent radiative recombination of electrons and holes at widely split dislocations in silicon. Phys Rev Lett 57:1472–1475
3. Sobolev NA (2010) Defect engineering in implantation technology of silicon light-emitting structures with dislocation-related luminescence. Semi-Conductors 44(1):1–23
4. Pavlyk BV, Kushkyk MO, Slobodzyan DP (2015) About the nature of electroluminescence centers in plastically deformed crystals of p-type silicon. J Nano Electron Phys 7(3):03043

5. Pavlyk BV, Kushkyk MO, Slobodzyan DP (2017) Origin of dislocation luminescence centers and their reorganization in p-type silicon crystal subjected to plastic deformation and high temperature annealing. Nanoscale Res Lett 12(358):1–8
6. Pavlyk BV, Kushlyk MO, Didyk RI et al (2013) Electrophysical characteristics of near-surface layers in p-Si crystals with sputtered Al films and subjected to elastic deformation. Ukr J Phys 58(8):742–747
7. Dzhafarov TD (1991) Radiation-stimulated diffusion in semiconductors. Energoatomizdat, Moscow
8. Vinetsky VL, Holodar GA (1979) Radiation physics of semiconductors. Naukova dumka, Kiev
9. Pavlyk BV, Lys RM, Didyk RI et al (2014) Investigation of conductibility changes of irradiated p-Si crystals by X-rays during the re-coverable deformation. Phys Chem Solid State 15(2):297–303
10. Lys RM, Pavlyk BV, Didyk RI et al (2017) Change in surface conductivity of elastically deformed p-Si crystals irradiated by X-rays. Nanoscale Res Lett 12(440):1–7
11. Nadtochiy VO (2006) Extended abstract of Cand. science. Dissertation, Kharkiv National University, Kharkiv
12. Pavlyk BV, Lys RM, Didyk RI et al (2015) Features of the uniaxial elastic deformation of X-ray-irradiated p-Si crystals. Semiconductors 49(5):638–643
13. Smyntyna VA, Kulinich OA, Glaubrman MA et al (2005) Simulation of the current transfer process in real metal-silicon structures with Schottky barrier. Phys Chem Solid State 6(4):656–660
14. Slobodzyan DP, Kushkyk MO, Pavlyk BV (2018) Electroluminescence energy efficiency of Si-structures with a different compound of nanoscale dislocation complexes. Appl Nanosci. https://doi.org/10.1007/s13204-018-0729-5

Structure and Optical Properties of CN_x–Ni and CN_x–$EuCl_3$ Core/Shell Nanostructured Films

A. Prudnikov, Yu. Pashkevich, K. Lamonova, P. Maksimchuk, O. Viagin, Yu. Malyukin and M. Pas'ko

Abstract The structural and optical properties of carbon nitride films doped with nickel or europium have been investigated. All films have been manufactured by the original reactive dc magnetron sputtering method with high pressure of the buffer gas and low energy plasma. It was shown that the formation or disintegration of the core (metal)-shell (carbon) structures in the Ni-C-N films can be achieved under some growth conditions. It was found that luminescent properties of the C-N-Eu films are strongly affected by the basic morphology of C-N films. Particularly the fullerene-like CN_x : $EuCl_3$ film demonstrates high luminescence efficiency at 620 nm band and Eu^{3+} concentration 4.5 at.%.

Keywords Core/shell structures · Magnetron sputtering · Carbon nitride · Nanostructures · Scanning probe techniques · Luminescent properties

1 Introduction

Hybrid nanomaterials with the metal encapsulated in a carbon cover (Me@C) attracts attention of researchers in recent years [1]. The carbon cover covering ferromagnetic nanowires, provides their protection against mechanical damages and also protects from oxidation. It is a reason, why the area of potential application of such materials is broadening [2]. There are two categories of areas of using a material of Me@C type: (1) Me@C structures can be applied as a separate functional element nano dimen-

A. Prudnikov (✉) · Yu. Pashkevich · K. Lamonova
O. O. Galkin Donetsk Institute for Physics and Engineering, National Academy of Sciences of Ukraine, Kiev 03680, Ukraine
e-mail: anatoliy.prudnikov@hotmail.com

P. Maksimchuk · O. Viagin · Yu. Malyukin
Institute for Scintillation Materials, National Academy of Sciences of Ukraine, Kharkiv 61001, Ukraine

M. Pas'ko
Donetsk National Medical University named after M. Gorky, Lyman 84404, Ukraine

© Springer Nature Singapore Pte Ltd. 2019

A. D. Pogrebnjak and V. Novosad (eds.), *Advances in Thin Films, Nanostructured Materials, and Coatings*, Lecture Notes in Mechanical Engineering,
https://doi.org/10.1007/978-981-13-6133-3_20

sional sources of electrons for electron microscopy and electron beam lithography, electrochemical gauges, and functional scanning probes [3]; (2) Me@C structures can be applied as a volume material manufacturing micro- and nanoelectromechanical devices [4], magnetic recorders, medical diagnostics and "transportation" of medical products into a human body [5], nanocontainers to storage of metal materials [6], etc.

There are many methods of obtaining the metal materials covered with a carbon cover. Among them: laser ablation, an arc method, chemical vapor deposition (CVD), physical vapor deposition (PVD), plasma enhanced chemical vapor deposition (PECVD) [7–9]. In these methods, the metal particles or clusters which are the nucleation centers of obtained structures, are entered into the formed carbon phase.

The listed methods of obtaining the Me@C structures require high temperatures. In present work, the method of obtaining similar materials without the high temperatures is proposed a method of magnetron deposition. Rather the high pressure of buffer gas also is the positive factor of the used method that leads to the formation of nanoclusters [10]. Use of a composite target allows obtaining the nanoclusters of "core-shell" type. This metal is a core, and carbon is a shell. Formation of nanocolumnar and fullerene-like structures on a substrate is possible using this method. Physical and chemical properties of these nanoclusters strongly depend on core and shell structure. The specified method allows controlling a chemical composition and the relative size of a core and a shell [11, 12].

The hybrid luminescent nanodiamond materials with phosphors formed in them exhibit unique physical properties and have been intensively studied recently also [13]. Due to the introduction of various impurities, it is possible to obtain luminescent micro- and nanocrystals of diamonds, which are promising for application in various fields of quantum photonics, computer science, optical microscopy, and sensors.

Of particular interest are nanometer diamonds, in which one or two luminescent centers are introduced. Such objects are promising as single-photon light sources necessary for the creation of quantum optical computers and cryptography. Unique and attractive from the point of view of the application of the properties of luminescent diamond microresonators (spherical UDD particles) are the photostability of optical parameters at room temperature, the unique mechanical properties and chemical resistance of the diamond matrix, which has optical transparency from ultraviolet to the infra-red areas and high refractive index, controlled brightness and large quantum luminescence yield (approaching unity), as well as short lifetimes of luminescence.

At the same time, it is necessary to notice, that the structural and phase state of nanoobjects with the structure of "metal core-carbon shell" is studied insufficiently.

The purpose of the present work is to obtain the hybrid Ni-C-N and Eu-C-N-films protected by a carbon cover, and to investigate their structure, optical properties, and modification at heating in the growing process.

2 Experimental Details and Films Preparation

Nanostructure hybrid Ni-C-N and Eu-C-N-films were grown by a method of magnetron sputtering of composite graphite-nickel and graphite-europium target onto cover and quartz glass substrates in an atmosphere of argon and nitrogen. A double disk structure was used as a target in which the nickel disk with apertures was imposed onto solid graphite disk (99.9% purity). Planar magnetron with the flat cathode and the ring anode created a plasma. Magnetron discharge power did not exceed 20 W. In a growth chamber, gas pressure was 26 Pa (150 mTorr). Series of films of C-N-Ni systems have been grown at a variation of the carbon concentration in a range of 10–50 at.% and substrate temperature T_S in a range of 30–300 °C. The quantitative analysis also has shown the presence of oxygen and nitrogen in films in the measure of 1–2 at.%. The growth time of films also varied over a wide range from 10 s (research of nanoclusters deposition process onto a substrate surface) up to one hour.

Substrates previously exposed to two-stage cleaning in chemically pure ethyl, isopropyl spirits, and isopropyl spirit vapors. Final cleaning of a substrate surface has been carried out directly ahead of deposition process by argon plasma of glow discharge.

We have already noticed that our method have a low temperature of plasma and rather high pressure of the buffer gas at obtaining of Ni@CN_x-nanoclusters.

The surface morphology of the films was analyzed by a JEOL ISM-6490 LV scanning electron microscope and also was studied by probe nanolaboratory Ntegra Aura atomic force microscope. The relative content of atoms in the samples was determined by energy dispersive X-ray (EDX) spectroscopy using an INCA Penta FETx3 (Oxford Instruments) attachment. The structure of samples was studied by X-ray diffraction (XRD) on a DRON-3 instrument using CoKα radiation.

Optical absorption spectra in the 190–1100 nm range were measured using the SPECORD 200 spectrophotometer (Analytik Jena, Germany). Luminescence spectra were obtained using the computer-controlled setup based on the grating monochromator. The samples were excited by the fourth harmonic (266 nm) of the YAG: Nd^{3+} pulsed laser (NL202 model, EKSPLA, Lithuania).

The laser beam was focused onto the sample by a quartz lens. The laser radiation was directed on the sample at an angle to the light axis of the monochromator, to minimize the light-exposing of the registration system. Luminescence of the sample was focused on the entrance slit of the MDR-23 monochromator using the capacitor, a spectral separation in the measured range reached 15 cm^{-1}. The beam inside the monochromator spread in a spectrum and fed onto the exit slit. Spectrum scanning on the output slit was carried out automatically during rotation of the spectral device dispersing element using a special mechanism and stepping motor.

The photomultiplier Hamamatsu R9110 operated in photon counting mode was used as a photodetector. The spectral resolution in the measured range was 15 cm^{-1}. An electrical signal from the PMT has been transmitted through the CAMAC system to a personal computer. Stepper motor control and luminescence spectra recording were carried out using the own development software Spectral measurement v.17. A resulting number of pulses on each wavelength were averaged by several measurements to minimize random errors.

3 Formation of Nanostructural Hybrid Ni@CN_x-Films on Growth Substrate Surface

Dense nanostructural hybrid Ni-C-N films on cover glass substrates at a temperature from 30 to 320 °C at constant growth time of 10 min (thickness ~100 nm) are grown up by a method of low-temperature magnetron deposition of Ni@CN_x-nanoclusters with the average size of a core in a range from 8 to 12 nm.

Figure 1 shows the dependence of the crystallites size D from substrate temperature for various samples. The first, which attracts attention, is the increase in crystallites size with increase in substrate temperature. The fact, that after heat treatment in growth process the size of grain D increases in comparison with an initial state of nanoclusters, obviously testifies that this growth occurs as a result of diffusion processes during temperature treatment of a substrate.

Figure 2 shows X-ray diffraction of nanostructural hybrid Ni-C-N-films obtained at various temperatures of a substrate and with various concentration of carbon in deposited nanoclusters. The concentration of carbon is determined in the synthesized films on cold substrates.

Analysis of the X-ray diffraction spectra of the investigated films has shown, that there is a set of lines which are close to the lines (111), (200), (220), and (113) of the nickel FCC-lattice. The difference consists only in some broadening and shift of lines towards smaller angles. Shift of Bragg's reflexes (Table 1) directly evidences about an increase in parameters of the FCC-lattice of nickel. It confirms a conclusion

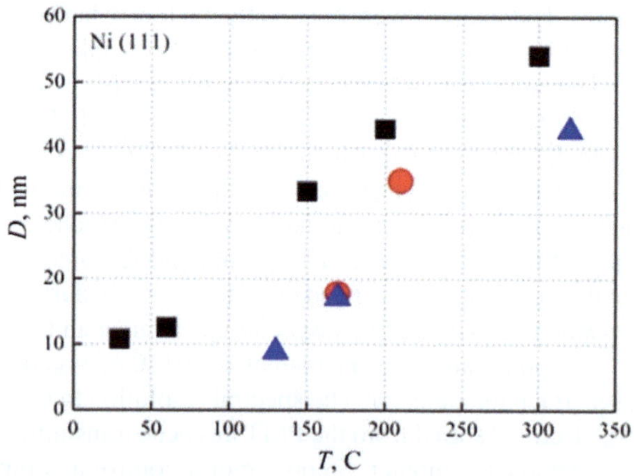

Fig. 1 Dependences of Ni crystallites size D from substrate temperature. Carbon concentration in deposited Ni@CN_x-nanoclusters, %: filled square—32, filled triangle—40, filled circle—61

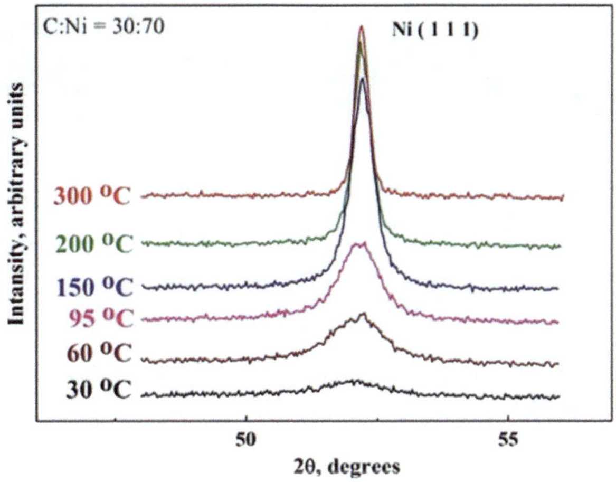

Fig. 2 X-ray diffraction of nanostructural hybrid Ni-C films, obtained at various substrate temperature. Carbon concentration in deposited nanoclusters is 30%

Table 1 Shift of Bragg's reflexes of nanostructural hybrid Ni-C-N-films

2θ, Ni(111) (°)	51.97	52.01	52.11	52.21	52.20	52.20
T (°C)	30	60	95	150	200	300

about the presence of a solid solution of carbon in the nickel matrix. Whereas the broadening of reflexes evidences about rather small size of crystallites from which films are constructed. In X-ray diffraction spectra the lines which are distinct from lines of nickel crystallite are not observed. The big shift of a X-ray lines observed in some samples, specifies in the big concentration of carbon in Ni@CN_x-nanoclusters, in the assumption of a solid solution of carbon in nickel.

So, in spite of the fact that in system Ni-C-N on the equilibrium phase diagram there is no appreciable solubility of carbon in a massive state (the maximum solubility of carbon in nickel does not exceed 2.7 at.%), in nanoclusters the solid solution Ni(C) with wide area of homogeneity is formed. Introduction of carbon atoms into a crystal lattice of nickel increases its parameter and, respectively, weakens exchange interactions.

Formation of supersaturated metastable solid solution Ni(C) and its subsequent disintegration, possibly, is controlled by kinetic parameters (diffusion) and thermodynamic factors (temperature and the nanoparticles size).

Figure 3 shows SEM-images of nanostructural hybrid Ni-C-N-films which have been grown up at various substrate temperatures. First of all occurrence of pores at temperature 270 °C. We assume that pores are formed at intensive evaporation of carbon into the atmosphere in the process of film growth. With the reduction of carbon concentration in films, considerable change of their structure is observed. At low substrate temperature (60 °C), the structure of films is amorphous. Nanoparticles which have the spherical form and have not the crystallographic facet are observed on a surface.

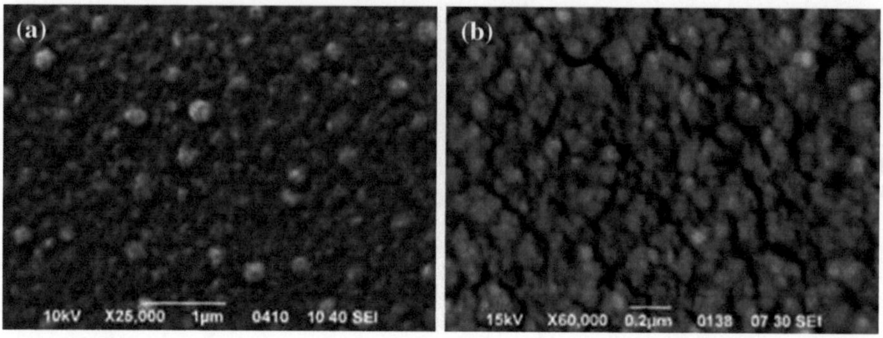

Fig. 3 SEM images of nanostructural hybrid Ni-C-N-films, grown at various substrate temperature: °C: **a** 60, **b** 270

They have a characteristic globular structure with the average size of elements ~10–25 nm with the association of single globules into aggregates up to 250 nm in size.

Thus, it is possible to conclude, that depending on growth parameters (concentration of carbon in nanoclusters, substrate temperature, flow clusterization, etc.) clusters provide processes of self-organization of substance on a substrate: there can be the centers of nucleation of nanocolumns, growing normally to the substrate surfaces, or to provide the globular character of films growth.

Nucleation processes study on a substrate surface shows that cluster character of films growth of Ni-C-N system takes place at magnetron sputtering. Clusters of the sputtered material of nanometer in size, which are already formed in plasma, are deposited on a substrate surface.

By the method of X-ray analysis, it is established, that Ni@C nanoclusters contain except nickel in a metal core (with the introduced nitrogen) in the form of a metastable supersaturated solid solution of Ni(C) with carbon covering. Ni(C) solid solution can break up by both the process of nanoclusters synthesis and by the subsequent temperature growth of nanostructures on a substrate. Composite of Ni-C nanostructures consist of Ni nanoparticles with a carbon cover. The carbon covering can be in an amorphous state or have a graphite-like structure.

The metastable solution of Ni(C) disintegrates under the fast increase in temperature. This process is accompanied by carbon diffusion, as the most mobile atom, from nickel crystallites into a particle surface, as well by the removal of carbon from the sample. Disintegration degree of the supersaturated solid solution depends on temperature and time at which solid solution is unstable.

4 Structural and Optical Properties of the Carbon Nitride Fullerene-Like Films Doped with $EuCl_3$

A series of samples of nanostructured nitride-carbon CN_x films and films supplemented with rare-earth Eu^{3+} with concentrations of dopant 4–5 at.% was synthesized by the method of reactive magnetron sputtering of graphite and graphite-europium target in a pure nitrogen atmosphere and a mixture of the argon-nitrogen atmosphere.

In Fig. 4 shows the luminescence spectra of fullerene-like films of carbon nitride with europium CN_x: $EuCl_3$ with concentrations of dopant: 4.5 at.% (#516). For comparison also shows the luminescence spectrum of compounds $EuCl_3$, the target of which was carried out of spraying europium.

It can be seen that the spectra of the samples have obvious differences in the splitting and intensity of the spectral lines compared to $EuCl_3$. This indicates that europium in films does not remain in the form of the $EuCl_3$ crystalline, but occupies a position with disturbed neighboring ordering, for example, on the surface or inside the nanostructures of carbon nitride (Fig. 5).

Comparison of the experimental electronic levels of energy of the ion Eu^{3+} of nanostructured carbon nitride films doped with europium CN_x: $EuCl_3$ (#516) with the luminescence spectrum from the $EuCl_3$ sample shows that the symmetry of the environment of the europium ion in the film is lower than the symmetry of the environment in a crystal of $EuCl_3$. In the crystal field of the $EuCl_3$ compound, the symmetry of the free Eu^{3+} ion decreases, thus degenerate terms split, forming singlet and doublet states [14].

The transitions between the terms $^5D_0 \rightarrow {}^7F_0$ and $^5D_0 \rightarrow {}^7F_1$ are observed in the region 16,800–17,300 cm^{-1}. Meanwhile, the transitions between the terms $^5D_0 \rightarrow {}^7F_0$ are forbidden, or have low intensity. The transitions between the levels $^5D_0 \rightarrow {}^7F_3$ obtained in the region 15,390–15,410 cm^{-1} are generally not observed in $EuCl_3$,

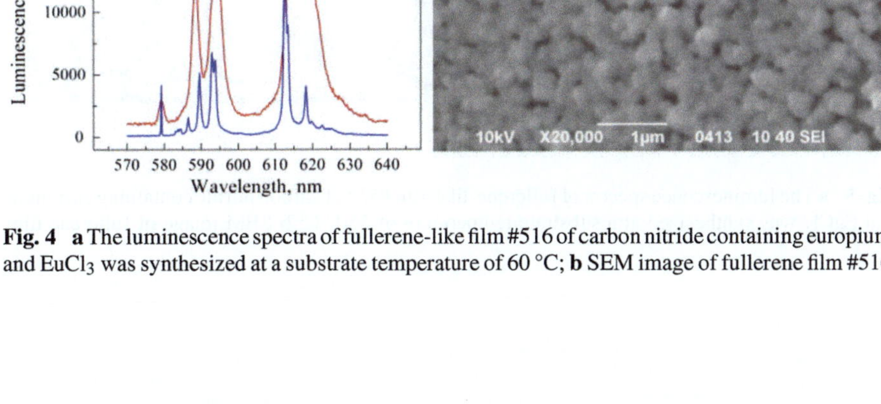

Fig. 4 **a** The luminescence spectra of fullerene-like film #516 of carbon nitride containing europium and $EuCl_3$ was synthesized at a substrate temperature of 60 °C; **b** SEM image of fullerene film #516

but they appear weakly in films of carbon nitride activated by europium. It can be argued that in the $EuCl_3$ group this transition group has practically zero intensity. The transitions between the levels $^5D_0 \rightarrow {}^7F_4$ are observed in the energy region 14,200–14,500 cm^{-1}. We note that intensive luminescence is observed in the energy range 15,680–16,330 cm^{-1}, which corresponds to the transitions $^5D_0 \rightarrow {}^7F_2$.

Radical changes in the luminescence spectra occur in the study of diamond-like films of carbon nitride doped with europium, CN_x: $EuCl_3$ (#549) (Fig. 6). In the luminescence spectrum, broad lines are observed instead of narrow lines. The luminescence spectrum of an amorphous diamond film of carbon monoxide doped with europium, CN_x: $EuCl_3$ (#549), and annealed at 225 °C during the growth process, contains broad, nonstructural bands in the wavelength regions of 580–630 and 680–720 nm (Fig. 6).

It is known that the intensity of radiation of a luminophore increases with an increase in the degree of crystallinity. Consequently, for a nanoparticle, the degree of a long-range order of a crystalline structure is not the main reason for increasing of the luminescence intensity. Another reason for changing of the luminescence intensity with the change in the sizes of nanostructured phosphors may be the change in the short-range order in the environment of europium cations. Quantitative short-range estimates can be determined by measuring the width at half-height (FWHM) of the most intense Stark luminescence line at 620 nm. The intensity of the emission band I_{620} is determined by the dipole-electric transition $^5D_0 \rightarrow {}^7F_2$, hypersensitive to the closest environment of the Eu^{3+} ion. The emission band 595 nm is determined by the dipole-magnetic transition $^5D_0 \rightarrow {}^7F_0$, insensitive to the nearest environment [15]. The ratio of the amplitude intensity of the emission bands I_{620}/I_{595} is used to explain the effect of the size of the CN-like luminescent particles: Eu^{3+} [16] on the

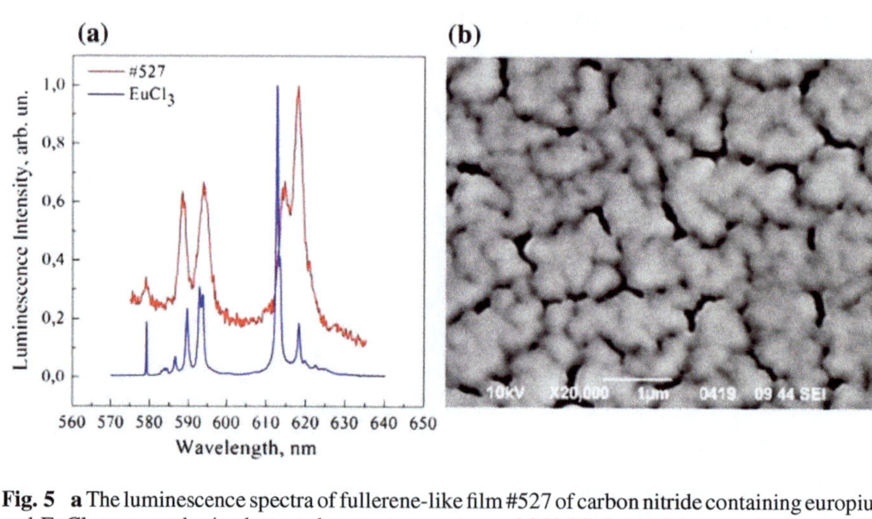

Fig. 5 **a** The luminescence spectra of fullerene-like film #527 of carbon nitride containing europium and $EuCl_3$ was synthesized at a substrate temperature of 250 °C; **b** SEM image of fullerene film #527

emission spectra. Our study reveals an increase in FWHM (samples with 4.5 at.%), indicating that there is an increase in the number of areas with broken long-range order around Eu^{3+} ions in the crystallites. In this case such samples show a lower intensity of luminescence. Increasing of the size of crystallites leads to an increase in the luminescence intensity of I_{620}. The competition of these effects determines the value of the integral efficiency of the luminescence.

5 Conclusions

A series of samples of nanostructured carbon nitride-CN_x films and films supplemented with Ni or rare-earth Eu^{3+} was synthesized by the method of reactive magnetron sputtering of graphite, graphite-nickel and graphite-europium target in a pure nitrogen atmosphere and a mixture of argon-nitrogen atmosphere. Methods of scanning electron microscopy and optical spectroscopy revealed that depending on the substrate temperature and the concentration of nitrogen different types of nanostructures are forming in the carbon nitride films CN_x. The mechanism of formation and disintegration of the core-shell structures in Ni-C-N films has been discussed. Investigation of the luminescence spectra of europium doped and undoped samples showed that films with europium on a glass substrate with a concentration of 4.5 at.% (#516) exhibited a luminescence with a spectrum characteristic for $EuCl_3$, but with frequency shift. The values of the frequency shifts and the modification of the spectra are determined by comparison with the luminescence spectrum of the initial compound $EuCl_3$. The morphological features of the films are clearly reflected in the luminescence spectra. So a thin structure of the Eu^{3+} spectrum is absent in films with

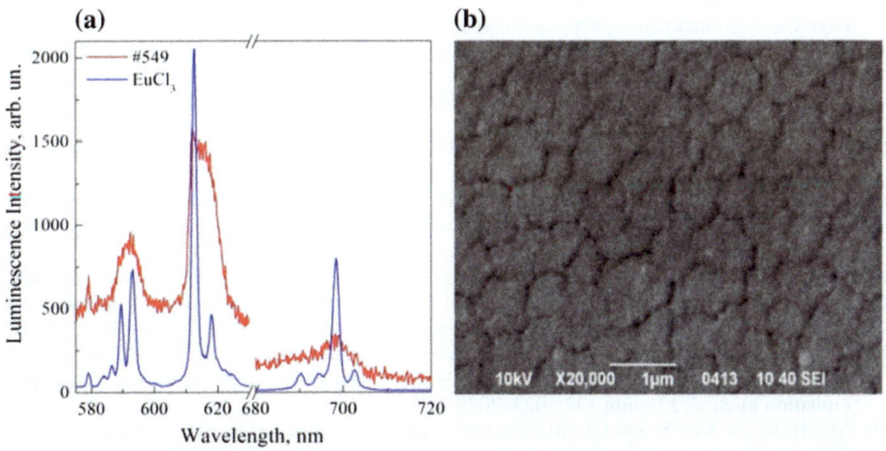

Fig. 6 **a** The luminescence spectra of diamond film #549 of carbon nitride containing europium and $EuCl_3$; **b** SEM image of diamond film #549

amorphous diamond-like phase. Whereas the fine structure of spectrum is preserved in the samples with fullerene-like phases. It is found that the phosphor films of CN_x: $EuCl_3$ in fullerene-like phase possesses abnormally high luminescence efficiency at 620 nm band and at Eu^{3+} concentration 4.5 at.%. The nature of an increase in the intensity of luminescence is associated with the structural rearrangement of nano-sized crystallites enriched with the activator Eu^{3+}. We conclude that it is caused by the formation of a hybrid ($EuCl_3$@CN_x) luminescent nanostructured material.

Acknowledgement This work was carried out within the framework of the project No. 53/18-N of the NAS of Ukraine Program "Fundamental issues of creation of new nanomaterials and nanotechnologies."

References

1. Sunny V, Kumar D, Yoshida Y et al (2010) Synthesis and properties of highly stable nickel/carbon core/shell nanostructures. Carbon 48:1643–1651
2. El Mel A, Gautron E, Angleraud B et al (2011) Synthesis of nickel-filled carbon nanotubes at 350 °C. Carbon 49:4595–4601
3. Melechko V, Merkulov V, McKnight T et al (2005) Vertically aligned carbon nanofibers and related structures: controlled synthesis and directed assembly. J Appl Phys 97:041301–041339
4. Vajtai R, Wei B, Zhang Z et al (2002) Building carbon nanotubes and their smart architectures. Smart Mater Struct 11(5):691–698
5. Sacanna S, Rossi L, Pine D (2012) Magnetic click colloidal assembly. J Am Chem Soc 134(14):6112–6115
6. Gy Kovács, Koós A, Bertoni G et al (2005) Structure and spectroscopic properties of C–Ni and CN_x–Ni nanocomposite films. J Appl Phys 98(3):034313–034316
7. Merkulov V, Melechko A, Guillorn M et al (2002) Simpson Effects of spatial separation on the growth of vertically aligned carbon nanofibers produced by plasma-enhanced chemical vapor deposition. Appl Phys Lett 80:476–478
8. Navas D, Hernandez-Velez M, Vazquez M et al (2007) Ordered Ni nanohole arrays with engineered geometrical aspects and magnetic anisotropy. Appl Phys Lett 90:192501–192503
9. Rossi G, Rapallo A, Mottet C et al (2011) Magic polyicosahedral core-shell clusters. Phys Rev Lett 3(10):105503–105504
10. Kashtanov P, Smirnov B, Hipple R (2007) Magnetron plasma and nanotechnology. Phys Usp 50:455–488
11. Ferlauto A, Alvarez F, Fonseca F et al (2004) Co-sputtered carbon-nickel nanocomposite thin films. J Metastable Nanocryst Mater 20:700–704
12. Galakhov V, Shkvarin A, Semenova A (2010) Characterization of carbon-encapsulated nickel and iron nanoparticles by means of X-ray absorption and photoelectron spectroscopy. J Phys Chem C 114:22413–22416
13. Aharonovich I, Greentree A, Prawer S (2011) Diamond photonics. Nat Photonics 5:397–405
14. Babkin R, Lamonova K, Prudnikov A et al (2017) Formation mechanism of luminescence spectra of carbon nitride films doped by europium chloride CN_x: $EuCl_3$. J Lumin 186:247–254
15. Gowd G, Patra M, Songara S et al (2012) Effect of doping concentration and annealing temperature on luminescence properties of Y_2O_3: Eu^{3+} nanophosphor prepared by colloidal precipitation method. J Lumin 132:2023–2029
16. Yoo H, Im W, Kim S et al (2010) Continuous nano-coating of Y_2O_3: EU^{3+} phosphor shell on SiO_2 core particles and its photoluminescence properties. J Lumin 130:153–156

Acoustic Activation of Radiation Defect Migration in Nanocrystalline Material Under Neutron or Ion Bombardment

A. I. Kalinichenko and V. E. Strel'nitskij

Abstract In the model of the nonlocal thermoelastic peak of low-energy ion, the acoustic action of low-energy (1/10 keV) ions or recoils on point defects in nanocrystalline material was investigated. The amplitude of the thermoacoustic pulse generated due to energy transfer from an incident ion to the target material is found using a spherical approximation of the thermoelastic peak. The spatial dependence of the amplitude of generated pulse adjusted for sound absorption follows a power law rather than the exponential one. Due to such long-range action, the acoustic pulses from the ions of Ti+ bombarding the nanocrystalline TiN target can activate migration of interstitial defects at distances up to 15 nm and considerably reduce the activation energy of migration process at distances up to 50 nm.

Keywords Nanocrystalline material · Ion bombardment · Thermoacoustic wave · Radiation defect migration

1 Introduction

The progress in the nuclear industry, nuclear power engineering, and space technology required the development of radiation-resistant materials and electronic equipment capable of operating under the influence of ionizing radiation. So, structural materials of nuclear reactor play an important role in ensuring the safety of nuclear power plants. The question of ensuring the stability of the structure of fuel elements, whose shells are exposed to the influence of high temperatures by the corrosive effect of the decay products of nuclear fuel and coolant, and also irradiation with neutrons, fission fragments, and γ radiation, is especially urgent. Possibilities of the traditional materials used in this field—austenitic stainless steels and zirconium compounds—have almost been exhausted, and the further development of nuclear power requires the development of new radiation-resistant materials. Such promising

A. I. Kalinichenko (✉) · V. E. Strel'nitskij
National Science Center "Kharkov Institute of Physics and Technology", Kharkiv, Ukraine
e-mail: aikalinichenko@kipt.kharkov.ua

© Springer Nature Singapore Pte Ltd. 2019

A. D. Pogrebnjak and V. Novosad (eds.), *Advances in Thin Films, Nanostructured Materials, and Coatings*, Lecture Notes in Mechanical Engineering,
https://doi.org/10.1007/978-981-13-6133-3_21

materials for nuclear reactors, in particular for fuel rod coatings, are nitrides of titanium, chromium, aluminum or their composites that have a nanocrystalline structure [1–3].

The results of the experiments show that the nanostructured nature of the material ensures its increased radiation resistance [4]. There are several explanations for this. First, the nanostructured material has a large integral area of grain boundaries and creates a large number of sinks for point defects induced by irradiation. Along with this, there is a reason to suppose that the energetic recoil ions produced by the interaction of neutrons with matter can accelerate the migration of defects. Experiments show that low-energy ions, which irradiate solid, or recoil ions that arise during neutron irradiation, affect the structure and properties at depths ≥ 10 nm, which are much larger than the ion range. One example of such a long-range action is the so-called radiation-stimulated diffusion of implanted ions during irradiation. The penetration depth of impurities for all the investigated cases was much higher than the projected mean free path of ions of given energy [5]. Another example of the long-range action of incident ions on the target structure is an increase in the transparency of carbon films when they are treated with Ar+ ions with an energy of E ~ 1.2 keV. The film thickness exceeded the ion mean free path in tens of times [6].

Earlier, it was already noted the possibility of generation by single neutrons of powerful thermoacoustic pulses associated with pulsed heating of small volumes of matter in neutron scattering [7]. However, the reasonable estimate of the magnitude of the emerging mechanical effects is impossible without analysis of the space-time characteristics of the region where energy transferred from neutron is localized.

Such an analysis can be carried out on the basis of the model of the nonlocal thermoelastic peak (NTP) of low-energy ion describing thermodynamics of "fast" processes in the interaction of the ion with solid and which made it possible to explain a number of effects observed during ion bombardment of solids, in particular, in the ion deposition of coatings [6, 8].

Indeed, the recoil, being essentially a heavy low-energy ion, is the cause of the emergence of the NTP—the overheated and overpressured region of the nanometer size. The spatiotemporal and thermodynamic characteristics of such a region are determined within the framework of the NTP model, which in turn allows one to investigate the thermo-mechanical and kinetic effects arising near the ion path.

We confine ourselves to the investigation of the acoustic action of neutrons of small and intermediate energies En $\leq$ 100 keV Such neutrons form a significant part of all reactor neutrons in PWRs. Elastic scattering of neutrons on nuclei prevails in this range of energies and for atomic weights A $\leq$ 60. The maximum energy of recoil nuclei (ions) is given by the expression:

$$E_m = \frac{4AE_n}{(A+1)^2}$$

In the case of neutron scattering by Ti or Cr atoms, neutrons with energies of En $\leq$ 100 keV generate recoil nuclei (i.e., Ti^+ and Cr^+ ions) with energy E $\leq$ 6 keV. Ions with such energies create the overheated regions in the TiN or CrN targets that can be considered in the NTP model.

The goal of this paper is the theoretical study of the radiation-acoustic mechanism for accelerating the migration of radiation defects in nanocrystalline TiN (CrN) coating when irradiated with low-energy (1–10 keV) Ti^+ (Cr^+) recoils or ions.

2 Mathematical Model

The thermodynamic conditions in the nanometer-sized region around the ion trajectory in different materials, where the ion energy loss is thermalized, were investigated [8, 9]. It was shown that substantial heating (more than 1000 K) and stress (more than 10 GPa) occurs in this region, which will be referred as the nonlocal thermoelastic peak (NTP) of the low-energy ion. According to SRIM2000 [10] simulation results, NTP of the low-energy ion can be approximated by a spherical region containing energy $\eta(E)E$ with its center in the middle of the projective range $l(E)$ and initial radius $R(E) = l(E)/2 + R_T$. Here $\eta(E)$ is the fraction of phonon loss of the ion, R_T is the radius of the " sphere of smearing" of the point thermal source at the ion-ion relaxation time τ.

Radius $R(E)$ and the fraction of the ion energy $\eta(E)$ in the NTP of the ion Ti^+ in TiN coating and ion Cr^+ in CrN coating are shown in Fig. 1.

At the normal incidence of the ion, the volume of the near-surface NTP is determined by the expression:

$$V = \frac{4\pi}{3}R^3 - \frac{\pi}{3}\left[R - \frac{l}{2}\right]^2\left[2R + \frac{l}{2}\right]$$

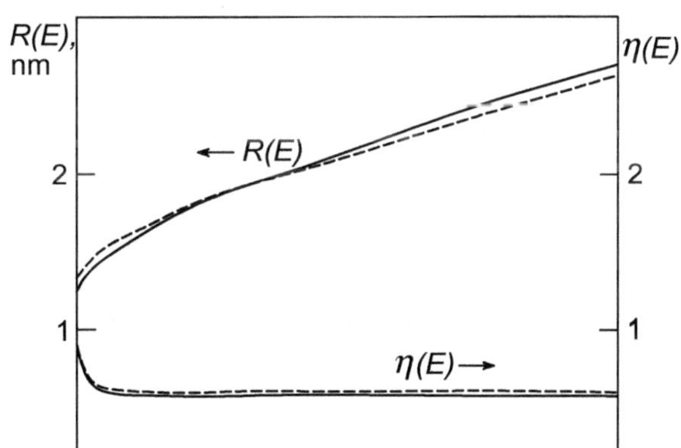

Fig. 1 Radius R of the NTPs and the fraction of phonon loss η of Ti^+ ion in TiN (solid curves) and Cr^+ ion in CrN (dashed curves) depending on ion energy E

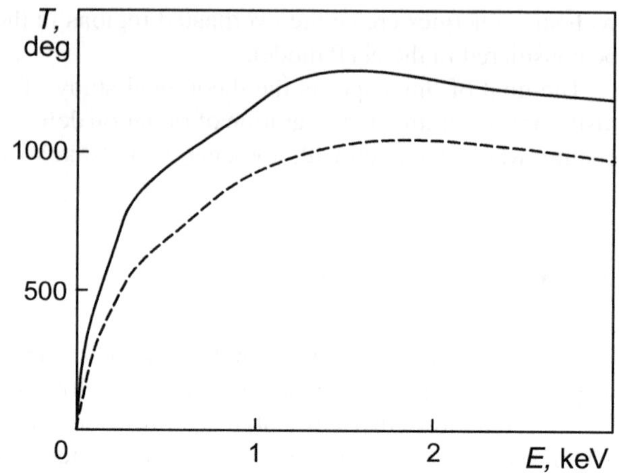

Fig. 2 Overheated temperatures in the NTPs of Ti^+ ion in TiN (solid curve) and Cr^+ ion in CrN (dashed curve) depending on ion energy

In case of the NTP in the boundless material we have $V = 4\pi R^3 / 3$.

The calculated functions $V(E)$ and $\eta(E)$ allow determining the average temperature increase in the NTP:

$$T(E) = \frac{\eta(E)E}{\rho C V(E)}$$

where ρ and C are the density and the specific heat of the coating, respectively. We note that at temperatures above room temperature, the temperature dependence of the heat capacity C can be neglected, making it equal to its high-temperature limit [11].

Figure 2 displays overheated temperature in the NTPs of Ti^+ ion in TiN (solid curve) and Cr^+ in CrN (dashed curve) depending on ion energy.

As can be seen from Fig. 2, Ti^+ (Cr^+) ions with energy 1 KeV $< E <$ 3 KeV create thermal peaks in the TiN (CrN) coatings with overheated temperature T ~ 1100–1200 K (T ~ 900–1000 K).

Pulsed heating of the NTP material leads to rapid thermal expansion and formation of variable thermoelastic stresses of ~10 GPa, which can affect the defect migration in the peak and its environs. The amplitude of thermoacoustic displacement $\vec{u}(r, t)$, which is a purely longitudinal wave, can be represented in the form $\vec{u}(\vec{r}, t) = \partial\Psi(\vec{r}, t) / \partial\vec{r}$ where the potential of acoustic displacements is found as a solution of the thermoacoustic equation [6–8]:

$$\frac{\partial^2\Psi(\vec{r}, t)}{\partial t^2} - s^2\Delta\Psi(\vec{r}, t) = -\frac{\alpha K}{\rho} T(\vec{r}, t) \tag{1}$$

Here s is the longitudinal velocity of sound, α is the volume coefficient of thermal expansion, and K is the bulk modulus of the material.

Estimates show that the heat transfer process is slow compared with the formation of the thermoacoustic pulse in the peak. This allows us to set $T\left(\vec{r}, t\right) = \varepsilon\left(\vec{r}, t\right)/\rho C$ in Eq. (1), where $\varepsilon(\vec{r}, t)$ is the density of thermal energy in the NTP.

Using the well-known Kirchhoff's formula with zero initial conditions, we obtain a solution of Eq. (1) in the form:

$$\Psi(\vec{r}_0, t) = -\frac{\Gamma}{4\pi\rho s^2}\iiint \frac{\varepsilon\left(\vec{r},\, t - |\vec{r}_0 - \vec{r}|/s\right)}{|\vec{r}_0 - \vec{r}|}\, dxdydz \tag{2}$$

where $\vec{r}_0$ is the radius vector from the center of the thermal field to the observation point, and $\Gamma = \alpha K/(\rho C)$ is the Grüneisen parameter? The integral is taken over the range of integration $|\vec{r}_0 - \vec{r}| \leq st$.

As an analytical approximation of the density of the absorbed energy, we choose the following spherically symmetric function:

$$\varepsilon(r, t) = \frac{E\eta(E)}{\left(\sqrt{\pi}R_{ef}\right)^3} e^{-(r/R_{ef})^2}\left(1 - e^{-(t/T_{ef})^2}\right) \tag{3}$$

Constants of R_{ef} and T_{ef} are related, respectively, to the radius R and the time τ of energy thermalization of the NTP. The analysis shows that we can put $R_e \approx 0.72R$, $T_e \approx 0.6\,\tau$.

We express the arising acoustic stresses σ through the potential of acoustic displacements $\Psi(r_0,\ t)$. Taking into account the spherical symmetry of the problem, we find the stress tensor component $\sigma \equiv -\sigma_{rr}$ as a function of the dimensionless variables $a = r_0/R_{ef}$, $b = st/R_{ef}$, and $f = (sT/R)^2$:

$$\sigma(a, b, f) = \frac{\rho s^2}{R^2}\left(-\frac{\partial^2\Psi(a, b, f)}{\partial b^2} + d\frac{1}{a}\frac{\partial\Psi(a, b, f)}{\partial a}\right)$$

where $d = 2(1 - 2\nu)/(1 - \nu)$, and ν is the Poisson ratio. The analytic expression for $\Psi(a, b, f)$ obtained from Eqs. (2), (3), is given in [8].

3 Calculation Results and Discussion

The calculation of thermoacoustic stresses was carried out at the values of the parameters $E = 2$ keV, $\eta = 0.57$ (0.60), $R = 2.34$ nm (2.2 nm), $\tau = 0.0775$ ps (0.11 ps), $s = 1.01 \times 10^3$ m/s (0.868×10^3 m/s), corresponding to the case of the thermoelastic peak of the Ti^+ (Cr^+) ion with energy of 2 keV in material TiN (CrN).

In Fig. 3 the spatial distribution of the thermoacoustic stresses generated by NTP of Ti^+ ion in infinite target TiN at different time moments are shown. The stress field is the superposition of the spherical bipolar acoustic wave diverging from NTP volume and the quasistatic stress remaining in NTP after leaving of the wave.

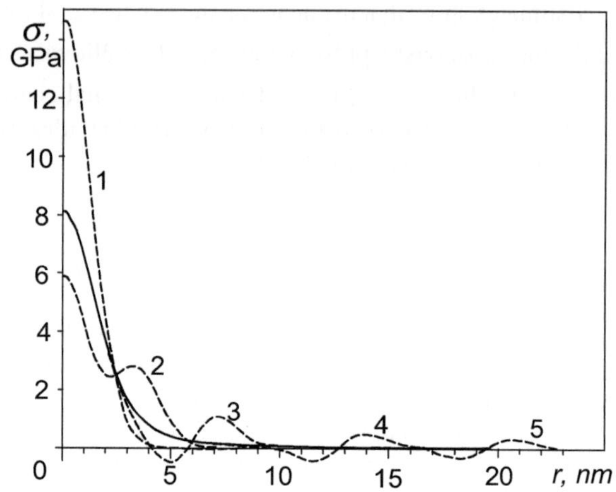

Fig. 3 The spatial distribution of thermoacoustic stresses generated by Ti^+ ion with energy E = 2 keV in TiN coating at successive times t = 0.085, 0.34, 0.68, 1.36, and 2.04 ps (dashed curves 1, 2, 3, 4, and 5, respectively). The solid curve corresponds to static stress distribution remaining in NTP

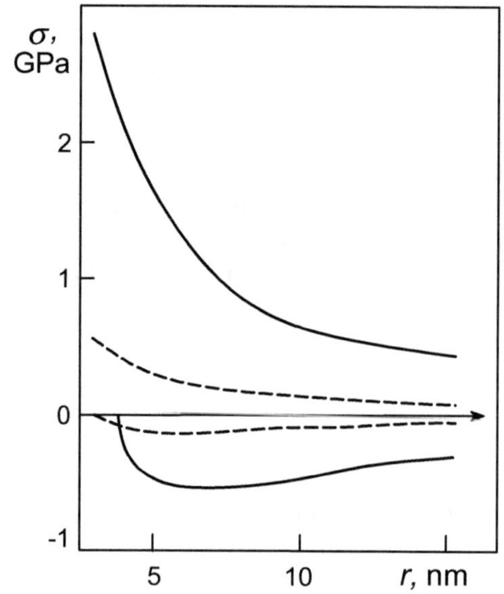

Fig. 4 Spatial distribution of the maximum values of compressive (positive) and tensile (negative) stresses in the diverging acoustic wave near the NTP of Ti^+ ion in TiN (solid curves) and near the NTP of Cr^+ ion in CrN (dashed curves)

The thermoacoustic stresses excited by the NTP of Cr^+ ion in CrN are much weaker than the analogous stresses in TiN. This is due to the small value of the Grüneisen parameter for CrN, in comparison with TiN. Figure 4 displays the spatial distribution of the maximum values of compressive and tensile stresses in a diverging acoustic wave near the peak of the Ti^+ ion in TiN (Cr^+ ion in CrN). As can be seen from Fig. 4, the stresses excited in the target CrN by Cr^+ ions are ~5 times lower than those excited in TiN by Ti^+ ions.

The thermoacoustic stress generated by the NTP is the spherically symmetric bipolar pulse with spatial extent $\lambda \sim 2R(E)$ diverging from the NTP with the velocity of longitudinal sound s and decreasing in the wave zone $r \gg R$ according to the law $\sim r^{-1}$. Such an attenuation law is valid only for distances $r < r_{at}$ where sound absorption for the boundary frequency ~s/R in the spectrum of the acoustic pulse can be neglected. If the absorption coefficient has the form $\beta(\omega) = \beta\omega^2$, then for the effective absorption length we obtain the estimate $r_{at} \sim R^2/\left(\beta s^2\right)$. Assuming $\beta = 2\ \text{m}^{-1}\ \text{GHz}^{-2}$ (typical value for solids like sapphire) we get $r_{at} = 15$ nm in the case of Ti^+ ion with energy 2 keV in TiN.

The analysis shows that in the region $r > r_{at}$ one can use the following expression [3], [8]:

$$\sigma_{at}(\vec{r}) \approx \frac{\Gamma E\eta(E)}{\sqrt{2\pi e}\beta s^2 r^2}$$

In the vicinity of the NTP, where the amplitude of the acoustic stress $\sigma_{at}(\vec{r})$ exceeds the threshold value of $\sigma_{is} = u/\Omega \sim 0.5$ GPa, it becomes possible to disrupt the interstitial defects. Here $u \sim (0.1 - 0.5)$ eV is the activation energy of the interstitials migration, $\Omega \sim 0.05\ \text{nm}^3$ is the defect volume. The boundary of the zone of activation of defects $\vec{r}_{is}$ can be found from the equation $\sigma_{at}(\vec{r}_{is}) = \sigma_{is}$. Thus, Ti^+ ion with the energy of 2 keV creates favorable conditions for "sweeping out" interstitial defects in the spherical zone around the NTP with radius $|\vec{r}_{is}| \sim 15$ nm. If TiN material of coating has a nanocrystalline structure with crystallite size $d < 2r_{is}$, then the effective "sweeping" of interstitial defects from the interior of crystallite to its boundary is possible.

In materials with relatively low coefficients of thermal expansion, as in the case of the CrN coating, the mechanism of thermoacoustic activation of defect migration is suppressed.

Acceleration of kinetic processes at large distances $r > r_{is}$ is also possible due to the decrease in the effective activation energy $U_{ef}(r) = U - \sigma(r)\Omega$. Estimates show that low-energy ions of Ti^+ can significantly accelerate the diffusion of defects and structural rearrangement in the target of TiN at depths of up to 50 nm.

4 Conclusions

1. Rapid heating of matter in the NTP of the heavy low-energy ion during ion-ion relaxation leads to the generation of the powerful thermoelastic pulse of subpicosecond duration, which has as static and waves components.
2. The acoustic pulse amplitude obeys the power law $\sigma(r) \sim r^{-1}$ at distances $2R < r < r_{at}$, and the power law $\sigma(r) \sim r^{-2}$ at distances $r_{at} < r$, so the acoustic pulse retains a significant value of ~1 GPa at distances exceeding the path length of the ion by tens of times. In the case of the NTP generated by Ti^+ ion with the energy of 2 keV in the TiN coating we obtain $r_{at} \sim 15$ nm.
3. The acoustic pulses from Ti^+ ion with energy 2 keV bombarding a nanocrystalline target of TiN can activate and "sweep out" interstitial defects at distances up to

15 nm, and accelerate their migration at distances up to 50 nm. Acoustic pulses from 2 keV Ti^+ ion bombarding the nanocrystalline TiN target can "sweep" interstitial defects from nanocrystallites of size $d \leq 30$ nm and accelerate their migration in larger nanocrystallites $d \leq 100$ nm.

4. In materials with low coefficients of thermal expansion, as in the case of the CrN coating, the mechanism of thermoacoustic activation of defect migration is ineffective.
5. The contribution to the acoustic activation effect of the impact component of the stress pulse associated with the transfer of momentum from the ion to the target material can be significant for materials consisting of heavy elements with atomic number A ~ 100, and also for ions or recoils with low energy E ~ 100 eV. However, this aspect of the acoustic activation problem requires additional analysis.

Acknowledgements This research was carried out within the framework of the joint scientific project of the National Academy of Sciences of Ukraine and the National Academy of Sciences of Belarus "Creation of radiation-resistant nanostructured metal and nitride coatings based on titanium, aluminum, and chromium by vacuum-arc deposition from filtered plasma," 2018.
(Reg. No. of Ukrainian projects: 02-03-18).

References

1. Pogrebnjak AD, Ivashchenko VI, Skrynskyy PL et al (2018) Experimental and theoretical studies of the physicochemical and mechanical properties of multi-layered TiN/SiC films: temperature effects on the nanocomposite structure. Compos B Eng 142:85–94
2. Pogrebnjak AD, Beresnev VM, Bondar OV et al (2018) Superhard CrN/MoN coatings with multilayer architecture. Mater Des 153:47–59
3. Abadias G (2008) Stress and preferred orientation in nitride-based PVD coatings. Surf Coat Technol 202(11):2223–2235
4. Komarov FF, Pogrebniak AD, Konstantinov SV (2015) Radiation resistance of high-entropy nanostructured (Ti, Hf, Zr, V, Nb)N coatings. Tech Phys 60(10):1519–1524. https://doi.org/10.1134/s1063784215100187
5. Safonov VI, Marchenko IG (2002) Ti and Al layers formation during low energy ion-plasma treatment of copper. Surf Coat Technol 158–159:105–107. https://doi.org/10.1016/s0257-8972(02)00228-1
6. Kalinichenko AI, Strel'nitskij VE (2007) Acoustic activation of structural rearrangement in carbon coating by ion bombardment. Surf Coat Technol 201:6090–6094. https://doi.org/10.1016/j.surfcoat.2006.08.127
7. Kalinichenko AI, Lazurik VT, Zalyubovsky II (2002) Introduction to radiation acoustics. In: The physics and technology of particle and photon beams, vol 9. Harwood Academic Publishers
8. Kalinichenko AI, Strel'nitskij VE (2003) The role of thermoelastic stresses in formation of diamond-like coating during amorphous carbon irradiation with low-energy ions C+ . Phys Chem Mater Process 2:5–10 (in Russian)
9. Kalinichenko AI, Perepelkin SS, Strel'nitskij VE (2006) Thermodynamic conditions of ta-C formation at implantation of noble-gas ions in carbon. Diamond Relat Mater 15:365–370. https://doi.org/10.1016/j.diamond.2005.10.022
10. Ziegler JF, Biersack JP, Littmark U (1996) The stopping and range of ions in solids. Pergamon Press, New York, p 297
11. Learned JG (1979) Acoustic radiation by charged atomic particles in liquids: an analysis. Phys Rev D 19:3293–3307. https://doi.org/10.1103/physrevd.19.3293

Conductive Polymer Nanocomposites for Novel Heating Elements

R. Kolisnyk, M. Korab, M. Iurzhenko, O. Masiuchok, A. Shadrin, Ye. Mamunya, S. Pruvost and V. Demchenko

Abstract Electrical and thermal properties of composites based on high-density polyethylene filled with nanofillers, namely carbon black (CB), carbon fiber (CF) and their mixture, were investigated under alternating current. The composite samples were prepared by thermal pressing from the mechanically mixed blend. High-density polyethylene (HDPE) was used as matrix. The nanofiller content was chosen to be 8 vol% for CF and 8, 12, 16, 30 vol% for CB. This content is higher loading than the percolation threshold and guarantees sufficient heating. The electrical resistivity of the abovementioned composites decreased significantly from 10^{12} Ω/cm to ~10^{1}–10^{2} Ω/cm with increasing of the CF/CB content. When a certain voltage was applied to the composites, the equilibrium temperature of the surface of the composite was reached within less than 100 s. The maximum surface temperature as the equilibrium state of the composite samples can be easily controlled by adjusting the composites filler ratio.

Keywords Heating element · Welding · Polymer nanocomposites

1 Introduction

In recent years, polymer nanocomposites reinforced with carbon-based nanofillers have drawn increasing interest from scientific and industrial points of view [1, 2]. These carbon-based nanofillers (carbon nanofiber and carbon black) have been applied to increase the mechanical properties and electrical conductivity of polymer composites. In particular, the carbon-based nanofillers are significantly effective for

R. Kolisnyk · M. Korab · M. Iurzhenko · O. Masiuchok · A. Shadrin · V. Demchenko (✉)
E.O. Paton Electric Welding Institute the NAS of Ukraine, Kiev, Ukraine
e-mail: dvaleriyl@ukr.net

Ye. Mamunya
Institute of Macromolecular Chemistry of the NAS of Ukraine, Kiev, Ukraine

S. Pruvost
INSA@LYON1, UMR CNRS 5223, Villeurbanne, France

© Springer Nature Singapore Pte Ltd. 2019
A. D. Pogrebnjak and V. Novosad (eds.), *Advances in Thin Films, Nanostructured Materials, and Coatings*, Lecture Notes in Mechanical Engineering,
https://doi.org/10.1007/978-981-13-6133-3_22

increasing the performance in the electrical conduction of nonconductive polymer matrix. It is expected that polymer nanocomposites with improved electrical conductivity can be also used for electric heating materials or devices, such as electrical resistors that convert electrical energy into thermal energy as heat. On the other hand, it is considered that polymer nanocomposites containing carbon nanofillers have other advantages such as lightweight, corrosion resistance, easy processing and lower manufacturing costs for electric heating applications.

The aim of the study is the creation of segregated polymer nanocomposites with defined electrical and thermal characteristics, needed mechanical parameters and required technological properties suitable for application in flexible energy-saving and low-energy devices (heating elements for climate clothes and heating elements for plastics welding). The present work involves the study of electrical and thermal process in polymer nanocomposites, depending on the nature of polymer/filler components; the hybrid nanocomposites with needed flexibility and strength can be formed.

2 Experimental Section

2.1 *Materials*

The samples were prepared using HDPE ($\rho = 0.95$ g/cm^3, $T_m = 120$ °C) as a polymer matrix to reach easy processability and carbon black P803 ($\rho = 1.8$ g/cm^3) and carbon fiber as a nanofiller, to have high electrical conductivity. The HDPE/CB, HDPE/CF, and HDPE/CF + CB blends ratio is chosen to be 8 vol% for CF and 8, 12, 16, 30 vol% for CB by volume, which is much higher than the percolation threshold. Mixture of HDPE and carbon black was blended under mechanical stirring, heated to 145 °C and moulded under 5 MPa pressure with simultaneously cooling. The thickness of the samples is 1.2–1.4 mm and 30 mm in diameter.

2.2 *Electrical Measurement*

To improve the quality of electrical measurements, two electrodes placed inside the samples before molding.

Heating properties were studied using an AC source and changes of electric current (I) with applied voltage for the composite samples were measured by employing multimeter. Electric heating behavior of the composite samples under a variety of applied voltages of 1–20 V was characterized with an infrared camera, and temperature changes were measured by placing thermocouple on the surface of the sample.

3 Results and Discussion

To investigate the influences of carbon black nanofiller on electrical properties of the nanocomposites, the resistivity of the HDPE + CB nanocomposites were measured and presented as a function of the CB content as shown in Fig. 1. Also, it is shown the model of structure evolution with increasing of CB content [3]. If the nanofiller covers the polymer particles surface with a uniform layer without aggregation, it can be achieved conductive nanocomposite with a low level of percolation threshold [4]. Figure 2 shows a TEM picture of carbon black nanoparticles [5].

Electric heating performance of the nanocomposite heating elements with a different nanofiller ratio of CF, CB and the nanocomposites of CF and CB were investigated by monitoring temperature changes as a function of time at different applied voltages of 1–20 V, as it is shown in Figs. 3, 4 and 5.

However, the temperature of the nanocomposite samples increased rapidly in time when the voltages above 10 V were applied. Maximum temperatures of the samples were then attained within around 100 s and remained unchanged over time. When the applied voltage was turned off after 900 s of heating, the temperature of the samples decreased to the room temperature within a minute.

It can be seen that the maximum temperature attained at a given applied voltage increases with rising of filler concentration and is higher for the HDPE + CB 12 vol% nanocomposite sample with higher content of carbon black comparing to HDPE + CB 8 vol%.

The temperature versus time curves in Figs. 3, 4 and 5 can be divided into three zones: the heating zone (0–200 s), the maximum temperature zone (200–800 s) and,

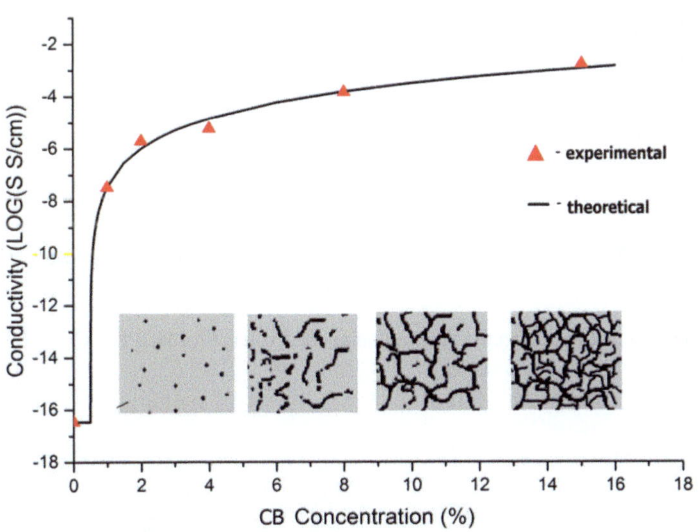

Fig. 1 Resistivity of the HDPE + CB nanocomposites as a function of the CB content

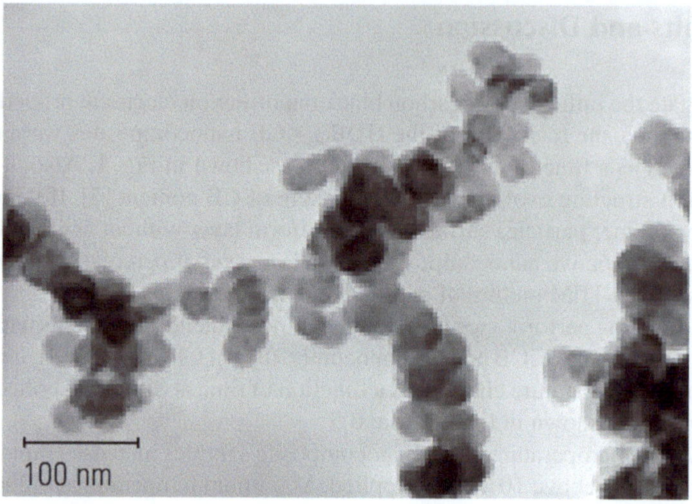

Fig. 2 TEM picture of carbon black showing the high level of aggregation [5]

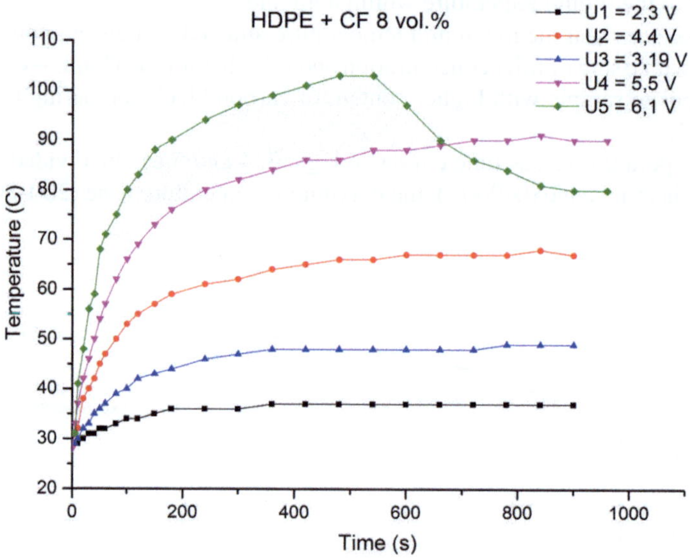

Fig. 3 Dependence of temperature changing on applied voltage for HDPE + CF 8 vol%

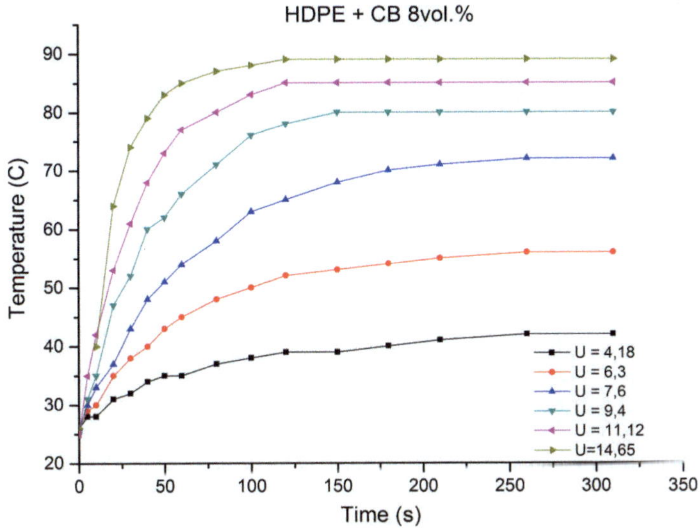

Fig. 4 Dependence of temperature changing on applied voltage for HDPE + CB 8 vol%

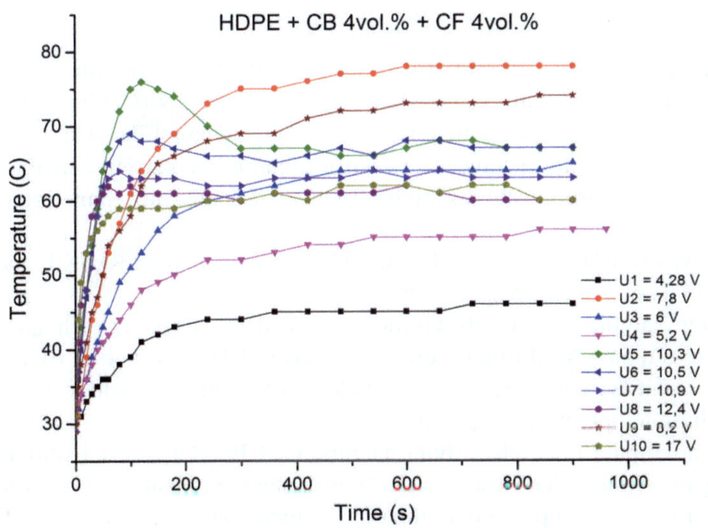

Fig. 5 Dependence of temperature changing on applied voltage for HDPE + HDPE + CB 4 vol% + CF 4 vol%

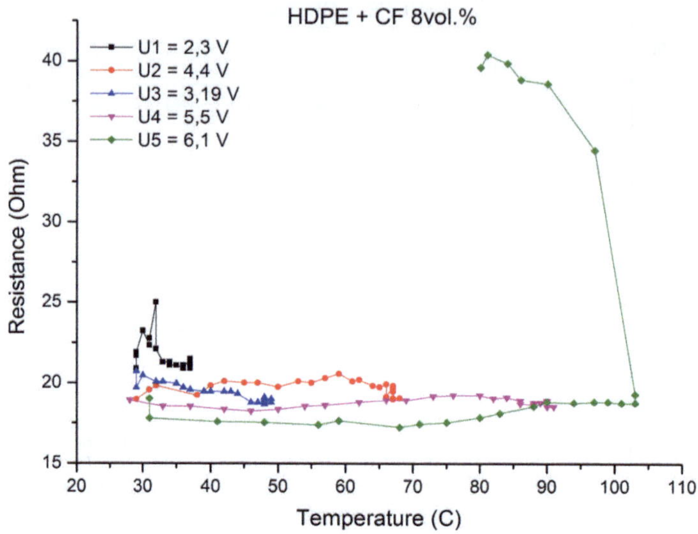

Fig. 6 Dependence of resistance changing on temperature for HDPE + CF 8 vol%

in some cases when the current still flows through the sample, the cooling zone (800–900 s) connected with positive temperature coefficient (PTC) effect owing to different thermal expansion coefficients of polymer and nanofiller. Therefore, during the electric heating, the distance between the nanofiller particles increased, thus increasing the nanocomposite resistivity because of decreasing tunneling of charge carriers.

The curves of PTC effect for CF, CB and the nanocomposites of CF and CB are shown in Figs. 6, 7 and 8, respectively.

However, samples filled with CF demonstrated good results of heating at low voltages (less than 5 V), further increasing leads to PTC effect (Fig. 6), that makes difficult to use it as a heating element for welding of the plastics, where it is necessary to obtain high temperature for polymer melting.

So we attempted to combine both nanofillers (CB + CF), which should lead to increasing of conductivity. However, this type of nanocomposites needs more voltage to achieve the same temperature comparing to nanocomposites with CB. The more voltage is applied, the more PTC effect is observed (Fig. 8).

The studies of heating properties showed that the best results have nanocomposites filled with CB as it has the most stable heating despite the PTC effect comparing to nanocomposites filled with CF and both CF and CB. It was decided to choose nanocomposites filled with CB for further researches.

Increasing the concentration of CB we tried to achieve sample melting point with minimum level of applied voltage. Thus, CB ratio was increased to 12 and 16 vol%, relatively. Temperature change as a function of time is shown in Figs. 9 and 10.

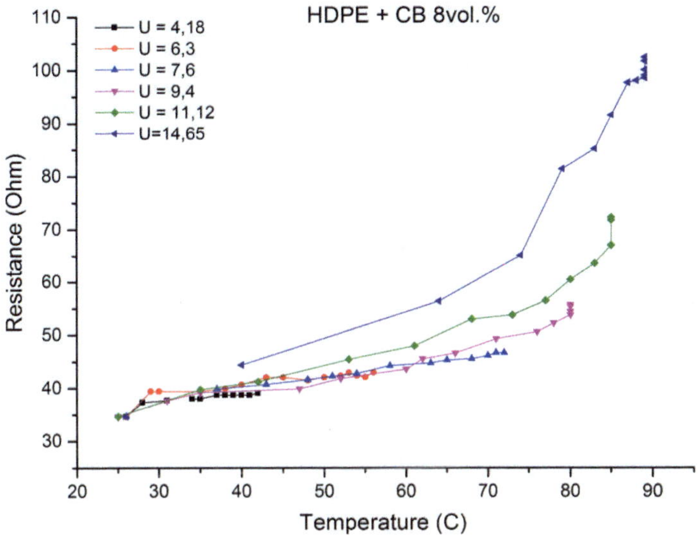

Fig. 7 Dependence of resistance changing on temperature for HDPE + CB 8 vol%

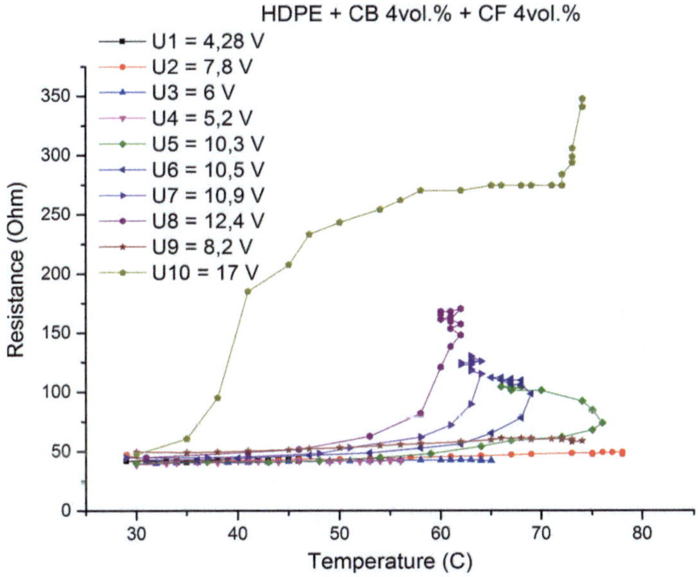

Fig. 8 Dependence of resistance changing on temperature for HDPE + CB 4 vol% + CF 4 vol%

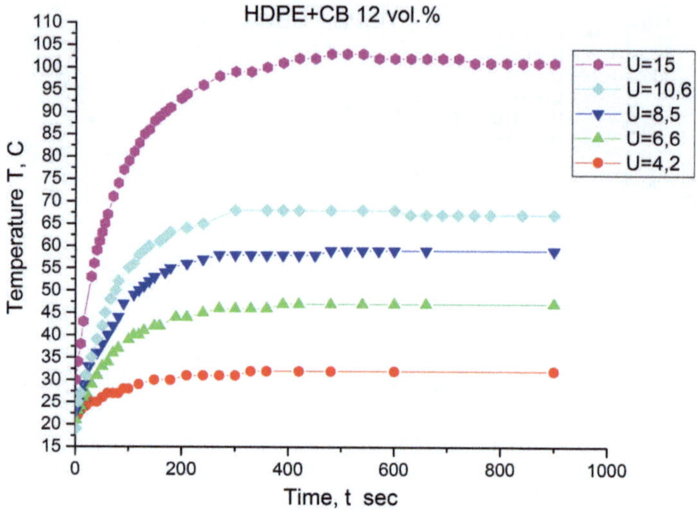

Fig. 9 Dependence of temperature changing on applied voltage + CB 12 vol%

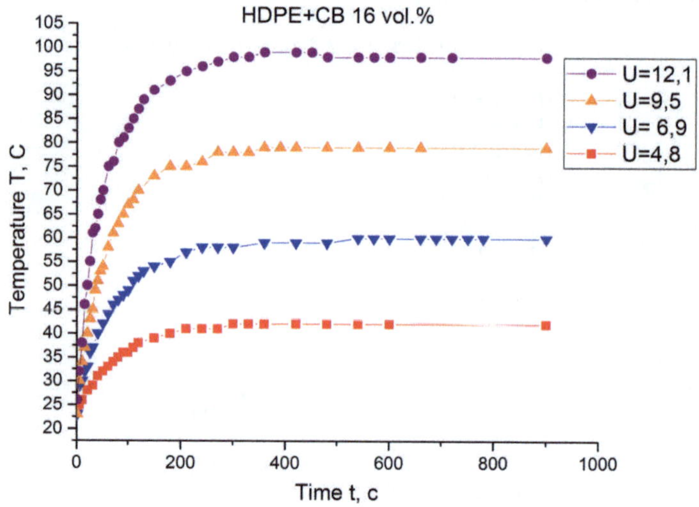

Fig. 10 Dependence of temperature changing on applied voltage of HDPE + CB 16 vol%

To estimate the heating efficiency of the developed heating elements, welding of T-joint of pure HDPE using heating element was carried out. Temperature distribution was characterized with an Infrared camera. Due to a big welding area of specimens, low heat generation could not provide high quality welded joint. The face of the adjusted HDPE specimen was heated only to the temperature approximately 75 °C, whereas to obtain the welded joint, the material should be heated to the temperature of 150 °C.

Fig. 11 Lap-welded joint after mechanical test

Fig. 12 Butt-welded joint after mechanical test

As nanocomposite with 16 vol%. did not show good results at welding, it was decided to increase CB ratio up to 30 vol% and to study the possibility of welding and to investigate effects on mechanical properties of the welded joint.

Lap-welded joint showed excellent results, as it was destroyed by basic material (Fig. 11). However, butt-welded joint ruptured in the heat-affected zone (HAZ) (Fig. 12), its mechanical strength was 15.9 MPa, while the strength of pure HDPE is approx. 18 MPa that says about the satisfactory quality of the welded joint. Regardless the fact that the maximum temperature, according to the infrared camera, along the fusion line was 120 °C, the temperature inside the welded joint was much higher than the melting point of HDPE, so that allowed to create a high quality welded joint.

Since the HAZ of the butt welded joint is less comparing to the same of the lap welded joint (Fig. 12), it was possible to observe that temperature on the edges of the heating element rose up to 250 °C, that is more than enough for melting of the surrounding material and for the welded joint formation.

The results allow to conclude that developed heating elements based on conductive polymer nanocomposites can be successfully used for welding of polymer materials.

The developed conductive polymer nanocomposites can be used not only for welding but also for production of filament for 3D printing. 3D printing is one of the additive technologies that allows to form 3D objects by adding layer-by-layer of material, in contrast to the conventional methods of products formation. There is some methods of creating of real objects from 3D models using additive technologies. One of the most common is FDM (Fused Deposition Modeling) or FFF (Fused Filament Fabrication) 3D plastic printing due to availability and usability.

FDM (FFF) modeling involves the creation of three-dimensional objects by applying successive melt-layers of thermoplastic materials, which are extruded from a temperature-controlled nozzle, according to a digital model. Thermoplastic materials for printing, such as ABS, PLA, HDPE, etc., are supplied in the form of coils of filaments.

One of the trends of the additive technologies is creation of functional materials and obtaining filament based on them for 3D printing. These materials, such as electrically active polymers allow not only prototyping but also manufacturing of the functional products.

In the present work, the same polymer nanocomposites (HDPE and carbon black P803) were used for producing conductive filament for FDM 3D printing with the necessary set of operational and electrical properties. As the result filament with the required diameter −1.75 mm based on the nanocomposite was formed using pressing and extrusion molding technique.

References

1. Hu Y, Shenderova OA, Hu Z, Padgett CW, Brenner DW (2006) Carbon nanostructures for advanced composites. Rep Prog Phys 69(6):1847–1895
2. Yang J, Liang J (2011) A modified model of electrical conduction for carbon black-polymer composites. Polym Int 60(5):738–742
3. Ye Mamunya, Iurzhenko M (eds) (2012) Advances in progressive thermoplastic and thermosetting polymers: perspectives and applications. TechnoPress editura, Iasi
4. Ye M, Iurzhenko M, Lebedev E, Levchenko V, Chervakov O, Matkovska O, Sverdlyakovska O (2013) Electroactive polymeric materials. Alpha Reklama, KyivSpecialty
5. Carbons for polymer. http://www.imerys-graphite-and-carbon.com

The Investigation of the Welding Process of Different-Type Polyethylenes

M. Kovalchuk, M. Iurzhenko, V. Demchenko and I. Senchenkov

Abstract The article presents the results of complex thermal and structural studies of two technical types of polyethylene (PE-80 and PE-100) and their welded joint. It was established by mathematical calculations as well as experimental investigations that during conventional heated tool butt welding recrystallization of polymers occurs resulting in the formation of crystalline areas with improved mechanical and thermal properties, due to the increased degree of their ordering.

Keywords Polyethylene · PE-80 · PE-100 · Modeling · Heated tool butt welding

1 Introduction

As it is known, polyethylene (PE) is one of the most widespread materials in the modern world, and polyethylene products occupy a great share at industrial and commercial markets [1]. For example, welding of such types of technical polyethylenes as PE-63, PE-80, and PE-100 is widely used in construction, particularly in installation and reparation of pipeline systems. The question of the possibility of old pipelines reparation using pipes made with polyethylene of modern types induces a great interest. The main technology that is used for the pipelines reparation is butt welding with the heated tool.

The technological process of the heated tool butt welding of polyethylene pipes (Fig. 1) includes five main stages: preliminary melting of pipes edges, their heating during the specified time, technological pause for removing the heated tool, upsetting under working pressure and following cooling of the formed joint. The main parameters of the technological process include temperature of the heated tool; melting time; heating time; duration of the technological pause and upsetting; pressure during melting, heating, and upsetting.

M. Kovalchuk · M. Iurzhenko · V. Demchenko (✉) · I. Senchenkov
E.O. Paton Electric Welding Institute of the NAS of Ukraine, Kiev, Ukraine
e-mail: dvaleriyl@ukr.net

© Springer Nature Singapore Pte Ltd. 2019

A. D. Pogrebnjak and V. Novosad (eds.), *Advances in Thin Films, Nanostructured Materials, and Coatings*, Lecture Notes in Mechanical Engineering,
https://doi.org/10.1007/978-981-13-6133-3_23

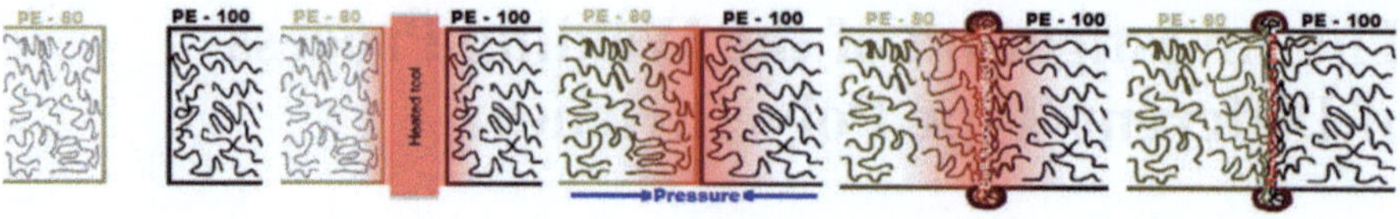

Fig. 1 Scheme of the heated tool butt welding of pipes of various types of polyethylene

Difficulties of reparation of plastic pipelines made with different types of technical polyethylene are their different structure, thermal, thermomechanical and rheological properties that follow in problems at welding process and further appearance of internal defects in their welded joints, which take a significant role in the duration of the pipeline lifetime.

2 Experimental Part

Polyethylene pipes produced from two types of high-density technical polyethylenes (HDPE) with different minimum required strength (MRS)—PE-80 ($MW_{bimodal}$ 300,000 g/mol, density 0.953 g/cm^3, MRS = 8 MPa), and PE-100 ($MW_{bimodal}$ 300,000 g/mol and density 0.960 g/cm^3, MRS = 10 MPa) as well as their welded joint were used for welding experiments, structure analysis and investigations of mechanical and thermal properties.

The welding experiments have been carried out with PE-80 and PE-100 pipes of 63 mm diameter and 6 mm wall thickness using conventional heated plate butt welding at SAT-1 welding device produced by Experimental Welding Equipment Factory of E.O. Paton Electric Welding Institute of the NAS of Ukraine under the following conditions: 220 °C welding temperature, 0.2 MPa welding pressure, and 60 s upset time. Technological pause was 3 s. The cooling time under pressure was 6 min.

Structure of PE (types PE-80 and PE-100) as well as of PE-80/PE-100 welded joint has been studied by means of wide-angle X-ray scattering (WAXS) using X-ray diffractometer DRON-4.07 (Burevestnik, Saint-Petersburg, Russia) with X-ray-optical scheme according to the Debye-Scherer method, using CuK_α-emission (λ = 0.154 nm), monochromated by Ni-filter. X-ray tube BSV27Cu working at U = 30 kV and I = 30 mA has been used like a source of characteristic X-ray radiation. X-ray measurements have been carried out by step-by-step scanning with scattering angles (2θ) from 2.6° to 40°, with exposure time 5 s at temperature $T = 20 \pm 2$ °C.

Thermal properties of the initial specimens and of the welded joint have been explored by means of differential scanning calorimetry (DSC) at DSC Q2000 device from TA Instruments (New Castle, DE, USA) in the dried air atmosphere in the temperature range from 40 to 200 °C with linear heating rate 20 °C/min. Specimen weight was between of 6 and 10 mg of each. Temperature measurement precision was ±0.01 °C, heat flow precision ±0.01 J/g.

Thermomechanical behavior and deformational characteristics (TMA) of the initial specimens and of the welded joint have been investigated at TMA Q400 EM device from TA Instruments (New Castle, DE, USA), in the dried air atmosphere with linear heating rate 10 °C/min in the temperature range from 30 to 250 °C. Measurements have been carried out in thermal expansion mode with use of quartz indenter of 2.8 ± 0.01 mm in diameter. Applied pressure to the specimen was permanent and equal to 10^{-1} MPa. Temperature measurement precision was ±0.01 °C, deformation control precision was ±0.01 μm. All the devices from TA Instruments have been certified according to the international standard ISO 9001:2000.

Mechanical properties (strength and elongation at break) of initial and welded specimens have been evaluated by means of the tensile axial test (according to DBN B.2.5-41 standard) with 50 mm/min tension rate at room temperature at FP-10 tension machine (Germany).

3 Research Results

Basing on the results of thermomechanical studies (Fig. 2) it is observed that the relative deformation curve of the PE-80/PE-100 welded joint material at temperatures above 140 °C is located between curves of the technical polyethylenes PE-80 and PE-100 [1]. However, as one can see in the insert of Fig. 2, in the temperature range 25–130 °C, the PE-80/PE-100 welded joint material has the maximum value of thermal expansion in comparison to individual polyethylenes PE-80 and PE-100. Such a situation can be explained by the fact that some internal stresses in the amorphous phase of polyethylenes arise during the welding process. Subsequently, heating leads to the increased molecular mobility and volume of the polymer in the welded joint due to the relaxation of these stressed areas.

For more detailed studying of the processes, which occur during welding and relaxation of polyethylene in the welded joint mathematical modeling was performed. The scheme of finite element grid of the simulated field of welding at edges of polyethylene pipes with an external diameter of 63 mm, the wall thickness of 5.8 mm, and heated tool, whose thickness is conventionally reduced to zero, is presented in Fig. 3.

It is noticeably that elements density of the grid increases according to logarithmic law with the distance from the conditional plane of the heated tool. The calculation results of the distribution of the temperature field at the key moments of the butt welding process at edges of polyethylene pipes are shown in Fig. 4. At the end of the heating stage, edges of the pipes have a maximum temperature for whole welding cycle, they are diverted from the working surfaces of the heated tool and remain in the air during the time interval of the technological pause. During this short period of time (in our case 3 s) an intense cooling of the heated edges of the pipes occurs until they become in contact with each other at the beginning of the upsetting process. The calculated temperature field in bulk of the walls of the pipe at the initial and final moments of the technological pause are shown in Figs. 4 and 5. A similar calculation

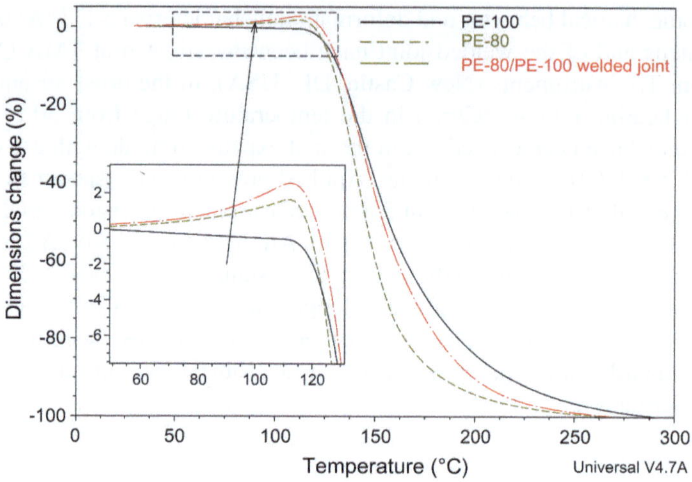

Fig. 2 Resulting plot of thermomechanical investigations of the technical polyethylenes PE-80, PE-100 and their welded joint PE-80/PE-100 [1]

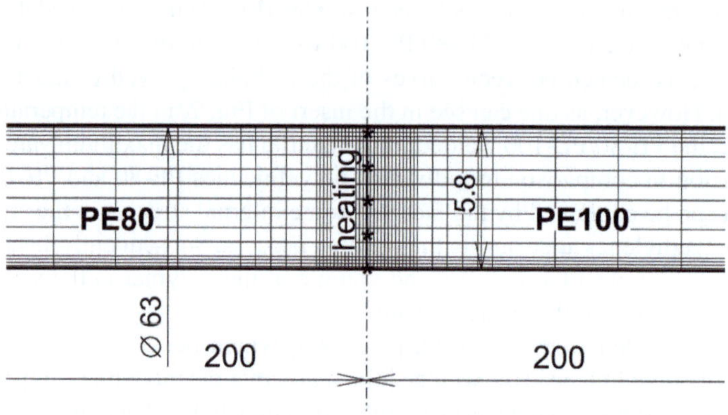

Fig. 3 The scheme of finite element grid of the simulated field of welding at edges of the polyethylene pipes

for the final moment of the welding process at the end of the cooling stage of the welded joint is shown in Fig. 6.

The alignment of temperature in the middle side of the pipe (r1 + r2)/2, namely the welding line, is shown in the temperature distribution graph (Fig. 7). It is also clearly seen that because of the difference in thermal conductivity of the materials PE-80 and PE-100, in bulk of the PE-100 pipe temperature is higher. Such difference in thermal properties of PE-80 and PE-100 definitely affects on thermomechanical and physical properties of the PE-80/PE-100 welded joint material.

Thus, since conventional tests of the welded joints of polyethylene pipes do not allow to evaluate adequately peculiarities of structure of the welded joint material

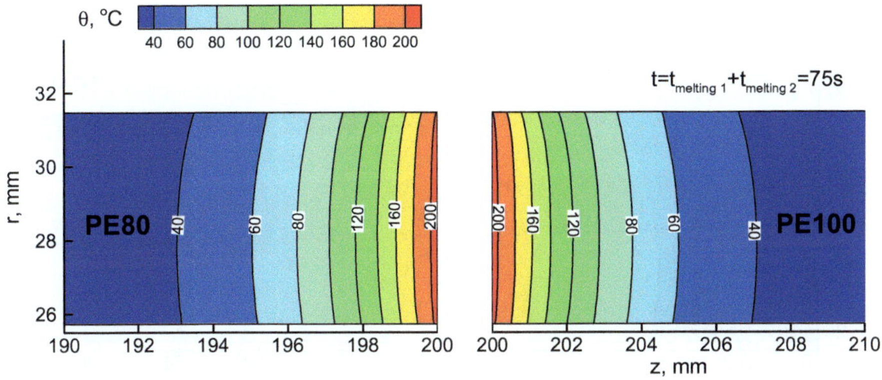

Fig. 4 Temperature field in bulk of the pipes walls at the initial moment of technological pause

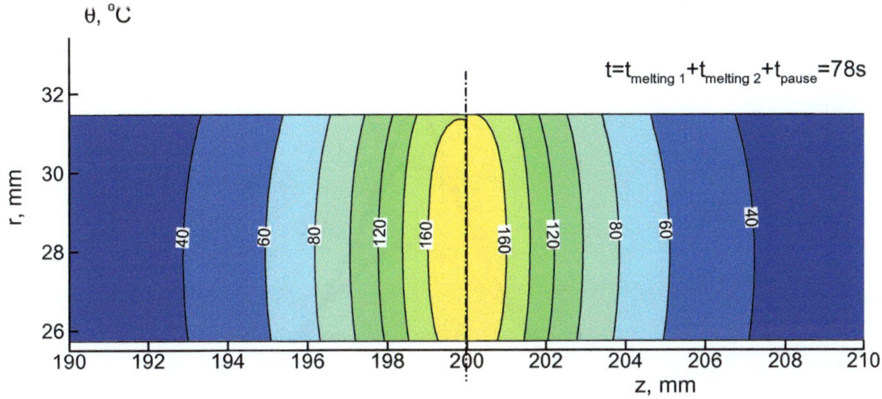

Fig. 5 Temperature field in bulk of the pipes walls at the final moment of technological pause

at macro and micro levels, complex structural investigations of the welded joint material were carried out using methods of the differential scanning calorimetry and the wide-angle X-ray scattering [1].

Proceeding data of the differential scanning calorimetry (Fig. 8) of all three samples, two minimums of the corresponding melting processes of the crystalline phase inside polyethylenes and their welded joint can be observed on curves [1]. Moreover, these two melting processes indicate polycrystallinity in all samples. The first minimum indicates the melting process of a crystalline fraction with the melting temperature from 117 to 125 °C. The second minimum corresponds to the melting of more ordered (better packed) crystallites with higher thermal stability and with the melting temperature from 133 fo 138 °C.

The increased melting temperatures of the welded joint material in comparison with the corresponding melting temperatures of both technical types of polyethylene

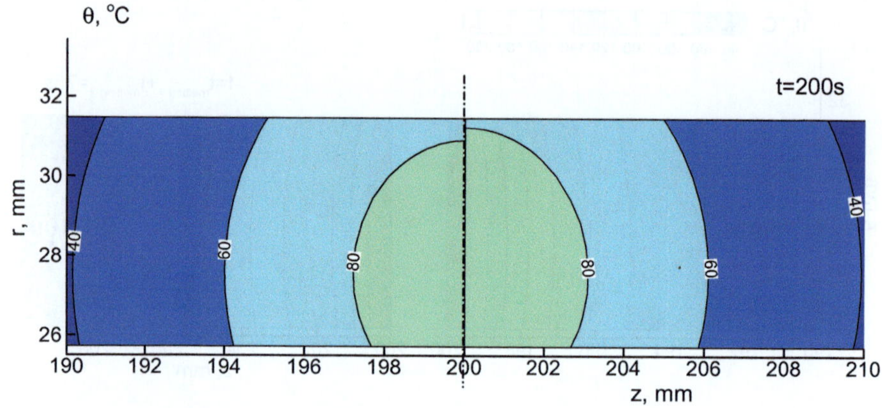

Fig. 6 Temperature field in bulk of the pipes walls at the final moment of the welding process at the end of the cooling stage of the welded joint

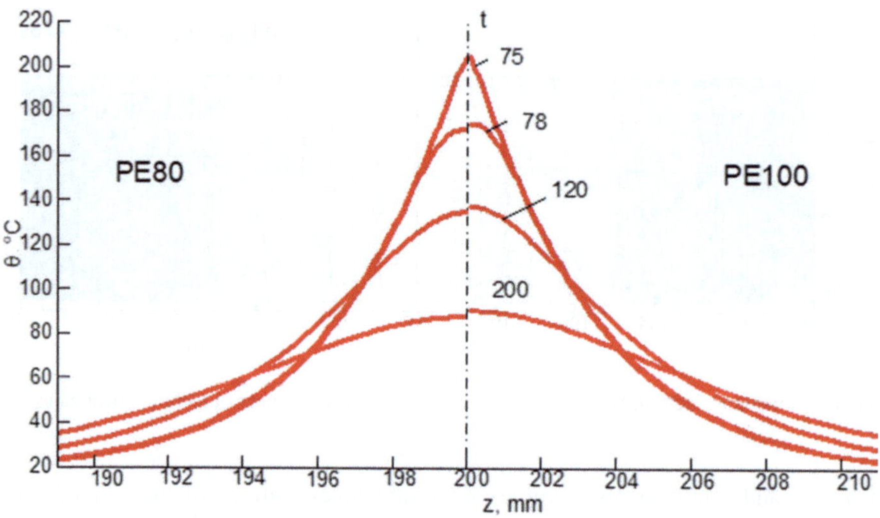

Fig. 7 Distribution of temperature along the welding line

may be explained by the fact that zones with higher thermal stability and, accordingly, crystallites with a higher degree of ordering have been formed in the welded joint.

Analysis of the diffraction curves of individual polyethylenes PE-80 and PE-100 and their welded joint PE-80/PE-100 (Fig. 9) showed that during welding of different-type polyethylenes structure of their welded joint becomes different from the structure of individual polyethylenes [1, 2]. That can be explained by melting of the crystalline phase with further recrystallization and simultaneous orientation of the crystallites in the welded joint under the action of the force field, which is applied during the welding process. This is evidenced from the increased intensity

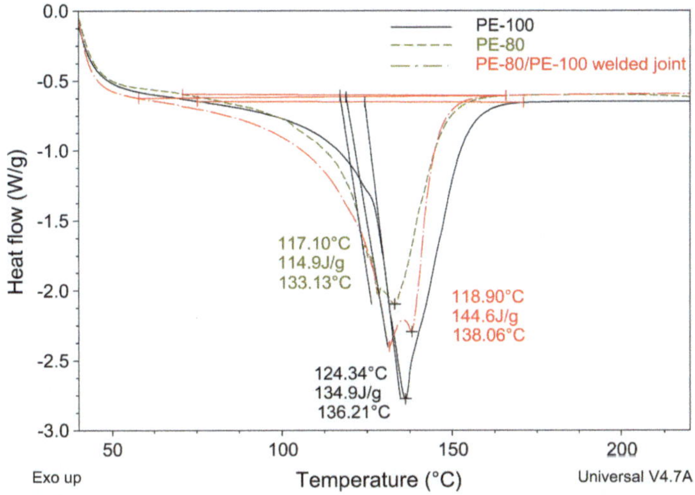

Fig. 8 Thermograms of the PE-80/PE-100 welded joint material and technical polyethylenes PE-80, PE-100 [1]

of diffraction maximum at the angular position 2θmax = 23.6° and the decreased intensity of the diffraction maximum at the angular position 2θmax = 21.2°.

The estimation of effective size of crystallites (L) of polyethylenes carried out by the Scherrer method:

$$L = K\lambda(\beta \cos \theta_{\max})^{-1} \quad (1)$$

where K—is constant, connected with the form of crystallites (if their form unknown K = 0.9), and β is the half-width (width at half height) of the diffraction maximum, showed that the mean value L ≈ 7.2 nm for both PE-80 and PE-100, and L ≈ 7.6 nm for PE-80/PE-100 welded joint material (for calculations the diffraction peaks at 2θmax = 21.2° and 23.6° were used). The values of the crystallites size were calculated from individual diffraction peaks for each sample (Table 1) [1].

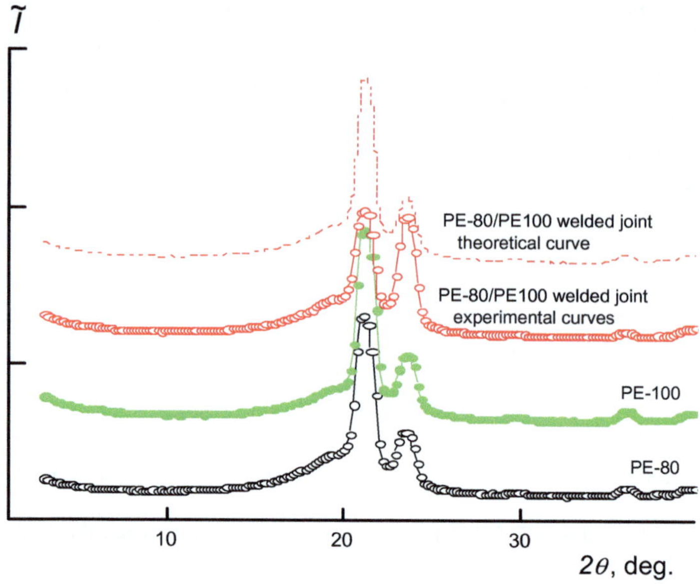

Fig. 9 Wide-angle X-ray diffraction patterns of the PE-80/PE-100 welded joint material and individual polyethylenes PE-80, PE-100 [1]

Table 1 Structural and mechanical characteristics of polyethylenes and of their welded joint

Sample	Degree of crystallinity (DSC), %	Degree of crystallinity (WAXS), %	Crystallites size L_1 ($2\theta =$ 23.6°) nm	Crystallites size L_2 ($2\theta =$ 23.6°) nm	Tensile strength, MPa	Relative tensile strenght, %
PE-80	42	56	7.2	7.2	19.6	100
PE-100	51	57	7.2	7.2	23.1	100
PE80/PE100 weld	53	66	7.2	8.0	Destroyed on the basic material	>100

4 Conclusions

It is established that during the process of formation of the welded joints of different-type polyethylenes a number of physical and chemical transformations of polymers, especially their crystalline phase, occurs. The main of them are recrystallization and orientation as a result of which the formation of a new structure of polyethylene with improved mechanical characteristics can be assumed.

References

1. Galchun A, Korab N, Kondratenko V et al (2015) Nanostructurisation and thermal properties of polyethylenes' welds. Nanoscale Res Lett 10:138
2. Demchenko V, Iurzhenko M, Shadrin A et al (2017) Relaxation behavior of polyethylene welded joints. Nanoscale Res Lett 12(1):280

Functional Selective Nanostructured Coatings Synthesized by Low-Temperature Ion-Plasma Method on Polymeric Substrates

Z. Tsybrii, F. Sizov, M. Vuichyk, K. Svezhentsova, E. Rudenko, I. Korotash and D. Polotskiy

Abstract The technology of low-temperature ion-plasma formation of the nanostructured films from the aluminum nitride on the various substrates, including the Mylar and Teflon flexible polymeric films was developed. The study of the structural and optical characteristics of such nanostructured coatings has shown the possibility to use them as the functional films with the selective transmissions and barriers in the infrared (IR) spectral region. Due to high thermal conductivity and specific optical properties (the Restrahlen band broadened from the wavelengths 10.3 μm up to almost 20 μm) and good transparency in visible and sub-THz spectral regions), the AlN thin polycrystalline films on polymeric substrates can be attractive for making the infrared radiation blocking filters from the background at 300 K. They can improve the performance of the sub-THz low-temperature detectors in the ground-based telescopes.

Keywords AN nanostructured coatings · IR selective filters · Polymeric substrates

1 Introduction

Recently, the study of the thin films process, particularly, in a number of group III metal nitrides (AlN, InN and GaN, etc.), have been attracting considerable attention of researchers as objects for the fundamental investigation and as the base materials for optoelectronics [1–3] too. The combination of materials properties has made it possible to process thin films for a variety of applications with the predefined parameters [4–6]. The aluminum nitride thin films have been the topic of research in the micro-electro-mechanical systems (MEMS) and in the semiconductor industry because of their good compatibility with a silicon substrate. AlN is typically used as a

---

Z. Tsybrii (✉) · F. Sizov · M. Vuichyk · K. Svezhentsova
V.E. Lashkaryov Institute of Semiconductor Physics NASU, Kiev, Ukraine
e-mail: tsybrii@isp.kiev.ua

E. Rudenko · I. Korotash · D. Polotskiy
Institute for Metal Physics NASU, Kiev, Ukraine

© Springer Nature Singapore Pte Ltd. 2019
A. D. Pogrebnjak and V. Novosad (eds.), *Advances in Thin Films, Nanostructured Materials, and Coatings*, Lecture Notes in Mechanical Engineering,
https://doi.org/10.1007/978-981-13-6133-3_24

gate dielectric for ultra large integrated devices or in GHz-band surface acoustic wave devices due to its strong piezoelectricity. It is used as an insulating layer responsible for the elimination of parasitic currents. It is mechanically strong, and have high thermal conductivity (up to 320 W/m K at 25 °C for single crystals [7]) that can be used for heat extraction. The AlN is thermally and chemically stable material, especially in an inert environment. The oxidation starts in it in the air at a temperature above 800 °C [8].

The decrease of the IR radiation influence is one of the needs in optical blocking for different applications. One of the examples of the IR optical blocking important applications due to AlN wide band-gap semiconductor can be the transparency in the microwave and THz spectral ranges in systems with cooled high sensitive detectors, e.g., for improving their performance in ground-based radio telescopes. These systems should provide completeness and quality of the processing of the received signals and suppression of increasing interference with other sources of noises [9]. Solving this problem also requires the creation of special protection systems against external parasitic thermal radiation, heat shocks and own thermal radiation of the antenna system and elements of the transmission path, which are at the ambient temperature ("warm part"). Achieve the necessary technical characteristics of the cryo-photoelectronic receiving system, while ensuring high reliability and resistance to extreme external factors, is possible by functional filtering devices, which should provide a high level of transmission with low losses of the informative signal in the microwave and sub-THz channels, and blocking elements from harmful infrared radiation. Realization of an upper limit system performance and improving the system noise imply the need of application of detector arrays or thermal protecting filters [10].

The IR signature suppression applications can be widened using thin flexible polymeric films (such as Teflon or Mylar) that are weakly transparent in defined bands of IR region. This is possible by developing technology for functional film coatings on flexible dielectric substrates that would provide the required spectral characteristics of selective radiation transmission and its suppression in desired spectral ranges.

Thin films of AlN (pure and oxidized) can be prepared by several techniques: chemical vapor deposition (CVD), molecular beam epitaxy (MBE), ion beam assisted deposition or direct current (DC) reactive magnetron sputtering [3]. Deposition of the AlN films at low temperature is necessary since a high-substrate temperature during film growth is not compatible with the processing steps of device fabrication or can destroy a substrate. Here we consider the technological method to realize the low-temperature ion-plasma synthesis of film coatings [11, 12]. This method has the advantages over other deposition methods due to the low growth temperatures on the substrate (30–300) °C, the ions flow formation of working gas of a helicon source with the high density in the substrate area of 5–10 mA/cm^2 and their high energies up to 100–150 eV. This method is also applicable to the polymeric substrates such as the Teflon or Mylar with a high degree of adhesion of the AlN films to the substrates. Utilizing the AlN optical properties (the Restrahlen band is at $\lambda \sim 11$–$16\ \mu$m [13, 14]), its high thermal conductivity ($\gamma \approx 134$–180 W/m K for polycrystalline films [15]) the composite structures AlN/polymeric substrate (Teflon, Mylar) can be applied for creation the large-area filters that blocks the IR radiation in the atmosphere transparency region $\lambda \sim 10$–$20\ \mu$m.

The aim of this investigation was the study of the morphological and optical characteristics of the AlN/(Teflon, Mylar) structures, which are formed by the hybrid helicon-arc ion-plasma deposition technique. It was supposed to ascertain the possibility of their using as the functional film coatings that can reduce the influence of IR background radiation (or radiation from a room temperature input window for high sensitive cooled detectors) in the spectral region $\lambda \sim 5$–$25\ \mu$m.

2 Technology of the Thin Films Growth

2.1 The Hybrid Helicon-Arc Ion-Plasma Deposition Technique

To obtain the thin film coatings, the ion-plasma deposition technology with magnetic filtration of plasma flow was used. The vacuum processing equipment consists of a helicon source of plasma, in which the helicon discharge was excited by a high frequency (HF) generator with a working frequency of 13.56 MHz, and the plasma-arc accelerators, integrated in one technological chamber, see Fig. 1 [11, 12]. The helicon source of plasma is electrodeless HF induction source with a magnetic field, that is capable to generate a low-temperature ($T_e = 2$–10 eV) dense ($n = 10^{11}$–10^{13} cm^{-3}) plasma in a wide range of operating gas pressures ($p = 0.07$–13.3 Pa) and the range of magnetic fields ($B = 0.001$–0.2 T). The principal feature of the helicon source is the presence in the discharge volume of the external magnetic field. In this case, it is possible to excite helicon waves in the plasma volume. The application of the magnetic field provides several advantages for practical purposes. First, it is possible to significantly increase, at the same input HF power, the density of the plasma and control its spatial distribution. Secondly, the magnetic field allows control effectively the intensity and energy of the ion flow that enters the treated surface. Finally, the stability of the discharge increases due to the growth of the resistance of the plasma load. A magnetic filtration allows to achieve a high content of the high-energy ionic component in the plasma flow directed to a substrate, as well as to separate the micro-droplet component. We have identified modes of operation of a helicon source (Fig. 2), where an additional magnetic field with high-efficiency helicon discharge occurs in the entire volume of the reactor chamber.

2.2 Samples Preparation

The specified configuration of a discharge plasma chamber, which functionally combines two sources of plasma—a helicon discharge and a plasma-arc accelerator, was

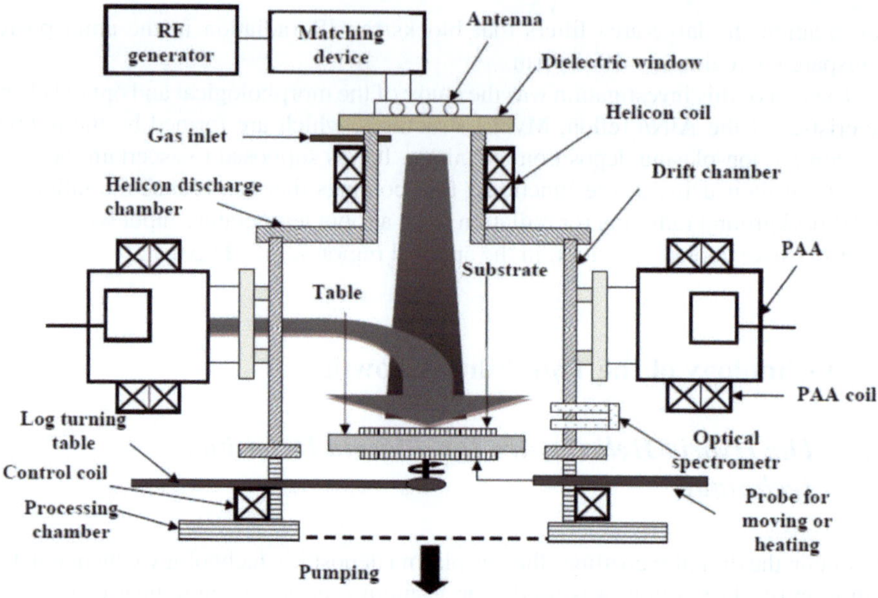

Fig. 1 The schematic of the hybrid helicon-arc ion-plasma reactor

Fig. 2 High-performance helicon ion-plasma discharge (the mode "Plasma column")

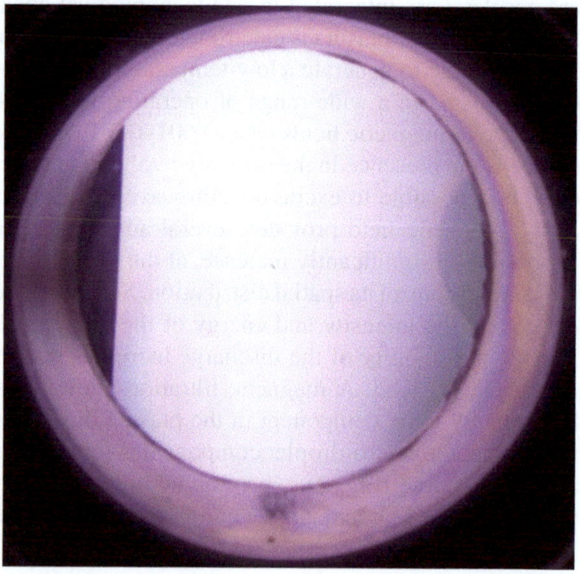

developed. It is possible to realize the new quality of the technological process of forming special film coatings on the dielectric flexible polymeric substrates. The formation of such coatings requires constant monitoring of the substrate surface at all stages of the process. Only in this case the reproducibility of results, the stability of the technological process and the high level of products can be provided. When applying the standard high vacuum equipment, the surface of the substrate always contains several adsorbed monolayers of the residual atmosphere component with the thickness of several nanometers. The helicon source of plasma provides at all stages of the technological process a constant effect on the substrate of argon ion flow. In this case, the density and ion flow energy are of a size, sufficient for effective removal from the substrate surface adsorbed light components of the residual medium. Due to the mechanism of selective spraying the basic technological operations are not blocked.

However, in the case of the plasma-arc method application for thin films depositing on polymeric substrates the important is the question of controlling and limiting the level of thermal heating of the substrate due to the plasma surface treatment. As it was established, the significant thermal stresses on the substrate, both at the processing stage in the plasma of the helicon source and during plasma-arc deposition, can lead to the destruction of the substrate or its significant damage. Therefore, various technological techniques for reducing the thermal load were used. Namely:

1. The breakdown deposition cycles, when the deposition was carried out in time interval t_1, and in the next interval t_2 the cooling of the substrate occurred.
2. The deposition on a substrate that rotated during plasma processing.
3. The deposition on a substrate that was rotating during plasma processing with the simultaneous scanning of the flux of precipitated particles by controlling the direction of deviation of the plasma flow by changing the magnetic field configuration in the drift chamber.

The application of the above-mentioned technological techniques allowed to reduce the level of thermal stresses on the substrate and to realize the deposition of films on polymeric substrates.

A single crystal silicon n-Si (100) and (111), as well as polymeric films of the Mylar and Teflon, were used as the substrates for the AlN films deposition. Combining the basic technological parameters, such as working gas pressure in the range of 0.4–4 Pa, the substrate temperature in the range of 20–400 °C, the current of cathode arc discharge, etc., the optimal growth regimes of AlN nanostructures thin films were selected.

The influence of the technological conditions of the formation of the AlN film structures on their quality was considered by using the atomic force microscopy (AFM). The surface morphology of the films on different substrates was investigated using the scanning probe microscope Nano Scope IIIa Dimension 3000 TM in the mode with the periodic contact. The measurements were carried out at the central area of samples by means of the serial silicon probes with the nominal tip radius of 10 nm.

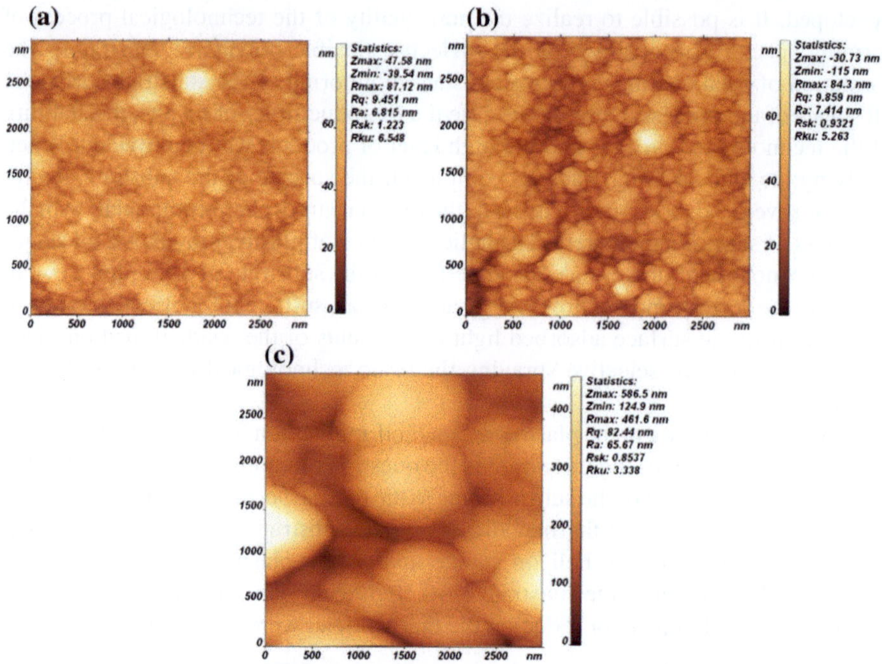

Fig. 3 The AFM images of the topology of AlN/Si (**a**), AlN/Mylar (**b**) and AlN/Teflon (**c**) films surfaces, synthesized at the simultaneous application of the helicon and arc plasma sources

The comparison of Fig. 3a–c shows that the samples of the AlN grown on silicon and polymeric substrates are characterized by different surface roughness and different grain sizes in spite of the same technological conditions of their deposition. The surface relief of the sample AlN/Si (Fig. 3a), where the roughness of the sample reached values of $R_a = 6.8$ nm in the area (3×3 μm), with the average grain size reaching 20–25 nm, was the most smooth. In the case of samples, AlN/Mylar and especially AlN/Teflon the developed relief is characterized by the presence of both shallow grains of nanometer size, and sufficiently large grains with dimensions of ~1 μm (Fig. 3b, c). This shape of the AlN films relief may be explained by the fact that before its growth, the flexible (in contrast to silicon) polymer substrates were subjected to ion cleaning. As a result, the initial relief of Mylar and Teflon substrates was modified, and the AlN film's morphology repeats the relief of the substrates.

3 The Optical and Structure Characterization

The sensitive radio receivers require the use of the cryogenic detectors [16]. The infrared filters are necessary to avoid the heat loading on the cold stages. The filters should be effective in blocking the 300 K blackbody radiation while remaining highly transparent at the radio frequencies of interest or must be transparent in the visible spectral region.

The background spectral radiant exitance (the flux power density at the given wavelength λ) taken here as the black-body radiation, is determined by Plank's formula (from the semi-sphere):

$$W(\lambda, T) \cdot \mathrm{d}\lambda = \frac{2\pi \cdot h \cdot c^2}{\lambda^5} \cdot \frac{1}{\exp(ch/k_B \cdot \lambda \cdot T_{BB}) - 1} \tag{1}$$

where c is the light speed, h is the Plank constant, k_B is the Boltzmann constant, T_{BB} is the black-body temperature, and W(λ,T) is the total power density fallen down at the input window from the hemi-sphere. For a black body at 300 K, the peak of the spectral radiant exitance determined by the Planck distribution (1) occurs at 9.66 μm (~18 THz). Because of the long tail at high radiation frequencies, a most part of a radiated energy lies above the maximum, squarely in the "mid-IR" range [17].

Teflon or its modification Zitex (the refractive index n ≈ 1.435, T = 300 K) and Mylar (refractive index at T = 300 K, n = 1.64–1.67) are often used as effective IR filters (see Fig. 4) due to the low transmission in the λ ~ 7.5–9.0 and 12.5–21 μm wavelength bands. But these polymeric films have a low thermal conductivity: down to γ ≈ 0.26 W/m K at 50 °C, slightly dependent on temperature for Teflon and γ ≈ 0.8 W/m K at $T = 300$ K (relatively strongly dependent on temperature) for Mylar. These low values of γ substantially suppress the heat transfer to heat sink from filters that are heated by radiation from the environment or the cryostat input window being at $T = 300$ K [17]. To reduce the influence of the radiation flows in the region of λ ~ 5–25 μm, in the filters that are based on the Teflon or Mylar films, there should be used coatings based on dielectric materials or wide-band semiconductors. The wide-band semiconductors have a low free-carrier absorption coefficient in the microwave region and low refractive indexes. They are composed of light atoms, which allows getting the Restrahlen band within the IR regions of interest. Among these coating films defined by the Restrahlen band, one can consider [17]:

1. MgO (Restrahlen band λ ≈ 14–26 μm)
2. Si/SiO_2 (Restrahlen band λ ≈ 7–9.5 μm)
3. GaN (Restrahlen band λ ≈ 15.88–17.57 μm)
4. Sapphire (Restrahlen band λ ≈ 13.65–18.62 μm)
5. BeO (Restrahlen band λ ≈ 9.0–14.5 μm)
6. AlN (Restrahlen band λ = 11–16 μm in amorphous films).

AlN was chosen as a coating material because of its high value of thermal conductivity, relatively wide Restrahlen band, non-toxicity (as, for example, in contrast to BeO).

The IR transmission spectra were measured at the radiation normal incidence at T = 300 K (see Fig. 4) using the Fourier Spectrometer "Spectrum BX-II." From their analysis, it is seen that the composite structures AlN/polymeric substrate are substantially blocking the IR radiation within the spectral range λ ≈ 5–25 μm.

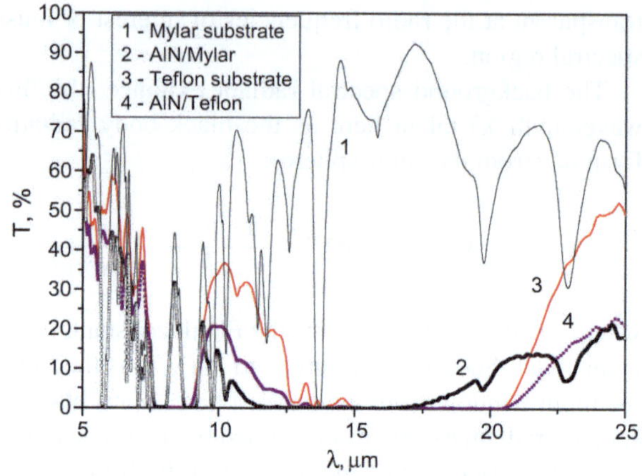

Fig. 4 The IR transmission spectra of polymeric substrates (Mylar with the thickness $d = 40$ μm (1) and Teflon with the thickness $d \approx 220$ μm (3)) and the AlN nanostructured layers ($d \sim 15$ μm) deposited on these substrates (2, 4)

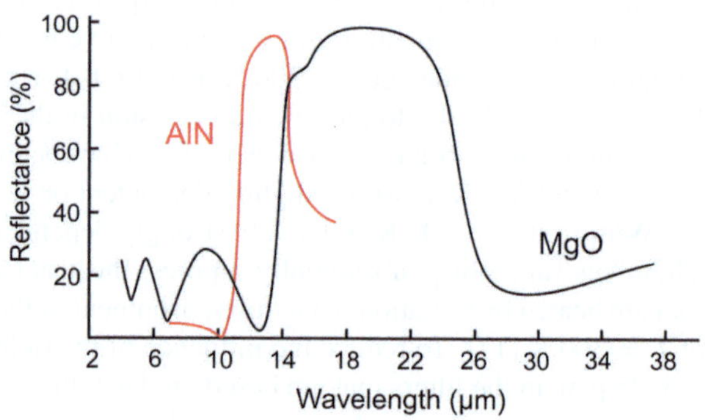

Fig. 5 The reflection spectra of the monocrystalline AlN and MgO layers [13, 18]

Taking into account the high AlN thermal conductivity, the composite structures of AlN on polymeric substrates can effectively withdraw the heat from the filters surfaces to their periphery (heat sink) if even absorb the radiation.

The high crystallinity AlN films within the Restrahlen band should reflect ~100% of radiation (Fig. 5). The IR reflectance spectra of the AlN/Si and AlN/Mylar films at 15° angles of incident radiation were studied and are presented in Fig. 6. From the analysis of the black-body spectral radiant exitance at $T = 300$ K (curve bb on Fig. 6) one can conclude, that for blocking the IR radiation influence, the spectral region from 5 to 25 μm can be mainly considered, where about of 82% of the black-body power density, as compared to in the whole spectrum, is concentrated.

Despite the AlN polycrystalline structure, the FTIR spectra of the AlN thin films exhibit the characteristic Reststrahlen band (between the longitudinal and transverse optical phonon frequencies) of the AlN layers (curves 3, 4 on Fig. 6). It concerns to the relatively thick AlN films, where the broadband of strong absorption related with the Restrahlen band is observed. For AlN/Si the short-wave edge of the Restrahlen band at λ ~ 10.3 μm in reflection spectra was observed (Fig. 6).

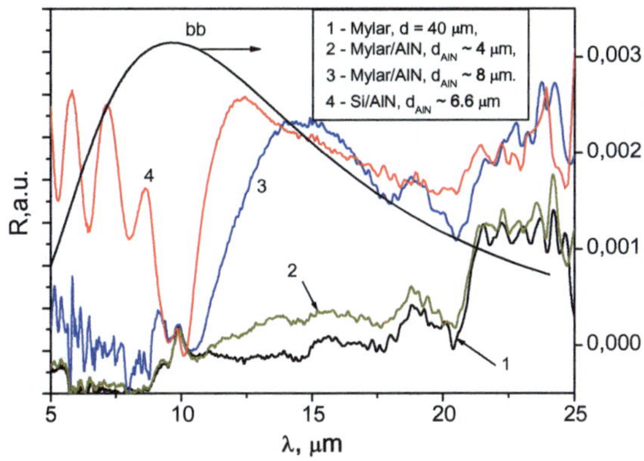

Fig. 6 The IR reflection spectra of the Mylar substrate (curve 1), Mylar films with AlN polycrystalline layers of different thickness (d ~ 4—curve 2, d ~ 8 μm—curve 3) and the AlN/Si structure with the thickness of AlN polycrystalline layer (d ~ 6.6 μm—curve 4) in the 5–25 μm IR spectral range at $T = 300$ K. The b.b. curve corresponds to the black-body spectral radiant exitance at $T = 300$ K (right scale)

As can be seen from Fig. 6, the long-wave edge of the Restrahlen band is shifted to the longer-wave region due to the influence of two factors: the polycrystalline structure of the AlN films, as well as the effect of substrates on the reflection spectrum. This band reflection spectrum is not so pronounced as in thick AlN crystalline films (see [13]), which is expected due to disordered tens-nanometer nano-size grains in these films with low crystallinity. As compare with the AlN/Mylar structures, from the Fig. 6 it is clearly seen the presence of the Restrahlen band ($\lambda \approx 10.6$–20 μm) on the AlN/Mylar structure with relatively thick (~8 μm) AlN film, while on the thinner structures the Restrahlen band is weakly observed. As it was pointed in [19], in the AlN/Si polycrystalline layers, the Restrahlen band is extended up to λ ~ 18.2 μm, which makes these layers (in the case of poor crystalline quality) attractive for making the IR radiation blocking filters from the background at $T \leq 300$ K. Another important advantage of the Restrahlen band broadening in the nanostructured polycrystalline AlN films is the possibility of blocking infrared radiation in the range $\lambda \approx 10$–20 μm by using one coating material, rather than two different (see Fig. 5).

The XRD pattern of one of the polycrystalline AlN film deposited onto *n*-Si (111) at $T = 300$ K is shown on Fig. 7. Because of relatively large background noise, it was difficult to precisely conclude that reflexes observed for AlN films (002), (101) and (100) are typical for hexagonal structure since they are weak. However, the noticeable signal at 38.03° can be related with the reflex of AlN (101) hexagonal structure. The reflex at 35° probably is caused by the availability of Al_2O_3 that could be formed because of oxygen presence on the walls of vacuum chamber, in the working gas and in the SiO_x surface layer. Oxygen impurity levels are an important factor in the growth of AlN films by any method [20]. Elemental aluminum reacts more favorably with oxygen than nitrogen: it is more possible to form Al_2O_3 by gaseous phase reaction of Al + (3/2)O_2 than AlN of Al + (1/2)N since $\Delta G(Al_2O_3) = -1480$ kJ/mol and

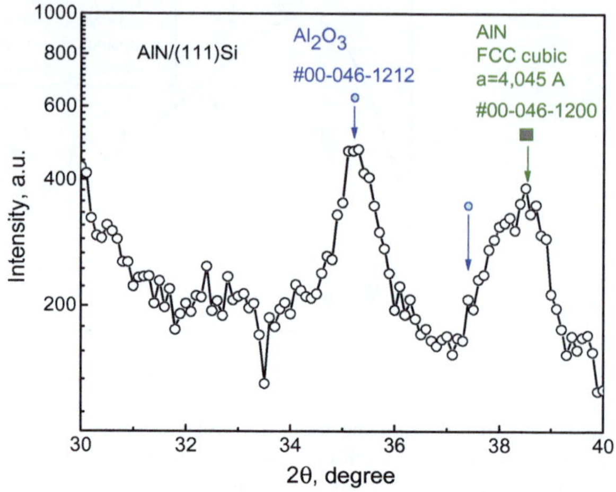

Fig. 7 XRD pattern of the AlN film deposited onto *n*-Si (111), $T = 300$ K. $Cu_{K\alpha 1}$ radiation, $\lambda = 0.15406$ nm (grazing incidence). Deposition temperature $T = 300$ K

$\Delta G(AlN) = -253$ kJ/mol, therefore, the existence of Al_2O_3 or even spinel AlNO phases in samples cannot be discarded [3]. The reflex at 38.5° is related with the availability in the AlN film crystallites with a cubic lattice, which is usually difficult for synthesizing.

The investigation of the composite structures AlN/polymeric substrate transparency in the sub-THz spectral regions by using quasi-optics attains $T \approx 79$–95%, and losses measured in the channels within the microwave region from 2 to 36 GHz are <0.06 dB. The results obtained argues that the composite structures "AlN film on the polymeric substrates" could have characteristics of a barrier filter in the mid-IR spectra and be transparent in the 2 to 270 GHz microwave region. They also are transparent in the visible region ($\tau \sim 50$–80% in dependence of the AlN films thickness).

4 The IR Blocking Properties

The obtained composite nanostructured AlN coatings on the Teflon and Mylar substrates can be efficiently used for decreasing the visibility of objects having the temperature $T \sim 300$ K and emitting in the IR range (blocking filters). Figure 8 shows the thermovision image obtained with using of the AlN/Mylar filter. From Fig. 8 it can be seen the part of woman's face masked behind the filter (Fig. 8a), as well as the transparency of this filter in visible spectral range (Fig. 8c). Figure 8b demonstrates the temperature distribution across the imaging (line 1 in Fig. 8a).

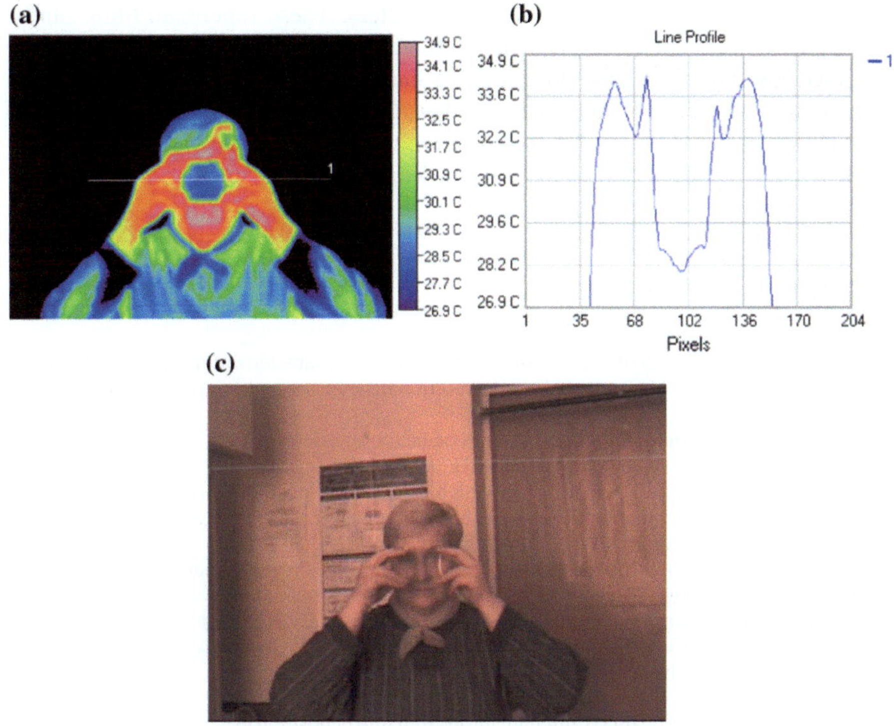

Fig. 8 The image of a woman obtained by thermovision camera (Mikron M7800) with the resolution of 60 mK in 8–14 μm IR spectral region: **a** the face is masked behind Mylar/AlN structure ($d_{Mylar} = 40$ μm, $d_{AlN} \sim 8$ μm); **b** the heat distribution along the line 1; **c** photo in the visible range

5 Conclusions

The technology of the low-temperature ion-plasma deposition of the AlN nanostructured thin films on different substrates, particularly on the Teflon and Mylar polymeric films, was developed. The essential difference of this method is the combined use of the helicon sources of low-temperature plasma and special elaborated the vacuum-arc sources, high deposition rate, and increased layer adhesion compared to other technologies, which allowed the formation of robust multicomponent functional nanostructures.

The morphological and optical characteristics of the AlN thin films on different substrates were studied. It was found that because of the Restrahlen band broadening in the nanostructured polycrystalline AlN films it is possible to sufficiently block the infrared radiation in the range $\lambda \approx 10$–20 μm by using single coating material. The "AlN film on Teflon or Mylar substrates" filters provide the radiation transmission ($\approx$85–95%) in wide microwave region 2–36 GHz and within 60–280 GHz. Therefore, diminishing the thermal noise in low-temperature microwave or sub-THz detector systems will increase their operational parameters. These functional films can be applied in cryogenic receiving systems for modern navigation networks, positioning and telecommunications (for protection of the cryoelements of detecting systems from harmful thermal IR radiation and thermal shock). The important feature of the AlN/Mylar structures is that they also are transparent in the visible region (τ ~ 50–80% in dependence of the AlN films thickness).

References

1. Moreira MA, Doi I, Souza JF et al (2011) Electrical characterization and morphological properties of AlN films prepared by dc reactive magnetron sputtering. Microelectron Eng 88(5):802–806
2. Morkoc H (2008) Handbook of nitride semiconductors and devices, vol 1: Materials Properties, Physics and Growth. Weinheim Wiley-VCH., Weinheim, Germany
3. Garcia-Mendez M (2012) Controlled growth of C-oriented AlN thin films: experimental deposition and characterization. In: Kolesnikov N, Borisenko E (eds) Modern aspects of bulk crystal and thin film preparation. InTech. https://doi.org/10.5772/1348
4. Ruterana P, Albrecht M, Neugebauer J (eds) (2003) Nitride semiconductors, handbook of materials and devices. Wiley-VCH Verlag GmbH&Co, Weinheim
5. Tansu N, Zhao H, Liu G et al (2010) III-Nitride Photonics. IEEE Photon J 2:241–248
6. Liang D, Quhe R, Chen Y et al (2017) Electronic and excitonic properties of two dimensional and bulk InN crystals. RSC Adv 7:42455
7. Chung DDL (2001) Materials for thermal conduction. Appl Therm Eng 21(16):1593–1605
8. Alrashdan MHS et al (2017) Aluminum nitride thin film deposition using dc sputtering. In: Proceedings of the 2014 IEEE international conference on semiconductor electronics (ICSE2014). https://doi.org/10.1109/SMELEC.2014.6920798
9. Balestraand F, Ghibaudo G (2001) Device and circuit cryogenic operation for low temperature electronics. Springer, Berlin
10. Graf UU, Honingh CE, Jacobs K et al (2015) Terahertz heterodyne array receivers for astronomy. J Infrared Milli Terahz Waves 36:896–921
11. Osipov L, Rudenko E, Semenyuk V et al (2010) Highly effective source of low temperature deposition of films and coatings. Nanoindustriya (Nanoindustry) 2:4–7 (in Russian)
12. Semenuk VF, Rudenko EM, Korotash IV et al (2011) Unified technological ion-plasma facility for formation of nanostructures. Metallofiz Noveishie Tekhnol 33(2):223–231 [In Russian]
13. Akasaki I, Hashimoto M (1967) Infrared lattice vibration of vapour-grown AlN. Sol St Comm 5:851–853
14. Kazan M, Pereira S, Correia MR et al (2009) Directional dependence of AlN intrinsic complex dielectric function, optical phonon lifetimes, and decay channels measured by polarized infrared reflectivity. J Appl Phys 106:023523. https://doi.org/10.1063/1.3177323
15. Park M-H, Kim S-H (2012) Thermal conductivity of AlN thin films deposited by RF magnetron sputtering. Sci Semicond Process 15:6–10
16. Sizov F (2018) THz radiation detectors: state-of-the-art. Semicond Sci Technol. https://doi.org/10.1088/1361-6641/aae473
17. Rudenko E, Tsybrii Z, Sizov F et al (2017) Infrared blocking, microwave and terahertz low-loss transmission AlN films grown on flexible polymeric substrates. J Appl Phys 121:135304-1–135304-8. https://doi.org/10.1063/1.4979858
18. Zhao S, Ribbing C-G (2010) Option for reststrahlen material in optical surfaces and filters. Chin Opt Lett 8:119–124
19. Antonova K, Szekeres A, Duta L et al (2016) Orientation of the nanocrystallites in AlN thin film determined by FTIR spectroscopy. J Phys. Conference Series 682:012024. https://doi.org/10.1088/1742-6596/682/1/012024

20. Jagannadham K, Sharma AK, Wei Q et al (1998) Structural characteristics of AlN films deposited by pulsed laser deposition and reactive magnetron sputtering: a comparative study. J Vac Sci Technol A 16(5):2804–2815

Estimating Qualitative Parameters of Aluminized Coating Obtained by Electric Spark Alloying Method

O. Gaponova, Cz. Kundera, G. Kirik, V. Tarelnyk, V. Martsynkovskyy, Ie. Konoplianchenko, M. Dovzhyk, A. Belous and O. Vasilenko

Abstract There are considered the features of the structural and phase state of the aluminized coatings obtained by the method of electric spark alloying (ESA) on the specimens made of 20 steel and 4 steel grades. It has been found out that with increasing discharge energy, there is increased the thickness and microhardness of the white and diffusion layers, as well as the surface roughness, and also there are changed the chemical and phase compositions. At low discharge energies, there is formed a layer predominantly consisting of α-Fe and aluminum oxides. It has been stated that increasing discharge energy results in obtaining the layer consisting of iron and aluminum intermetallics and free aluminum as well. In comparison with 20 steel, at electric spark alloying of 40 steel, there is increased the depth of the zone of increased hardness and microhardness thereof. In order to reduce the roughness and increase the continuity of the coatings obtained, it is recommended to conduct the electric spark alloying process applying the same electrode (aluminum), but at low discharge energies (Wp = 0.52 J). The comparative studies of the heat resistance of the aluminized coatings, which had been obtained with the use of the classic technology, that is, in aluminum melt, and by the ESA method with the use of an aluminum electrode, showed that electric spark coatings were characterized by a higher heat resistance. The results of the study make it possible to recommend the ESA technology with the use of an aluminum electrode in order to increase steel resistance to oxidation at elevated temperatures.

Keywords Electric spark alloying · Aluminizing · Microstructure · Coating · Surface · X-ray diffraction analysis · X-ray spectral analysis · Microhardness · Roughness · Heat resistance

O. Gaponova
Sumy State University, Sumy, Ukraine

Cz. Kundera
Kielce University of Technology, Kielce, Poland

G. Kirik · V. Tarelnyk (✉) · V. Martsynkovskyy · Ie. Konoplianchenko · M. Dovzhyk · A. Belous · O. Vasilenko
Sumy National Agrarian University, Sumy, Ukraine
e-mail: tarelnik@i.ua

© Springer Nature Singapore Pte Ltd. 2019
A. D. Pogrebnjak and V. Novosad (eds.), *Advances in Thin Films, Nanostructured Materials, and Coatings*, Lecture Notes in Mechanical Engineering,
https://doi.org/10.1007/978-981-13-6133-3_25

1 Introduction

Recently, to improve the environmental safety of machine-building enterprises, there have been developed and implemented a lot of environmentally friendly and resource-saving technological processes that allow increasing the resource and reliability, ensuring the operability of parts and apparatus under harsh operating conditions. In many events, changing physical and chemical properties of the surface layers of structural materials and products is a sufficient and cost-effective way to improve their performance properties, since the weakest element in the system of 'material-working medium' is the surface of the material. From this it follows that the development of the methods and technologies for applying protective coatings onto the surface of materials remains one of the most important technical problems to be solved. The need in the coatings for the parts operating at elevated temperatures can be justified by the fact that in some cases, it is impossible to provide the required increase in physical and chemical properties of parts and assembly units even when using new materials having an improved complex of chemical, physical, mechanical and other properties.

2 Analysis of Main Achievements and Publications

The authors of work [1] propose to divide the methods for controlling surface properties into the main groups as follows: applying coatings [2–6], alloying surfaces [7–10] and modifying structures of surface layers [11, 12]. The choice of the surface treatment is always based on providing a complete set of requirements for the surface of the product operating in the intended operating conditions [13–15].

The perspective methods for strengthening and modifying surfaces are considered the methods based on processing surfaces by concentrated flows of energy and matter (CPE). The method of electric spark alloying (ESA) relates to the number of such modern CPE methods for processing metallic surfaces, which make it possible to obtain surface structures having unique physical, mechanical and tribological properties. The ESA advantages are characterized by ecological safety of the process, high strength of bond between an alloyed layer and base material, the possibility of applying any conductive material onto the surface being processed, low energy consumption for processing procedures, and the simplicity of performing technological operations [5, 6, 16]. While possessing a wide range of possibilities to form certain structures, phase and chemical compositions in the surfaces being processed, the electric spark alloying process allows improving their operational properties.

The work is aimed at the following: to analyze the peculiarities of the process for forming the structures of the surface layers of carbon steels after aluminizing them by the method of electric spark alloying (ESA); to study the influence of the energy parameters of aluminizing process by the ESA method on the qualitative parameters of the layers obtained; to evaluate the heat resistance feature of the aluminized coatings obtained by the ESA method.

3 Methodology of Research

The studies were carried out on the specimens made of the high quality structural steels of 20 and 40 grades having dimensions of 15 mm × 15 mm × 8 mm. The ESA method was carried out using the unit of Элитрон-52A (Elitron-52A) equipped with the hand held vibrator; the rod of СвА99 (ГОСТ 7871–75) (SvA99 (GOST 7871-75) aluminum wire with a diameter of 4 mm and a length of 45 mm (Table 1) was used as an alloying electrode. The ESA modes are presented in Table 1.

Table 1 ESA modes

Mode number	Energy of discharge, W_p, J	Productivity, cm^2/min
1	0.52	1.0–1.3
2	1.30	1.3–1.5
3	2.60	1.5–2.0
4	4.60	2.0–2.5
5	6.80	2.5–3.0

To conduct the metallographic analysis of the coatings, there was used the optical microscope of МИМ-7 (MIM-7) type, the electron microscope of JSM 7100f (JEOL Ltd., Japan) type, and the ПМТ-3 (PMT-3) apparatus for durametric studies.

X-ray studies were carried out in Cu-Kα radiation using the instrument of Diffractometer PROTO AXRD type. To carry out the local X-ray-micro-spectral analysis of the coatings obtained, there was used the electron microscope equipped with the X-ray spectral micro-analyzer of ISIS 300 (Oxford Instruments, Great Britain) type. The surface roughness after the ESA process was studied by reading and processing the profilograms on the profilograph-profilometer of model 201.

4 The Study Results

The microstructure of the aluminum coating on 20 steel, as depended on the discharge energy values, is shown in Fig. 1. Metallographic analysis of the coatings obtained showed that the microstructure consists of 3 zones, namely:

(1) The white layer that cannot be etched with the use of ordinary reagents;
(2) The diffusion zone;
(3) The base metal having a ferrite-pearlite structure.

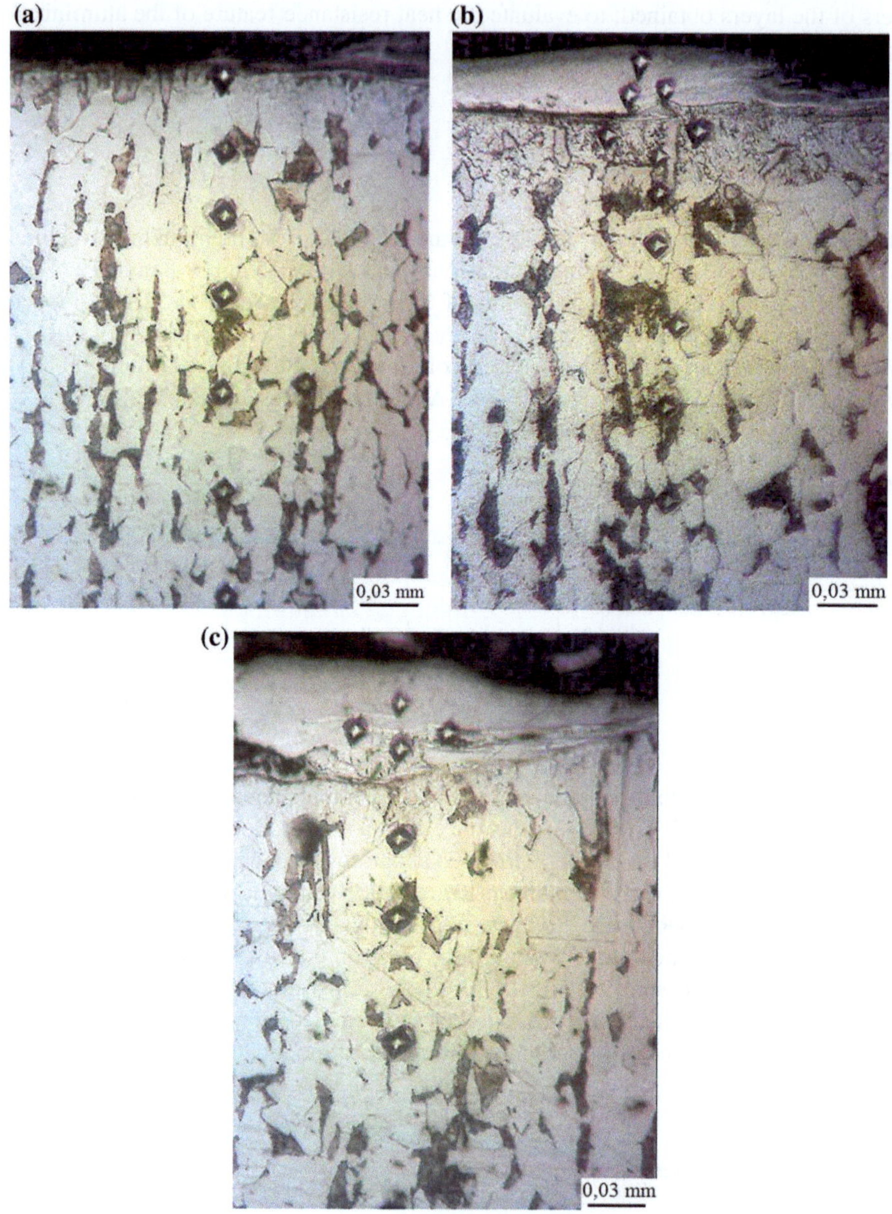

Fig. 1 Cross section microstructure of the specimens made of 20 steel with aluminum coating obtained by the ESA method: **a** $W_p = 0.52$ J, **b** $W_p = 1.30$ J, **c** $W_p = 2.6$ J

Table 2 Qualitative parameters of aluminized coatings obtained by the ESA method

W_p, J	Thickness, μm	Microhardness, H_μ, MPa	Roughness, μm	Continuity of white layer, %				
	White layer	Transition zone	White layer	Transition zone	R_a	R_z	R_{max}	
20 Steel								
0.52	10–12	20–30	2000 ± 70	1900 ± 50	1.3	2.3	9.3	60
1.30	30–50	30–40	2050 ± 70	1850 ± 80	1.9	6.2	21.6	80
2.60	40–50	30–50	2700 ± 70	2000 ± 200	3.3	9.3	23.2	85
4.60	50–70	40–60	5010 ± 90	2250 ± 200	6.2	16.3	40.6	95
6.80	Up to 70	110–130	7270 ± 50	2370 ± 70	9.0	18.1	58.3	100
40 Steel								
0.52	10–15	10–20	2350 ± 50	2000 ± 50	1.6	3.0	8.1	50
2.60	30–70	30–70	3500 ± 50	4500 ± 50	1.9	4.1	11.6	70
6.80	60–130	130–150	7400 ± 70	2390 ± 70	8.1	17.3	49.0	100

In Table 2, there are shown the parameters of the microstructure zones being formed as functions of the discharge energy values.

In the course of the ESA process, according to mode 1 (Table 1), there is formed a thin layer of a transition zone having a thickness of 20–30 μm and continuity that tends to 100%. There are met separate sections of white layers (up to 60%) of 10–12 mm thick. With increasing discharge energy, there are increased the values of thickness of the white layer and the transition zone. At Wp = 1.30 J, the thicknesses of the white layer and the transition zone are 30–50 and 30–40 μm, and at Wp = 4.60 J, they are 50–70 and 40–60 μm, respectively (Table 2). With increasing Wp, the continuity of the surface layer increases. Thus, at Wp = 2.60 J (mode 3), the continuity of the diffusion layer tends to 100%, and the white layer tends to 85%.

In Fig. 2, there are shown the results of the durametric analysis of the coating microhardness distribution as deepening from the surface. From the figure, it can be seen that the maximum hardness is achieved in the very surface portion of the specimen layer, and then it has smoothly been decreasing to the microhardness of the substrate, that is, to 1600–1700 MPa. The microhardness values of the coating zones are determined using the energy parameters of the ESA process: the greater the energy of the discharge, the higher the hardness of the white layer and, accordingly, the transition zone. Such a change in the microhardness is apparently due to the diffusion of aluminum into the substrate, and it is also caused by a change in the phase composition of the layer.

According to the Fe–Al [17] state diagram, in the aluminized layer obtained by the ESA method, there is a possibility of forming intermetallic compounds (this fact is confirmed by the high level of the coating hardness). Also, the fact of formation of nitrides and oxides could not be excluded [18], since the coating had been applied

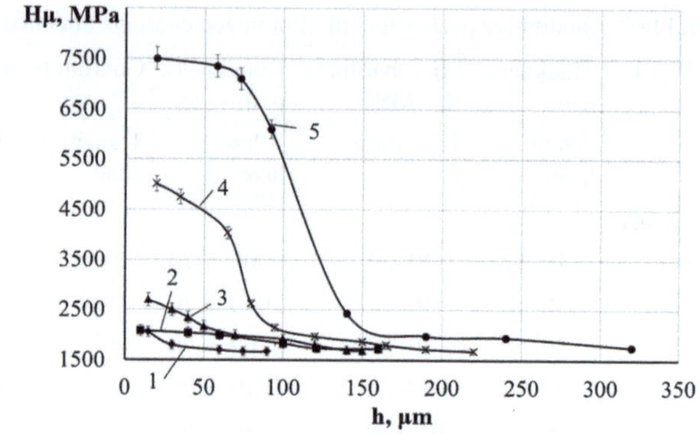

Fig. 2 Dependences of the microhardness (Hμ) values for the specimens made of 20 steel on the values for the layer (h) depth (height): 1—$W_p = 0.52$ J, 2—$W_p = 1.30$ J, 3—$W_p = 2.6$ J, 4—$W_p = 4.6$ J, 5—$W_p = 6.8$ J

in the air atmosphere. There is also very likely the availability of small amounts of pure electrode materials.

The phase compositions of the coatings formed after ESA process were determined by X-ray diffraction analysis. Diffractograms (Fig. 3) were taken from the surfaces of the specimens at applying the aluminum coating being obtained under mode 1 (Table 1). They indicate the presence of the diffraction maxima of two solid solutions based on the bcc phase structure—α-Fe and α-Fe′ (space group 229) and aluminum oxide γ-$Al_{2,67}O_4$ (space group 227). As shown in [19], the α-Fe and α-Fe′ phases differ only in the lattice period, which fact, apparently, is a consequence of the macrostresses arising as a result of accelerated cooling after the ESA process.

With increasing the discharge energy values, there are appeared the new phase components. If under mode 1 (Tables 1 and 3), there is only a solid solution of aluminum in iron and a small amount of aluminum oxide, since ESA was carried out in the air atmosphere, then under mode 2, there are provided conditions for creating a phase with the monoclinic structure of Fe_4Al_{13} (space group 12) and pure aluminum

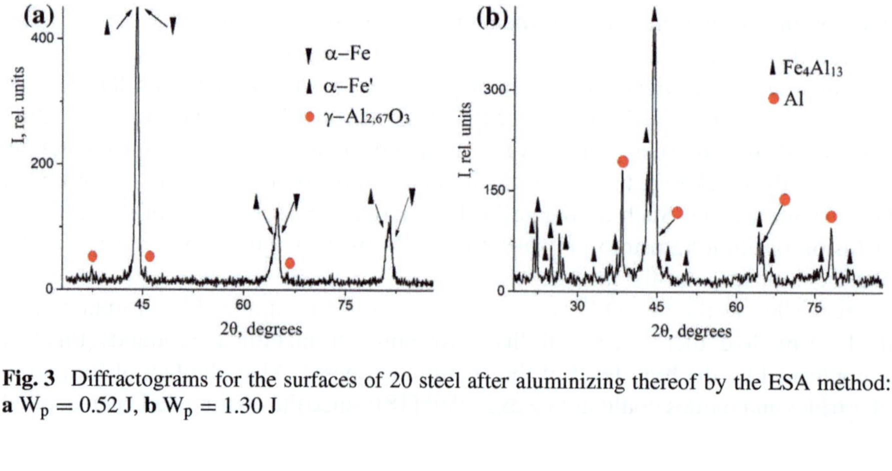

Fig. 3 Diffractograms for the surfaces of 20 steel after aluminizing thereof by the ESA method: **a** $W_p = 0.52$ J, **b** $W_p = 1.30$ J

Table 3 Parameters of crystal lattices for phases and quantitative phase analysis after aluminizing

Phase number	Phase	Parameters of crystal lattices, nm	Phase contents, % (wt.)
1	α-Fe	a = 0.2887	36
	α-Fe′	a = 0.2907	47
	γ-$Al_{2,67}O_4$	a = 0.7980	17
2	Al	a = 0.4056	19
	Fe_4Al_{13}	a = 1.5403, b = 0.8134, c = 1.2473	81

(space group 225). Apparently, the presence of intermetallic phases contributes to increasing the coating hardness and discharge energy values (Fig. 2).

As the authors of works [20, 21], using the pairs of Mo–Fe and Cu–Fe, we have demonstrated that increasing the discharge energy values under the condition, when aluminum interacts with iron (steel) in the air atmosphere, results in increasing the thickness of the coating and the diffusion zone being formed. In accordance with the samples obtained in modes 1 and 2, the thickness of the diffusion zone of aluminum in iron makes up 34 and 50 μm, respectively (Fig. 4). We herein note that the amount of aluminum in iron also increases: at Wp = 1.30 J, on the surface of the specimen, there is formed a thin layer (up to 4 μm) containing ~1.5–2 times more aluminum. The presence of free aluminum in the layer is confirmed by X-ray diffraction data (Table 3).

Investigating the roughness of the surface layer of 20 steel after aluminizing by the ESA method showed that the roughness of the surface increased with increasing the discharge energy values (Table 2): Ra = 1.3 μm at Wp = 0.52 J and Ra = 3.3 μm at Wp = 2, 60 J. The further increase in the discharge energy up to 6.8 J is accompanied by a significant increase in the surface roughness: Rmax = 58.305 μm, Ra = 9.039 μm and Rz = 18.142 μm (Fig. 5).

The metallographic analysis of the aluminized coatings on 40 steel showed that there was observed a formation of 3 zones just like on 20 steel (Fig. 6). It should be noted that under the same conditions of the ESA process, the thickness values of the white layer and the diffusion zone are greater on 40 steel, (Table 2). In addition, those have the higher microhardness values (Fig. 7). Thus, in the course of ESA processing 40 steel in accordance with mode 3, the thickness of the white layer makes up 30–70 μm, Hμ = 3500 ± 50 MPa; and the same for 20 steel are 40–50 μm, Hμ = 2700 ± 70 MPa, correspondingly. With increasing the discharge energy values, there is increased the continuity of the white layer, and the transition layer tends to 100% (Table 2).

With increasing the discharge energy values at the ESA process, the depth of the transition zone increases. At Wp = 0.52 J, it is not clearly expressed, whereas at Wp = 2.60, J it is about 30–40 μm and characterized by an increased hardness (~4500 MPa).

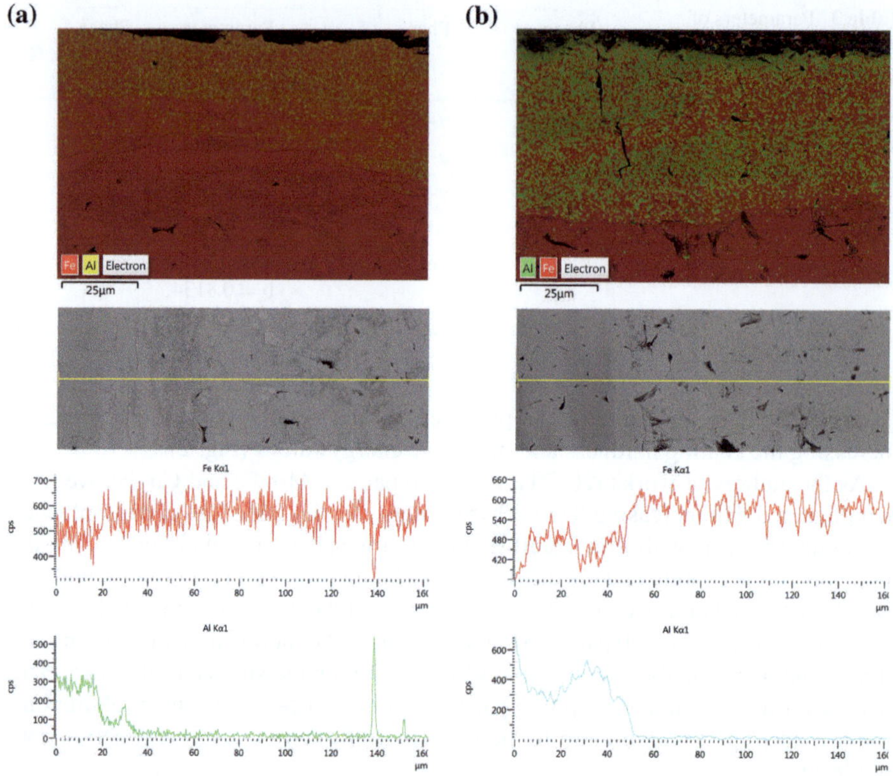

Fig. 4 Concentration curves for distributing iron and aluminum in the coating obtained by the ESA method: **a** $W_p = 0.52$ J, **b** $W_p = 1.30$ J

Increasing the hardness of the transition zone can be caused by phase transformations occurring, when the steel is heated above critical temperatures, and by accelerated cooling in the air atmosphere. In the microstructure, there is clearly traced a region of incomplete re-crystallization: in the course of the ESA process, the pre-eutectoid 40 steel is heated up at the intercritical temperature range (interval) of A1–A3 (730–755 °C), and after cooling, a large grain of ferrite is retained in this section, with the pearlite component being crushed. The procedure of accelerated cooling after the ESA process results in some strengthening the surface layer of the steel and increasing hardness thereof. Just like on 20 steel, the formations of the intermetallic compounds, nitrides and oxides were possible in the surface layer of 40 steel since the ESA process had been carried out in the air atmosphere, which fact also could cause increasing the microhardness of the coating and the transition zone.

On analyzing the roughness of the surface layer of the specimens made of 40 steel after aluminizing them by the ESA method, it has been found out that the surface roughness increased with increasing the discharge energy values. At Wp = 0.52 J,

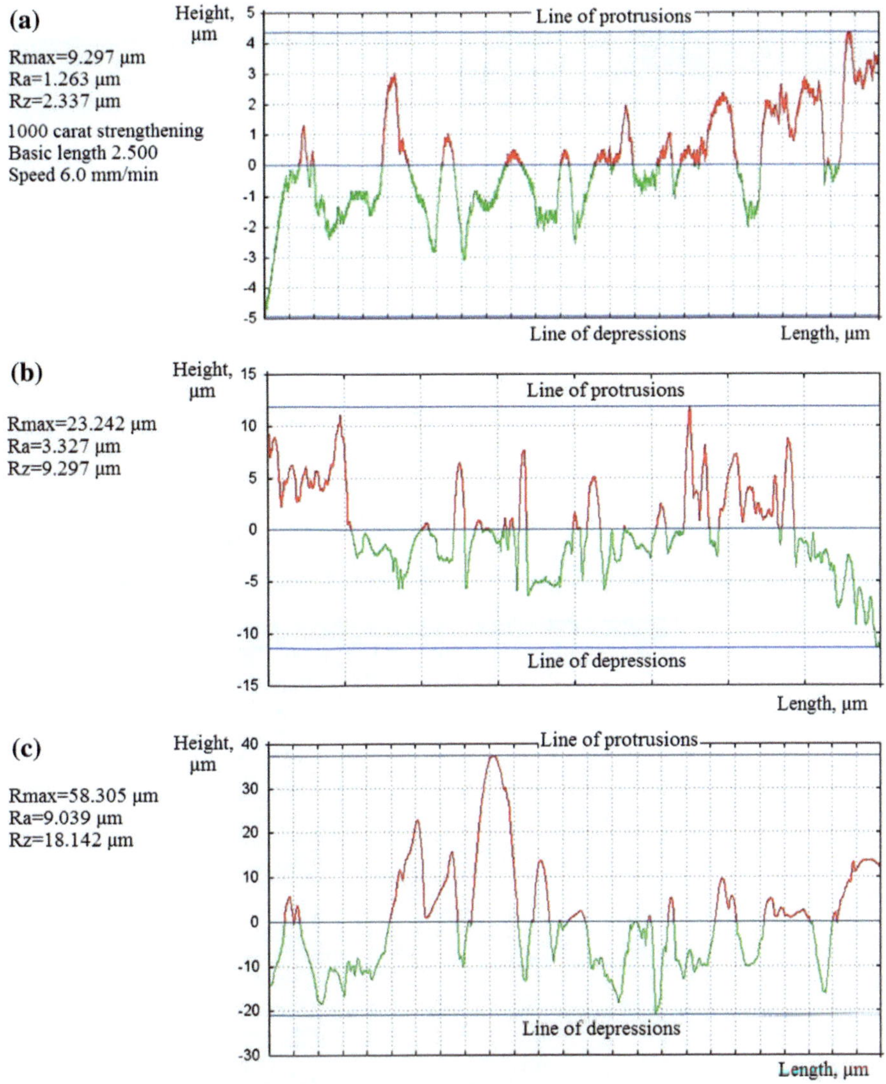

Fig. 5 Profilograms of surface roughness for the specimens made of 20 steel after aluminizing them by the ESA method: **a** $W_p = 0.52$ J, **b** $W_p = 2.6$ J, **c** $W_p = 6.8$ J

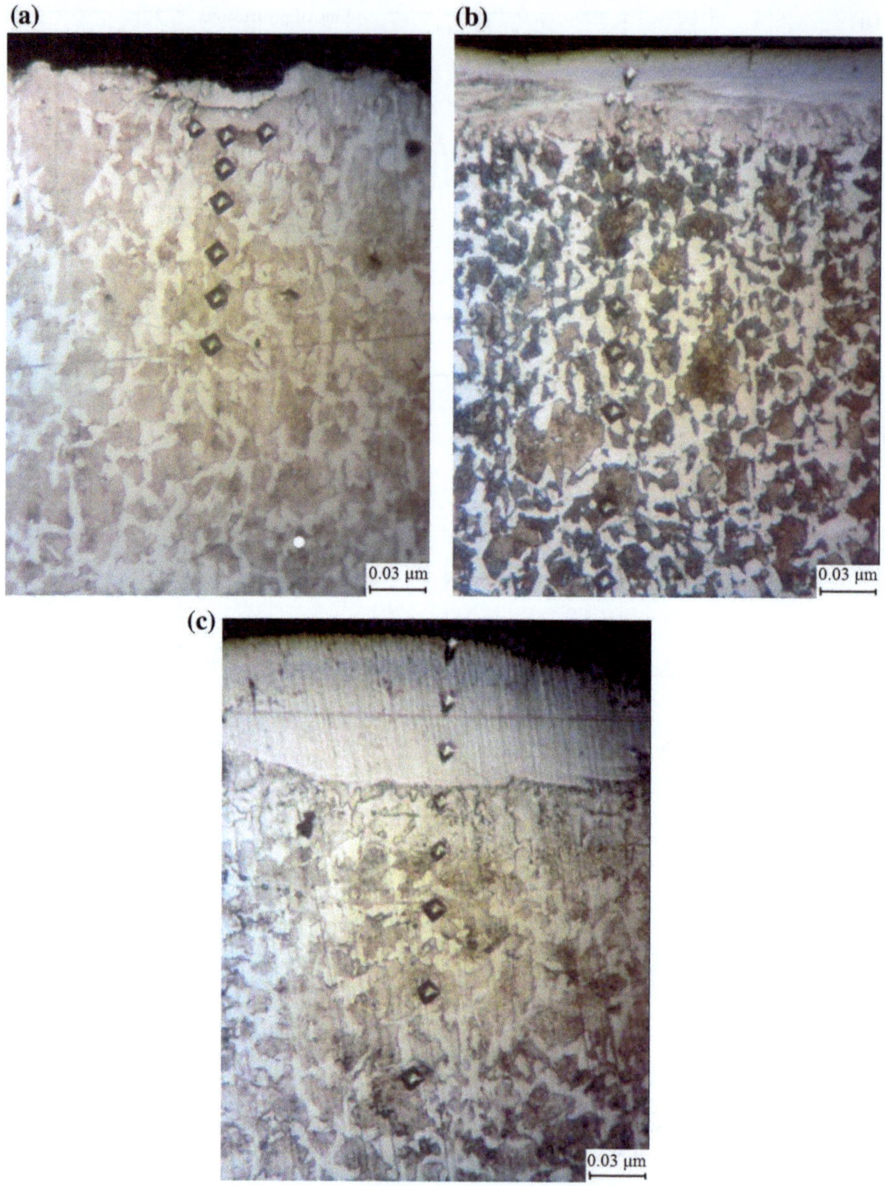

Fig. 6 The microstructure of the cross section of the specimens made of 40 steel after aluminizing them by the ESA process: **a** $W_p = 0.52$ J, **b** $W_p = 2.6$ J, **c** $W_p = 6.8$ J

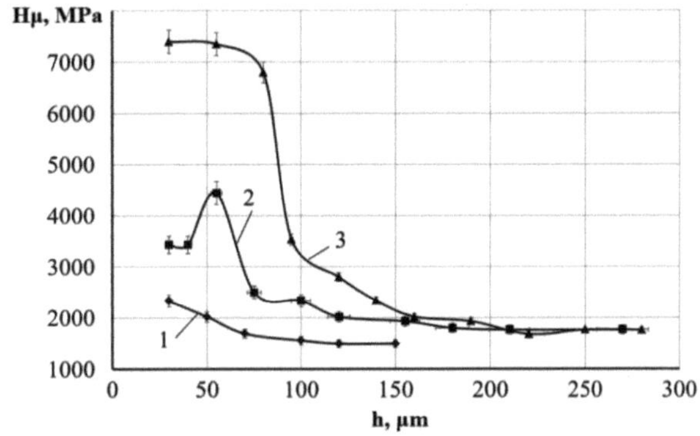

Fig. 7 Dependences of the microhardness (Hμ) values for the specimens made of 40 steel on the values for the layer (h) depth: 1—Wp = 0.52 J, 2—Wp = 2.6 J, 3—Wp = 6.8 J

the maximum surface roughness is Rmax = 11.555 μm, and the surface roughness mean values were respectively Ra = 1.853 μm and Rz = 4.144 μm (Fig. 8).

Thus, at ESA processing the steels of 20 and 40 grades with the use of an aluminum electrode, with increasing discharge energy, there are increased such qualitative parameters of the surface layer as roughness, thickness, microhardness of the white layer and the transition zone (Table 2). The white layer continuity at Wp = 0.52 J is low and makes 50–60%, and further, with a subsequent increase in the discharge energy to Wp = 6.8 J, it increases to 100%.

To reduce the surface roughness after ESA processing with an aluminum electrode, we herein propose to carry out subsequent processing with the same electrode (aluminum), but at lower discharge energies. In this event, the electric discharge flows between the roughness protrusion tip and the aluminum electrode, as a result of which fact the tip of the protrusion collapses, and the roughness of the surface decreases. Previous studies have shown (Patent RU No. 2468899, Patent UA No. 101715) that such an ESA technology is rather effective.

In Fig. 9, there are shown the results of measuring the roughness of the surface of the specimen made of 20 steel, which was aluminized by the ESA method first with the discharge energy of Wp = 2.6 J and the productivity of 1.8 cm^2/min, and then with Wp = 1.3 J and the productivity of 0.8 cm^2/min. From the data obtained, it is clearly seen that the surface roughness has really decreased: Rmax = 9.297 μm, Ra = 1.263 μm and Rz = 2.337 μm.

5 Estimating Heat Resistance Characteristic of Aluminized Surface Layers Obtained by the ESA Method

It is known that the aluminized coatings provide for iron-carbon alloys with the increased heat resistance values. In this connection, the actual technical problem is to study the heat resistance of the aluminized coatings obtained by the ESA method. 20 Steel was used as a substrate, the ESA layers were produced in two passes, first

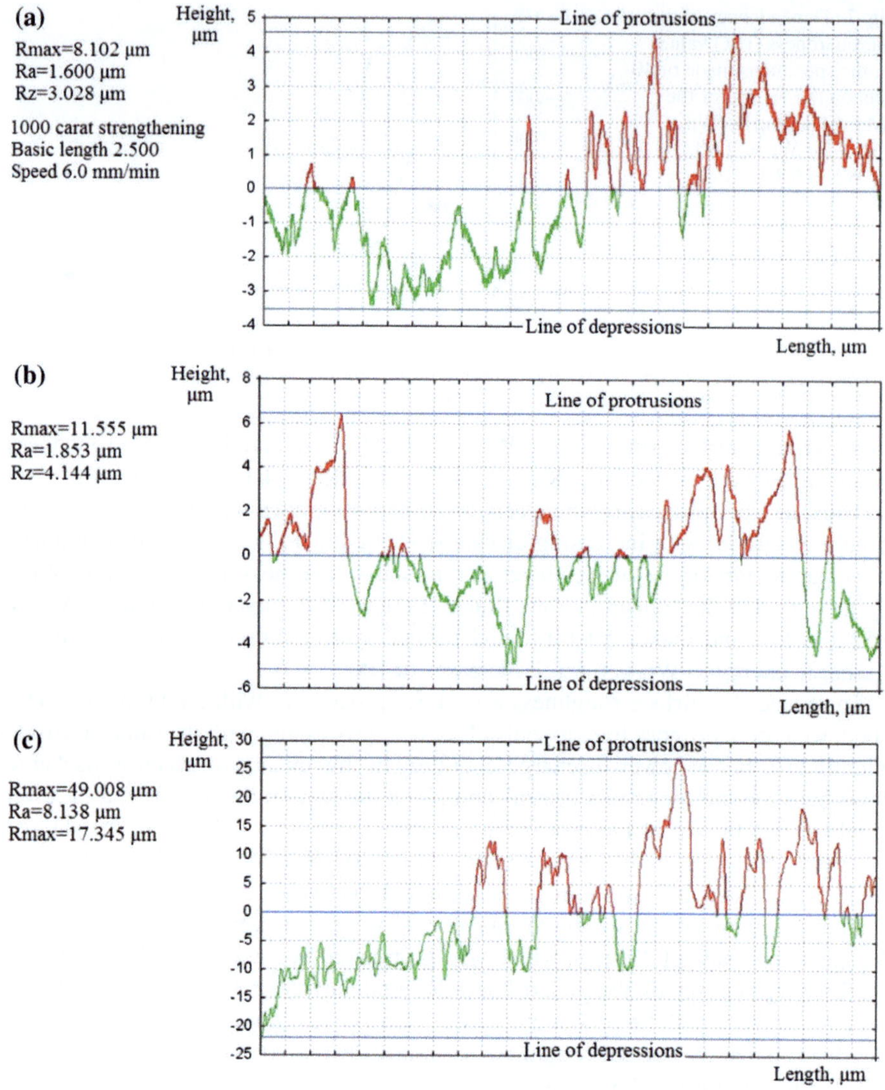

Fig. 8 Profilograms of the surface roughness values of the specimens made of 40 steel after aluminizing them by the ESA method: **a** $W_p = 0.52$ J, **b** $W_p = 2.6$ J, **c** $W_p = 6.8$ J

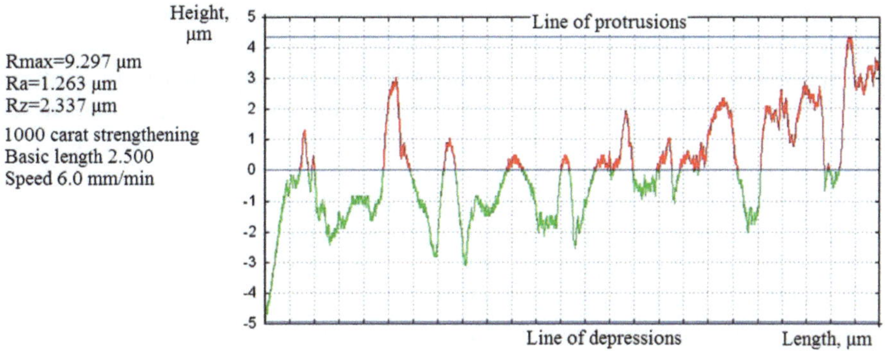

Fig. 9 Profilogram of the surface roughness of the specimen made of 20 steel that was successively aluminized by the ESA method with the aluminum electrode at the discharge energy values first of $W_p = 2.6$ and then of $W_p = 1.3$ J

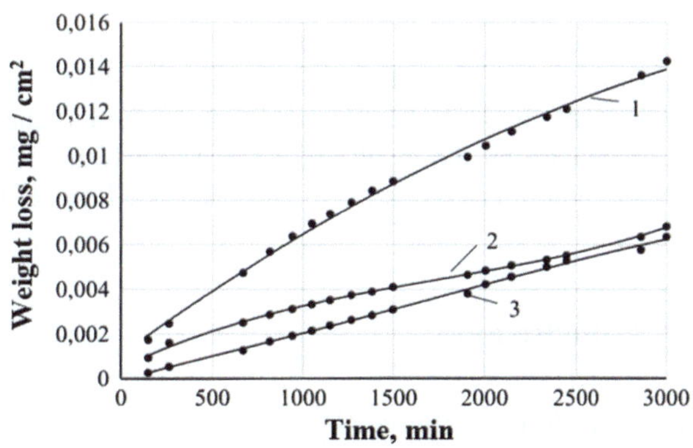

Fig. 10 Heat resistance of aluminized coatings on 20 steel (the tests in the air atmosphere at 980 °C, for 50 h): 1—uncoated, 2—aluminized by the ESA method; 3—aluminized in aluminum melt

at the discharge energy of Wp = 2.6 J and the productivity of 1.8 cm^2/min, and then, at Wp = 1.3 J and the productivity of 0.8 cm^2/min. The studies have shown that after such a treatment, there are formed the high-quality coatings (100% continuity) having low roughness (Table 2, Fig. 9).

To obtain a comparative evaluation of the heat resistance (Δg-weight loss per unit of the surface area of the test specimens, mg/cm^2), there were investigated the uncoated specimens made of 20 steel and treated having applied the classic aluminizing technology (aluminizing in aluminum melt, described in detail in Ref. [22]). The results of the tests, which had been carried out in the air atmosphere at the temperature of 980 °C for 50 h, showed that ESA coatings were characterized by the high heat resistance values being even slightly higher than after the classical melt aluminizing technology, which fact would make it possible to recommend the ESA technology with an aluminum electrode for the purpose of increasing the resistance of the steels to oxidation at elevated temperatures (Fig. 10).

There were carried out metallographic studies of specimen oxidation nature after heat-resistance test. The oxide inclusions were observed on all the specimens. The specimens without protective coatings were especially intensively oxidized (Fig. 11a). After processing by the method of liquid aluminizing, in the surface layer, there was located a zone of aluminum remained after terminating the aluminizing process (Fig. 11b, c). After the heat-resistance test (Fig. 11b), there were observed the results of the process of oxidation of the metal under the coating, as evidenced by the presence of an oxidized layer that separates the surface layer from the base metal. Obviously, this type of destruction of the coating and, consequently, oxidation of the base metal would result in decreasing the mass of the specimens in the course of high-temperature tests (Fig. 10) [23–26].

There is observed decreasing the thickness of the white layer for the specimens with the aluminized coating obtained by the ESA method (Fig. 11d), because at the high-temperature tests, the surface layer breaks down, and in this event, the loss of mass is associated with physical and chemical processes occurring predominantly in the surface layer, and not in the base metal [27–30]. In addition, despite the oxidation process, the surface layer retains an increased hardness (Fig. 12).

Thus, the aluminized coatings obtained in aluminum melt and done with the use of the ESA method retain increased hardness, protect the base metal against oxidation, as evidenced by the results of the tests for heat resistance (Fig. 10), as well as a smaller amount of oxides in the near-surface layer of the base metal (Fig. 11).

6 Conclusions

There are described the peculiarities of the structure formation at aluminizing 20 steel and 40 steel under various ESA modes. It is shown that the structure of a layer obtained consists of three sections, namely, a white layer, diffusion zone, and base metal. With increasing discharge energy, there are increased such qualitative parameters of the surface layer as thickness, microhardness of white layer and transition zone, as well as roughness. At $Wp = 0.52$ J, the continuity of the white layer is low making up 50–60%, and with increasing the discharge energy to $Wp = 6.8$ J, the continuity increases to 100%. In the course of the ESA process, increasing the discharge energy results in changing the chemical and phase composition of the layer. Thus, at low discharge energies, there is formed a layer predominantly consisting of α-Fe and aluminum oxides. According to the data of the local X-ray-micro-spectral analysis, it was stated that increasing discharge energy results in obtaining the layer consisting of iron and aluminum intermetallics, as well as of free aluminum.

The comparative analysis of the influence of the substrate on the qualitative parameters of the surface layer during the process of aluminizing by the ESA method showed that, when 20 steel having been replaced by 40 steel, there was increased the thickness of the white layer and the transition zone, namely, the depth of the strengthened zone, as well as its microhardness. The roughness of the surface practically remained unchanged.

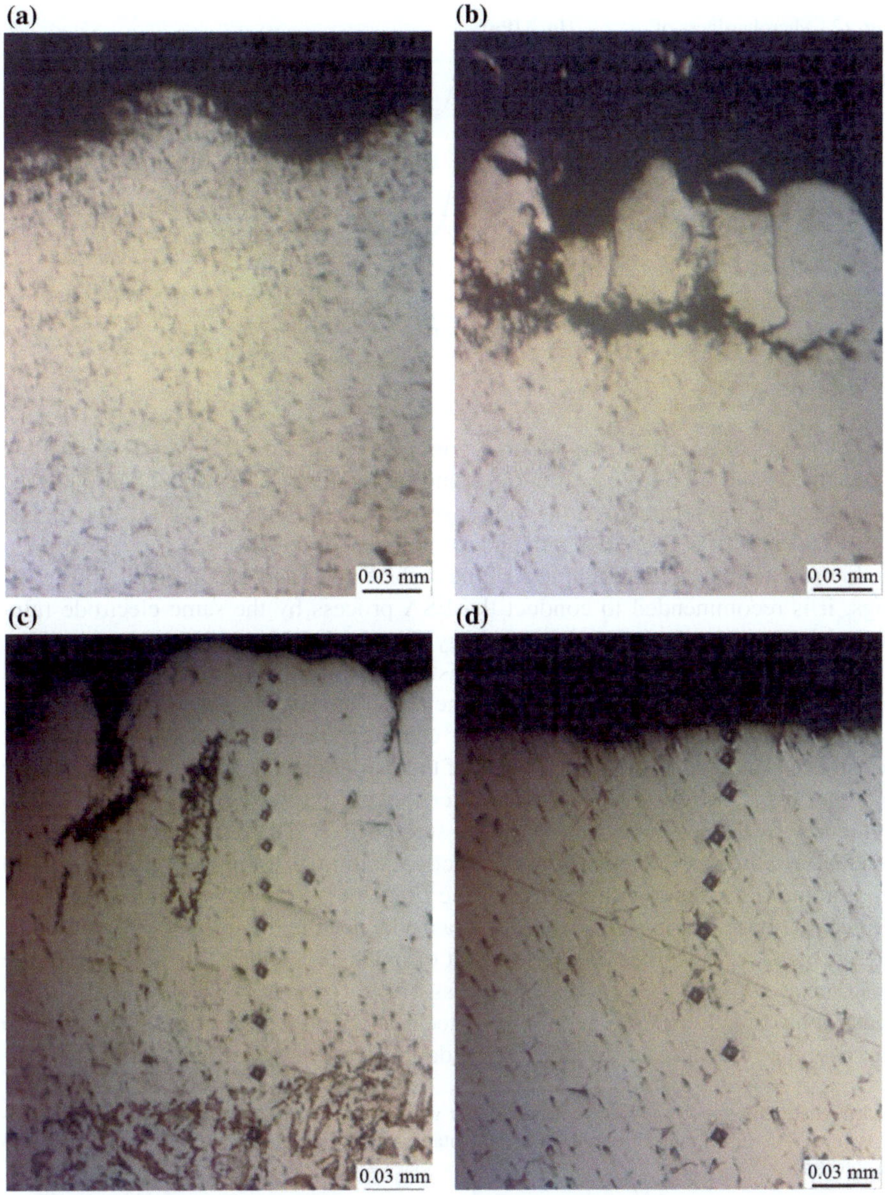

Fig. 11 Microstructures of aluminized coatings on 20 steel after the heat resistance test (as tested in the air atmosphere at 980 °C, for 50 h): **a** uncoated (non-etched) thin section; **b** aluminized in aluminum melt (non-etched) thin section; **c** aluminized in aluminum melt (after etching in 3–5% HNO_3 solution); **c** aluminized by the ESA method (after etching in 3–5% HNO_3 solution)

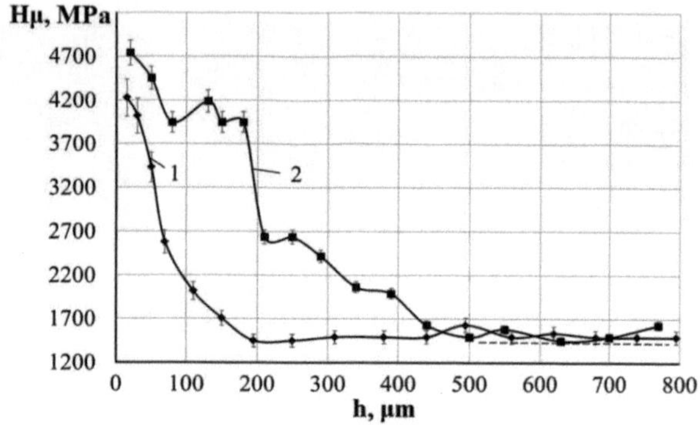

Fig. 12 Microhardness of aluminized coatings on 20 steel after the heat resistance test (as tested in the air atmosphere at 980 °C, for 50 h): 1—aluminized by the ESA method; 2—aluminized in aluminum melt

For practical use, it is possible to recommend aluminizing by the ESA method under modes 4 and 5 (Wp = 4.6–6.8 J and productivity 2.0–3.0 cm^2/min), which would ensure the formation of a white layer with the thickness of 70–130 μm, Hμ = 5000–7500 MPa, Ra = 6–9 μm and a continuity of 95–100%.

In order to reduce the roughness of the surface layer and obtain continuous coatings, it is recommended to conduct the ESA process by the same electrode (aluminum), but with a lower discharge energy (Patents UA No. 119316, No. 119707).

The comparative studies of the heat resistance of the aluminized coatings obtained by the classic technology (in aluminum melt) and the ESA method by an aluminum electrode showed that the electric spark coatings are characterized by high heat resistance. The metallographic analysis of the oxidation state of the specimens after the heat resistance test indicates that after the test the base metal is oxidized, as evidenced by the presence of oxides in the surface layer. There is especially intensive oxidation of the specimens without a protective coating. Aluminum coatings obtained in aluminum melt and with the use of the ESA method retain increased hardness, protect the base metal against oxidation, as evidenced by the results of the tests for heat resistance, as well as a lower content of oxides in the near-surface layer of the base metal and also the sufficient hardness of the coating. The results of the study make it possible to recommend the ESA technology with an aluminum electrode to increase the resistance of the steels to oxidation at elevated temperatures.

Acknowledgements This work was carried out within the framework of the project No. 53/18-N of the NAS of Ukraine Program "Fundamental issues of creation of new nanomaterials and nanotechnologies".

References

1. Yushchenko KA, Borisov YS, Kuznetsov VD, Korzh VM (2007) Surface engineering. Naukova dumka, Kyiv
2. Martsinkovsky V, Yurko V, Tarelnik V et al (2012) Designing radial sliding bearing equipped with hydrostatically suspended. Procedia Eng 39:157–167. https://doi.org/10.1016/j.proeng.

2012.07.020
3. Tarel'nik VB, Martsinkovskii VS, Zhukov AN (2017) Increase in the Reliability and durability of metal impulse seals. Part 2*. Chem Pet Eng 53:266–272. https://doi.org/10.1007/s10556-017-0333-7
4. Tarelnyk V, Martsynkovskyy V (2014) Upgrading of pump and compressor rotor shafts using combined technology of electroerosive alloying. Appl Mech Mater 630:397–412. https://doi.org/10.4028/www.scientific.net/AMM.630.397
5. Tarel'nik VB, Paustovskii AV, Tkachenko YG et al (2017) Electric-spark coatings on a steel base and contact surface for optimizing the working characteristics of babbitt friction bearings. Surf Eng Appl Electrochem 53:285–294. https://doi.org/10.3103/S1068375517030140
6. Tarelnyk V, Martsynkovskyy V, Dziuba A (2014) New method of friction assemblies reliability and endurance improvement. Appl Mech Mater 630:388–396. https://doi.org/10.4028/www.scientific.net/AMM.630.388
7. Antoszewski B (2014) Influence of laser surface texturing on scuffing resistance of sliding pairs. Adv Mater Res 874:51–55
8. Antoszewski B, Tarelnyk V (2014) Laser texturing of sliding surfaces of bearings and pump seals. Appl Mech Mater 630:301–307. https://doi.org/10.4028/www.scientific.net/AMM.630.301
9. Donets SY, Klepikov VF, Lytvynenko VV et al (2015) Aluminum surface coating of copper using high-current electron beam. Probl Atom Sci Tech 4(98):302–305
10. Tarelnyk V et al (2017) New method for strengthening surfaces of heat treated steel parts. In: IOP conference series: materials science and engineering, vol 233. IOP Science, p 012048. https://doi.org/10.1088/1757-899X/233/1/012048
11. Pogrebnjak AD, Beresnev VM, Bondar OV et al (2018) Specific features of microstructure and properties of multielement nitride coatings based on TiZrNbAlYCr. Tech Phys Lett 44:98–101
12. Pogrebnjak OD, Dyadyura KO, Gaponova KP (2015) Features of thermodynamic processes on contact surfaces of multicomponent nanocomposite coatings with hierarchical and adaptive behavior. Metallofiz Noveishie Tekhnol 37:899–919
13. Smyrnova KV, Pogrebnjak AD, Beresnev VM et al (2018) Microstructure and physical-mechanical properties of (TiAlSiY)N nanostructured coatings under different energy conditions. Met Mater Int 24(5):1024–1035
14. Pogrebnjak AD, Ivashchenko VI, Skrynskyy PL et al (2018) Experimental and theoretical studies of the physicochemical and mechanical properties of multi-layered TiN/SiC films: temperature effects on the nanocomposite structure. Compos B Eng 142:85–94
15. Pogrebnjak AD, Beresnev VM, Smyrnova KV et al (2018) The influence of nitrogen pressure on the fabrication of the two-phase superhard nanocomposite (TiZrNbAlYCr)N coatings. Mater Lett 211:316–318
16. Gitlevich AE, Mikhailov VV, Nya Parkansky, Gitlevich AE, Revutsky VM (1985) Electric spark alloying of metal surfaces. Shtintsa, Chisinau
17. Matysik P, Jóźwiak S, Czujko T (2015) Characterization of low-symmetry structures from phase equilibrium of Fe–Al System—microstructures and mechanical properties. Materials 8(3):914–931. https://doi.org/10.3390/ma8030914
18. Mulin YI, Verkhoturov AD (1999) Electric spark alloying of working surfaces of apparatus and machine parts with electrode materials obtained from mineral raw materials. Dal'nauka, Vladivostok
19. Kirik GV, Gaponova OP, Tarelnyk VB (2018) Quality analysis of aluminized surface layers produced by electric spark deposition. Powder Metall Met Ceram 56(11–12):688–696
20. Mazanko VF et al (2015) Interaction of aluminum with iron and air gases at electric spark alloying. In: Collection of materials of the 11th international conference "Interaction of Radiations with Solid Body", Minsk, 23–25 Sept 2015
21. Gertsriken DS et al (2010) Interaction of nickel and molybdenum with air gases as influenced by spark discharges. In: collection of materials of the 50th intern. Scientific symposium "Actual Problems of Strength". Vitebsk State Technology University, Vitebsk, 27 Sept–1 Oct 2010

22. Minkevich AN (1965) Chemical and thermal treatment of metals and alloys. Monograph. Mechanical Engineering, Moscow
23. Kobets AG, Horodek PR, Lonin YF et al (2015) Melting effect on high current electron beam on aluminum alloy 1933. Surf Eng Appl Electrochem 51(5):478–482. https://doi.org/10.3103/S1068375515050075
24. Panda A, Dyadyura K, Valíček J et al (2017) Manufacturing technology of composite materials—principles of modification of polymer composite materials technology based on polytetrafluoroethylene. Materials 10(4):377. https://doi.org/10.3390/ma10040377
25. Ivanov V, Mital D, Karpus V et al (2017) Numerical simulation of the system "fixture—workpiece" for lever machining. Int J Adv Manuf Technol 91:79–90. https://doi.org/10.1007/s00170-016-9701-2
26. Karpus VE, Ivanov VA (2012) Choice of the optimal configuration of modular reusable fixtures. Russ Engin Res 32(3):213–219. https://doi.org/10.3103/S1068798X12030124
27. Korczak A, Martsynkovskyy V, Peczkis G et al (2012) A diagnosis of the phenomenon of flow as an inspiration to inventions in the domain of constructing hydraulic machines. Procedia Eng 39:286–302. https://doi.org/10.1016/j.proeng.2012.07.035
28. Vanyeyev S, Getalo V (2014) Jet-reactive turbine: experimental researches and calculations by means of softwares. Appl Mech Mater 630:66–71. https://doi.org/10.4028/www.scientific.net/AMM.630.66
29. Kravchenko YO, Coy LE, Peplińska B et al (2018) Nano-multilayered coatings of (TiAlSiY)N/MeN (Me = Mo, Cr and Zr): influence of composition of the alternating layer on their structural and mechanical properties. J Alloy Compd 767:483–495
30. Pogrebnjak AD, Beresnev VM, Bondar OV et al (2018) Superhard CrN/MoN coatings with multilayer architecture. Mater Des 153:47–59

Thin ZnO:Al and CdS Films' Optical Properties

I. P. Koziarskyi, E. V. Maistruk and D. P. Koziarskyi

Abstract Optical properties of thin ZnO:Al and CdS films produced by RF magnetron sputtering are studied. Transmission and reflection coefficients are studied. The optical band-gap width depending on substrate temperature is estimated.

Keywords ZnO:Al · CdS · Thin films · Band-gap · Sputtering · Evaporation

1 Introduction

X-ray and gamma detectors are an important element today in a number of astronomical telescopes and other areas of life and medicine. CdTe and CdZnTe (CZT) cathodes are conductive materials for room temperature detectors for semiconductor radiation designed to meet the needs of many of these applications. Long-term stability is very important for most of these needs. However, one critical problem facing this technology is the instability of the detectors over time due to the lack of optimal interconnection between the metal contacts and the semiconductor material.

To solve this problem in the application of a nuclear radiation detector, it is proposed to use a transparent conductive oxide (TCO) to replace the commonly used metal contacts. These transparent conductive oxides are in essence degenerate semiconductors, where the Fermi level (E_f) lies inside and above the conduction band (E_c), as well as for metals. These strongly doped degenerative semiconductors have more metallic properties than semiconductor properties with respect to their electrical resistance. In this study, we chose ZnO alloyed with aluminum (AZO) as a promising material for contacting with CdZnTe. Air defense on the basis of alloyed (Al) ZnO can be grown with different specific resistance. In addition, AZO offers a number of advantages over traditional metallic contacts, which are now used for radiation detectors. Due to the oxide nature of the contact, the interface provides better chemical adhesion to the surface of CdZnTe and increases the stability of

I. P. Koziarskyi (✉) · E. V. Maistruk · D. P. Koziarskyi
Chernivtsi National University, Chernivtsi 58012, Ukraine
e-mail: i.koziarskyi@chnu.edu.ua

© Springer Nature Singapore Pte Ltd. 2019

A. D. Pogrebnjak and V. Novosad (eds.), *Advances in Thin Films, Nanostructured Materials, and Coatings*, Lecture Notes in Mechanical Engineering,
https://doi.org/10.1007/978-981-13-6133-3_26

the interface electrode/CdZnTe. The coefficient of thermal expansion ZnO exactly corresponds to the level of CdZnTe, thereby reducing the mechanical stresses on the interface. In addition, the hardness of ZnO is 8–20 times higher than that of conventional metals.

In addition, it was interesting to explore the properties of thin CdS films with the prospect of creating ZnO:Al/CdS/CdZnTe hetero structures and studying their properties.

2 Experiment

Solid solutions of $Cd_{1-x}Zn_xTe$ with zinc content $x = 0.2$ were grown by vertical Bridgman method at a low pressure of cadmium vapor in an ampule ($P_{Cd} \approx 0.02$ atm). Crystals were of p-type conductivity and were characterized by a small resistance of $R \approx 10^2\ \Omega$.

Preparatory synthesized alloys were used to produce targets for high-frequency magnetron dispersion. The obtained powder was pressed into the aluminum cup by a hydraulic press.

In order to prevent cup material dispersion, the inner diameter of the cup was much bigger than the magnetron's erosion zone (disperse zone) diameter.

UVN-70 unit was used for high-frequency magnetron dispersion method of thin layers production [1–4]. The unit was equipped with the magnetron. Turbo-molecular pump TMP-500 was used instead of the original high vacuum pump.

In order to keep the temperature of lining in the range from room temperature to 600 °C a stainless steel oven with 100 mm diameter was constructed (Fig. 1). Low resistance is a special feature of this oven, as it is made of 0.8–0.9 mm diameter nichrome wire. This feature is necessary to avoid electric break down in the oven's

Fig. 1 Sputtering process

power circuits. Power for the oven is provided by mean of laboratory stabilized switched-mode power supplies BVP 900–1500 W with high currents (10 s A).

The temperature was controlled with chromel-alumel thermocouple.

3 Results and Discussion

To research optical properties of thin layers ZnO:Al and CdS those thin layers were sputtered on glass lining by high-frequency magnetron dispersion method. Glass of standard size was used as a lining. In order to obtain high-quality thin layers, the sputtering was performed multiple times in different modes (different magnetron power P, W; time of sputtering t, min; lining temperature T, °C) researching their properties in different modes. To calculate optical coefficient was used method based on an independent reading of reflectivity and transmittance [5–8].

At a normal incidence of the beam dependence of incidence coefficient (R) on refraction (n) and absorption (k) indices is described by the following relation:

$$R = \frac{(n-1)^2 + k^2}{(n+1)^2 - k^2} \tag{1}$$

In semiconductor crystals the domain of energies of incident electromagnetic radiation, in which absorption can be neglected, is in most cases small; however the condition $n^2 \gg k^2$ is satisfied on much larger domain, thus refraction index was being determined on the basis of the value of reflection coefficient of non-polarized radiation (at the incidence angle tending to a normal one) by the formula (2):

$$R = \frac{(n-1)^2}{(n+1)^2} \tag{2}$$

that gave a rather reliable value of refraction index (n) on a wide interval of photons energies (at least for $\hbar\omega < E_g$).

If the condition $n^2 \gg k^2$ is satisfied, transmission ratio for each of the investigated samples with the corresponding thickness d (at the absence of interference) can be represented by formula (3):

$$t = \frac{(1-R)^2[1 + (\lambda\alpha/4\pi n)^2]}{e^{\alpha d} - R^2 e^{-\alpha d}} \tag{3}$$

As $n^2 \gg k^2$, $(\alpha\lambda/4\pi n) < 1$, then in the domain of change of transmission from $(1 - R)/(1 + R)$ to 10% the absorption coefficient can be calculated by means of the formula:

$$t = \frac{(1-R)^2 e^{-\alpha d}}{1 - R^2 e^{-2\alpha d}} \tag{4}$$

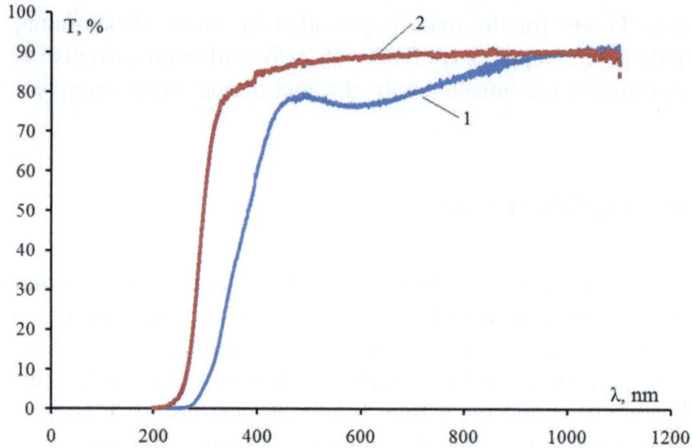

Fig. 2 The transmission spectra of thin films ZnO:Al: 1—P = 180 W, t = 60 min, T_S = 300 °C; 2—P = 150 W, t = 30 min, T_S = 300 °C

At that

$$\alpha = \frac{1}{d}\ln\left[\frac{(1-R)^2}{2t} + \sqrt{\frac{(1-R)^4}{4t^2} + R^2}\right] \tag{5}$$

When transmission T < 10% (at $n^2 \gg k^2$), for calculating absorption coefficient (α) the following expression is obtained from formula (4):

$$\alpha = \frac{1}{d}\ln\frac{(1-R)^2}{t} \tag{6}$$

Thus, the value of the absorption coefficient (α) was obtained for each crystal on the basis of independent measurements of reflection and transmission coefficients.

Errors in absorption coefficient calculations in many articles were analyzed, from which main error is calculating α is made while reading t and R and while measuring sample thickness. Considering these errors brings total value error calculating absorption coefficient to 15%.

Optical coefficients were researches with spectrophotometer SF-2000 in range from 0.2 to 1.1 μm. Figures 2 and 3 illustrate transmission spectra of thin layers ZnO:Al and CdS accordingly obtained in different modes. As we can see from the figure the longer the time of sputtering the further is a shift of the edge of its own absorption into the long wavelength range, which points out on the decrease of the optical band gap. Transmittance absolute value is decreasing, which means thin layers' transmittance is getting worse. This can be explained by increased layer thickness that comes from a decrease of interference on Fig. 2.

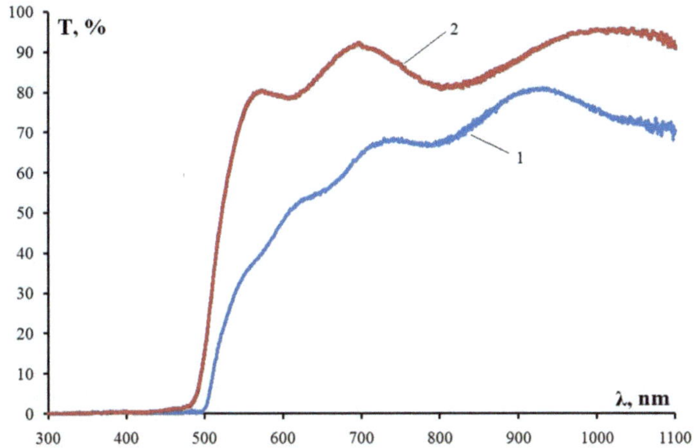

Fig. 3 The transmission spectra of thin films CdS: 1—P = 190 W, t = 80 min, T_S = 330 °C; 2—P = 190 W, t = 30 min, T_S = 330 °C

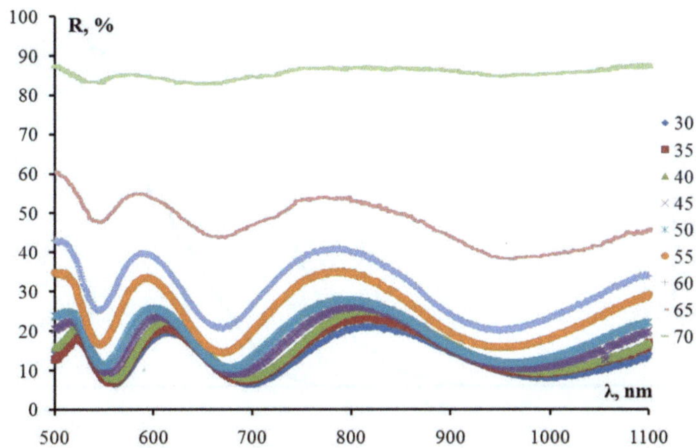

Fig. 4 Dependance R from wave length for CdS thin layer at different incident angles of electromagnetic wave

For R reading Pike console was used in the wavelength range λ = 0.4–1.1 μm. The angle of incidence at Pike console can be changed in the range 30°–80°. Spectral function of the coefficient of reflectivity was measured at a different angle of incidence for thin layer CdS and is shown in Fig. 4.

Researches on the dependence of R from the angle of incidence (Fig. 5) has shown that R in the range of angle of incidence from 30° to 45° barely change so that we can say R = R(0°) = R(30°).

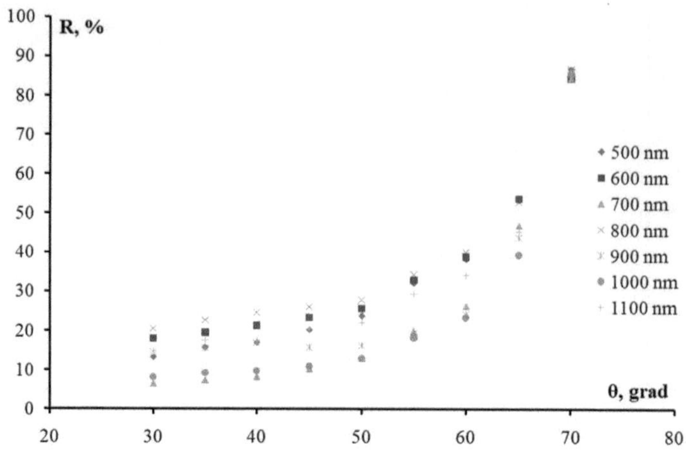

Fig. 5 Dependence R from incident angle for thin layers CdS at different wave lengths

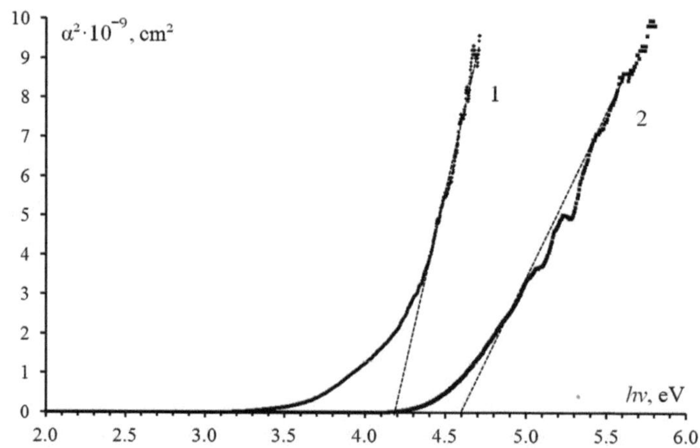

Fig. 6 Dependance α^2 from the energy of incident electromagnetic radiation at T = 300 K for thin layers ZnO:Al: 1—P = 180 W, t = 60 min, T_S = 300 °C; 2—P = 150 W, t = 30 min, T_S = 300 °C

Obtained dependencies $\alpha = f(h\nu)$ were calculated in dependencies $\alpha^2 = f(h\nu)$. According dependencies were made for thin layers ZnO:Al, CdS (Figs. 6 and 7).

Dependencies $\alpha^2 = f(h\nu)$ (Figs. 6 and 7) illustrate some straight sections, which points on existence in researched crystals direct permissible interband optical transitions. By extrapolation of these straight sections, $\alpha^2 = f(h\nu)$ to $\alpha^2 = 0$ value of optical band gap (E_g^{op}) for each thin layer was calculated (Table 1).

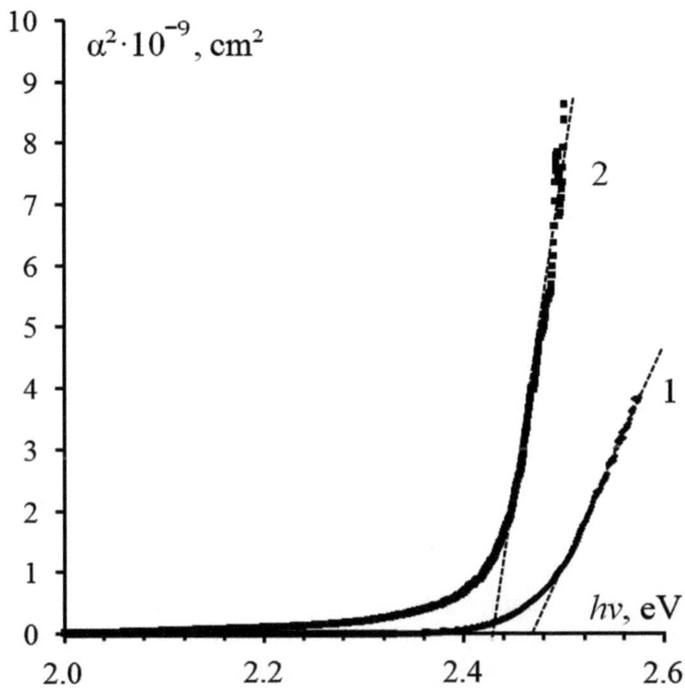

Fig. 7 Dependance α^2 from the energy of incident electromagnetic radiation at T = 300 K for thin layers CdS: 1—P = 190 W, t = 30 min, T_S = 330 °C; 2—P = 190 W, t = 80 min, T_S = 330 °C

Table 1 Optical band gap (E_g^{op}) for each thin film

Film	Mode	E_g, eV
ZnO:Al	P = 180 W, t = 60 min, T_S = 300 °C	4.2
	P = 150 W, t = 30 min, T_S = 300 °C	4.6
CdS	P = 190 W, t = 80 min, T_S = 330 °C	2.43
	P = 190 W, t = 30 min, T_S = 330 °C	2.47

4 Volt-Amps Diagram (CVD) of Heterostructure ZnO:Al/CdS/$Cd_{1-x}Zn_xTe$

Solid solutions containing zinc $x = 0.2$ were produced by vertical Bridgman's method [9–13] in ampule with low pressure cadmium steam ($P_{Cd} \approx 0.02$ Atm).Those p-type conductivity crystals were low resistance with $R \approx 10^2\ \Omega$.

$Cd_{1-x}Zn_xTe$ surface, which was a lining for the thin layer CdS, was polished with a gradual decrease of powder size from 28 to 5 μm. After that samples were polished on silk and batting with diamond paste and rinsed with alcohol. CdS thin layer were produced by magnetron sputtering of target CdS in UVN-70 unit with magnetron

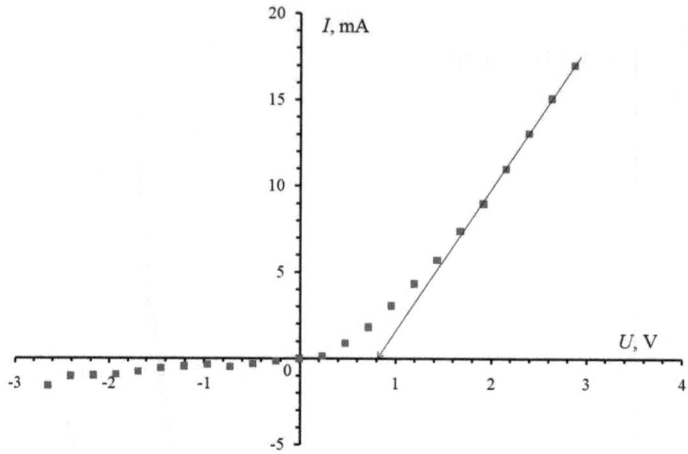

Fig. 8 Anisotropic heterostructure's ZnO:Al/CdS/$Cd_{1-x}Zn_xTe$ volt-amps diagram at room temperature

power 190 W in argon gas atmosphere. The prepared lining of $Cd_{1-x}Zn_xTe$ (diameter 1 cm) was placed above the magnetron. A thermocouple in vacuum camera provided lining's temperature control at 330 °C. CdS sputtering lasted 30 min.

Next, transparent conductive "window" ZnO:Al (98:2) was applied on CdS/$Cd_{1-x}Zn_xTe$ structure by reactive magnetron sputtering method with magnetron power 150 W in the atmosphere of argon and oxygen mixture. CdS/$Cd_{1-x}Zn_xTe$ temperature was locked at 300 °C. Sputtering lasted 30 min.

Anisotropic heterostructure's ZnO:Al/CdS/$Cd_{1-x}Zn_xTe$ volt-amps diagram was investigated by measuring unit Agilent 34410A (digital multimeter)/Keysight B2985A (picoammeter). Based on a received volt-amps diagram by extrapolation of linear part/straight section of it (Fig. 8) height of the potential barrier $\varphi = 0.8$ was received and straightening factor k = 12 was calculated at $U_{+.} = |U_{-.}| = 2.5$ V ($k = I_{+.}/I_{-.}$) at room temperature.

Using formula $R_S = \Delta U/\Delta I$ in the range of straight section part volt-amp diagram $R_S = 8.38$ kΩ was calculated.

5 Conclusions

1. Thin Layers of ZnO:Al, CdS were obtained by magnetron sputtering method at different sputtering modes, and optical properties of these thin layers were researched.
2. Researched best sputtering modes to obtain thin layers with high transparency and high optical band gap for ZnO:Al (P = 150 W, t = 30 min, $T_S = 300$ °C) $E_g = 4.6$ eV; CdS (P = 190 W, t = 30 min, $T_S = 330$ °C) $E_g = 2.47$ eV.

3. Volt-amp diagram of anisotropic heterostructure n-ZnO:Al/n-CdS/p-$Cd_{1-x}Zn_xTe$ shows its promising use in solar power: $\varphi = 0.8$ V, $k = 12$ ($U = 2.5$ V), $R_S = 8.38$ kΩ.

References

1. Maistruk EV et al (2017) Optical properties of thin films CZTSe produced by RF magnetron sputtering and thermal evaporation. In: Pogrebnjak AD (ed) Proceedings of the 2017 IEEE 7th international conference on nanomaterials: applications & properties (NAP-2017), Odessa, Ukraine, Sept 2017, vol CFP17F65-ART. IEEE, Sumy, p 01PCSI28. https://doi.org/10.1109/NAP.2017.8190163
2. Maistruk EV et al (2018) Structure and optical properties of thin films CZTS obtained by the RF magnetron sputtering. In: Angelsky OV (ed) Proc. SPIE. Thirteenth international conference on correlation optics, Chernivtsi, Ukraine, Jan 2018, vol 10612. SPIE, Bellingham, p 1061215. https://doi.org/10.1117/12.2304801
3. Koziarskyi IP et al (2018) The influence of selenium on the optical properties of thin films KZTS. In: Angelsky OV (ed) Proc. SPIE. Thirteenth international conference on correlation optics, Chernivtsi, Ukraine, Jan 2018, vol 10612. SPIE, Bellingham, p 1061218. https://doi.org/10.1117/12.2304879
4. Maistruk EV et al (2018) Optical properties of thin films cadmium chalcogenide obtained by the RF magnetron sputtering. In: Angelsky OV (ed) Proc. SPIE. Thirteenth international conference on correlation optics, Chernivtsi, Ukraine, Jan 2018, vol 10612. SPIE, Bellingham, p 1061212. https://doi.org/10.1117/12.2304328
5. Maistruk EV, Mar'yanchuk PD, Solovan MN et al (2017) Optical properties of thin Cu_2ZnSnS_4, films produced by RF magnetron sputtering. Opt Spectrosc 123(1):38–43. https://doi.org/10.1134/S0030400X17070153
6. Koziarskyi IP et al (2011) Temperature dependence of optical properties $(3HgSe)_{0.5}(In_2Se_3)_{0.5}$, doped with Mn or Fe. In: Angelsky OV (ed) Proc. SPIE. Tenth international conference on correlation optics, Chernivtsi, Ukraine, Nov 2011, vol 8338. SPIE, Bellingham, p 833814. https://doi.org/10.1117/12.917802
7. Koziarskyi IP et al (2013) Optical filters on the base of $(3HgSe)_{0.5}(In_2Se_3)_{0.5}$, doped with Mn or Fe. In: Angelsky OV (ed) Proc. SPIE. Eleventh international conference on correlation optics, Chernivtsi, Ukraine, Dec 2013, vol 9066. SPIE, Bellingham, p 90661E. https://doi.org/10.1117/12.2053538
8. Koziarskyi IP, Marianchuk PD, Maistruk EV (2011) Optical properties of $(3HgSe)_{0.5}(In_2Se_3)_{0.5}$ crystals doped with Mn or Fe. Ukr J Phys Opt 12:137–142. https://doi.org/10.3116/16091833/12/3/137/2011
9. Mar'yanchuk PD, Koziarskyi IP (2010) Magnetic and band parameters of $(3HgS)_{1-x}(Al_2S_3)_x$ ($x = 0.5$) crystals doped with manganese. Russ Phys J 53:60–66. https://doi.org/10.1007/s11182-010-9387-0
10. Maryanchuk PD, Koziarskyi IP (2012) Magnetic, electrical, and optical properties of $(HgS)_{1.5}(Al_2S_3)_{0.5}$ ⟨Mn⟩ and $(HgS)_{1.5}(In_2S_3)_{0.5}$ ⟨Mn⟩ crystals. Inorg Mater 48:655–661. https://doi.org/10.1134/S0020168512070096
11. Koziarskyi IP, Abashin SL, Maistruk EV et al (2015) Surface morphology and composition of crystals of indium and mercury selenides doped with 3d metals. J Synch Invest 9:415–419. https://doi.org/10.1134/S1027451015020329
12. Koziarskyi IP, Maistruk EV, Koziarskyi DP et al (2014) Charge transport and mechanisms of electron scattering in $(HgSe)_3(In_2Se_3)$ crystals doped with 3d transition metals. Inorg Mater 50:447–451. https://doi.org/10.1134/S0020168514050070
13. Kovalyuk TT, Maryanchuk PD, Maistruk EV et al (2014) Physical properties of Hg_{1-x-y} Cd x EuySe crystals. Inorg Mater 50:241–245. https://doi.org/10.1134/S002016851403008X

Excitation of Vortex-Antivortex Pairs in Thin Superconducting Films and Superlattices

P. Mikheenko

Abstract Direct imaging of the accumulation of magnetic flux and antiflux resulted from the excitation of vortex-antivortex pairs inside thin superconducting films is reported. Thin film superconductors, like $YBa_2Cu_3O_7/PrBa_2Cu_3O_7$ superlattices or NbN films grown by pulsed laser deposition, were used in experiments. The superlattices provide enhanced pinning for vortices, facilitating imaging of accumulated flux, and feature nanoscale fractures ideal for excitation of vortex-antivortex pairs. The idea of the experiment is to record images using a specific magneto-optical mode that allows distinguishing between positive and negative magnetic field in the sample. Two types of flux-antiflux patterns are observed. In one type, flux and antiflux are entering on permanent defects, like nano-fractures formed in the process of film deposition. In another type, flux and antiflux patterns are formed by application of a strong localized magnetic field and not are linked to permanent defects. In the first type, the amount of flux and antiflux entering superconductor from the defect is strongly affected by an external magnetic field. In the second type, the permanently frozen pattern is hardly influenced by the field. An unusual and dramatic effect of the excitation of flux-antiflux dendritic avalanches is also reported.

Keywords Superconductivity · Thin films · Vortex-antivortex pairs

1 Introduction

Superconducting thin films have wide range of applications in electronics. They are used, for example, in thin-film SQUD-on-tip microscope [1], and thin films of topological superconductors can solve existing problems in quantum computing [2]. Based on thin-film technology, pulse laser-deposited superconducting tapes of second generation are already in trial use in cable and magnets applications [3].

P. Mikheenko (✉)
Department of Physics, University of Oslo, P.O.Box 1048, Blindern, 0316 Oslo, Norway
e-mail: pavlo.mikheenko@fys.uio.no

© Springer Nature Singapore Pte Ltd. 2019

A. D. Pogrebnjak and V. Novosad (eds.), *Advances in Thin Films, Nanostructured Materials, and Coatings*, Lecture Notes in Mechanical Engineering,
https://doi.org/10.1007/978-981-13-6133-3_27

Advanced preparation of thin films is based on nano-techniques designed to improve their critical current density, especially in strong magnetic fields [4]. Still, physics of magnetic behavior of superconducting films is not fully understood and continues to surprise by unusual effects. One of these effects is the generation of vortex-antivortex pairs inside superconducting films.

One can expect that applying a magnetic field of the certain direction perpendicular to superconducting film would introduce magnetic flux of the same direction entering from a sample edge. Such entrance is well described theoretically for most of important sample's geometries, starting with simplest disc geometry [5], and confirmed in multiple experiments. However, presence of defects or areas of suppressed superconductivity would lead to somewhat unexpected effect of the excitation of magnetic field loops with equal amount of quantized flux and antiflux in the plane of the sample. The current flowing in the film would increase size of these loops moving vortices and antivortices in opposite direction. If pinning is strong, vortices will be trapped on pinning centers creating localized areas with equal, and opposite in direction, amount of magnetic flux. If pinning is weak and the dissipation of the energy during the flux motion is high, it could lead to thermomagnetic instability.

The mechanism of excitation of vortex-antivortex pairs was described as early as in 1983 in [6] explaining formation of localized resistive domains on artificial areas of suppressed superconductivity inside the film. A modern demonstration of this effect with low-temperature scanning tunneling microscope can be found in [7]. This technique, however, is not ideal for samples with large in-plane dimensions. Another technique, namely magneto-optical imaging (MOI) [8–10] is more suitable for this. MOI was already used to image accumulation of large amount of flux and antiflux on artificially produced slits [11]. The application of this technique to samples with naturally formed nano-fractures and point-like defects is described in the following sections.

2 Experimental

The main sample used in experiments was a $YBa_2Cu_3O_7$/$PrBa_2Cu_3O_7$ (YBCO/PrBCO) superlattice prepared by pulsed laser deposition on a $SrTiO_3$ substrate. In this superlattice, 48 layers of $YBa_2Cu_3O_7$ (YBCO) of the thickness of about 60 nm were separated by incomplete, less than one-unit-cell thick, layers of $PrBa_2Cu_3O_7$ (PrBCO) used to nucleate defects for increasing pinning in YBCO [4, 12]. The details of the film growth and its behavior in magnetic field are described in [13]. The specific feature of the film is formation of nano-fractures leading to channeled magnetic flux flow [13].

In MOI technique, a Bi-substituted iron garnet indicator film was placed on top of the film, and together they were attached to cold finger of an optical cryostat. Measurements of YBCO/PrBCO superlattice were performed using liquid nitrogen or liquid helium as coolant. Another sample was a NbN film with critical temperature of about 16 K. MOI images were recorded using a field-cooled procedure or slightly uncrossed polarizer and analyzer to distinguish between positive and negative magnetic fields.

3 Results and Discussion

An example of magnetic flux entering YBCO/PrBCO superlattice on internal defects as described above is shown in Fig. 1. There are several nano-fractures in the superlattice. When the external field is increased, the superconductor is filled with positive (white) and negative (dark) magnetic field. It is important to note that fractures themselves are so narrow (less than one micron) that they are not resolved in imaging. What is seen is magnetic flux in superconductor close to the fractures. With generation of vortex-antivortex pairs, amount of positive magnetic flux trapped in superconductor is equal to amount of negative flux. This is qualitatively demonstrated by Fig. 2, which contains three-dimensional images of the areas selected by red frames in Fig. 1. In these plots, the intensity of the light is plotted along the z-axis.

Magneto-optical image in Fig. 1 was recorded at polarizer and analyzer uncrossed by 5°, which is suitable to distinguish between positive (white) and negative (dark) magnetic fields. The temperature of the record was 3.7 K, and the applied magnetic field was 4.3 mT.

The areas of positive and negative flux in the figure have specific dumb-bell shapes directed along a line. However, in some cases, positive and negative parts are at 90°, as it is marked by a red arrow in the figure or just one part could be present if another is erased by a closely-spaced singularity as in a feature to the bottom-left from the upper red frame. To the right of the same frame, another configuration is seen as the central dark area surrounded by the two bright islands.

Such wealth of configurations reflects length and positions of nano-fractures in the sample and also the flow of supercurrent circulating in the film in the presence

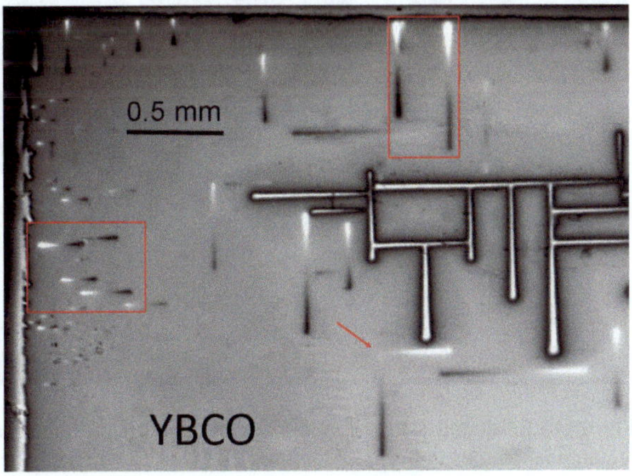

Fig. 1 Magnetic flux pattern in superlattice composed of 48 YBCO layers. The temperature of the sample at the record was 3.7 K. A magnetic field of 4.3 mT was applied after zero-field cooling. The polarizer and analyzer are uncrossed by 5° to distinguish between positive (white) and negative (dark) magnetic fields

of magnetic field. If current flows perpendicular to a nano-fracture, it starts bending around it and drags excited vortices and antivortices in different parts, first along the fracture, and further perpendicular to it into a superconductor, where they become trapped on pinning centers.

If two perpendicular fractures are connected in a point, the distribution of flux will be at the right angle, like it is pointed to by the red arrow in Fig. 1. For an isolated linear nano-fracture with the perpendicular current flow, the largest current density is close to the edges, and the lowest is in the middle. Therefore, no separated flux in a superconductor is seen in the middle, and its amount gradually increases towards the edges. The distribution of flux around nano-fractures is strongly affected by the applied field. At higher than in Fig. 1 field, patterns of trapped flux start merging with the corresponding annihilation of flux and antiflux, and they are overlapped by magnetic flux advancing from the edges.

Completely different pattern could appear when superconducting film is exposed to localized magnetic field, strong enough to suppress the superconductivity. One of the examples of such a pattern is shown in Fig. 3. It is a stripy line with periodical repetition of the fragments of positive and negative magnetic flux, seen on a background of weak horizontal and vertical lines. These lines also appear along with a network of nano-fractures. The sample in this figure is another YBCO/PrBCO superlattice with four, much thicker than those in the superlattice in Figs. 1 and 2, layers of YBCO of 750 nm.

The stripy line in Fig. 3 reminds a track left in a superconductor by a particle. In fact, it was first treated like the track of high-energy particle. If such particle enters superlattice, it starts destroying superconductivity inside the film. In a very short time, superconductivity recovers trapping positive and negative magnetic flux formed during the suppression of superconductivity. This process periodically repeats along the track while the particle moves inside the film, resulting in a stripy pattern that can be seen by magneto-optical imaging. Period of this pattern is defined by

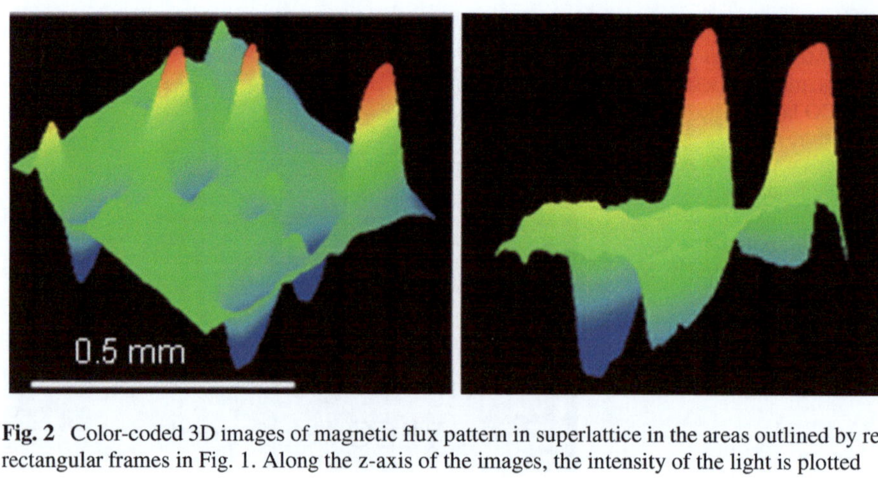

Fig. 2 Color-coded 3D images of magnetic flux pattern in superlattice in the areas outlined by red rectangular frames in Fig. 1. Along the z-axis of the images, the intensity of the light is plotted

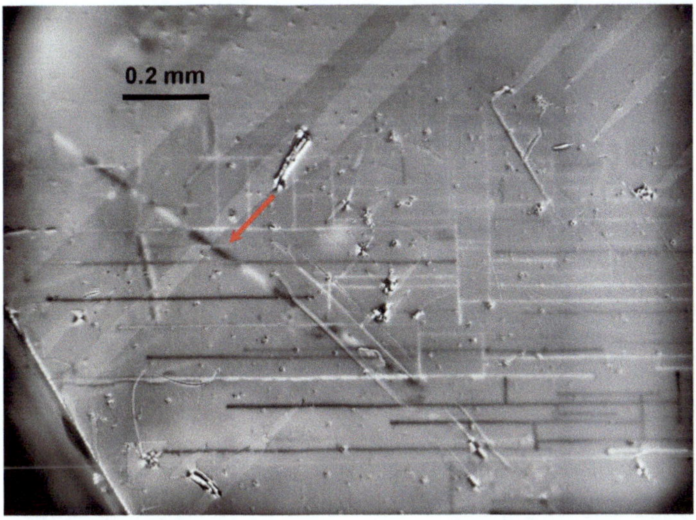

Fig. 3 A stripy pattern (marked with red arrow) with periodical repetition of the fragments of positive and negative magnetic flux, on the background of weak horizontal and vertical lines, which belong to a network of nano-fractures in a superlattice with four layers of YBCO of a thickness of 750 nm. The image was recorded at 68 K after application and removal of a very small magnetic field below 1 mT

the speed of the particle and by relaxation time of superconductor. Knowing the volume of the film with suppressed superconductivity allows the calculating amount of energy deposited into the material. The calculation gives realistic agreement with a recovery time of superconductor (about 2 ns) and reveals a very high amount of energy deposited into a superconductor, of about 160 GeV, which, judging by track appearance, is likely to correspond to a dark-matter particle. The idea was abandoned when it was found that the position of the track exactly coincides with the position of the edge of MOI indicator film that previously has been positioned there before being shifted to a different place during the orientation of the sample.

Similar to the accumulations of positive and negative flux in Figs. 1 and 2, there is a balance between flux and antiflux along the track in Fig. 3, but the flux here behaves differently from the patterns in the first two figures, in that it virtually insensitive to the external magnetic field. For example, Fig. 4 shows the same part of the sample as in Fig. 3, but in a higher magnetic field of 6.8 mT. The background is completely changed, while stripy pattern remains seemingly the same in the places, where it is not erased by magnetic flux advancing from the nano-fractures.

Such insensitivity to a magnetic field could be explained by the fact that after removal of the edge of MOI film, superconductivity fully recovers below the stripy pattern and keeps the frozen flux, whose density is too low to significantly affect the local critical current density. In nano-fractures, however, local critical current density is always zero.

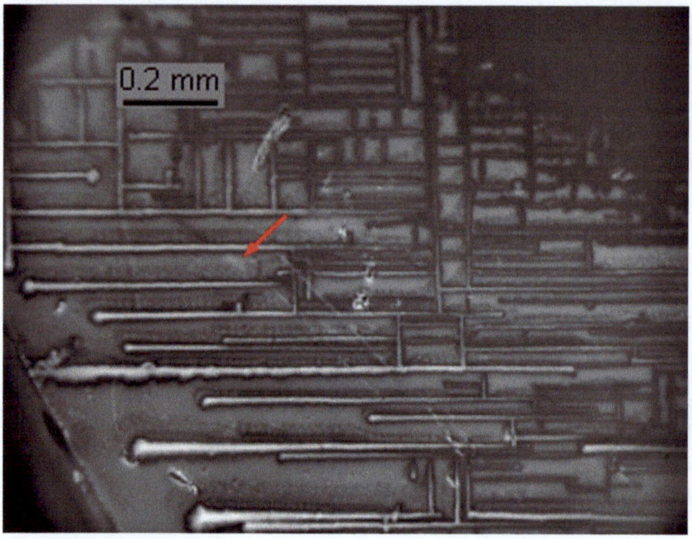

Fig. 4 MOI image of the same sample as in Fig. 3. The image is recorded at 68 K with application of magnetic field of 6.8 mT

At high field, or even at zero field after the application of the high field, the excitation of vortex-antivortex pairs seems to have little influence on global distribution of magnetic flux or energy stored in the superconductor. An example of one of the distributions of trapped magnetic flux in main superlattice composed of 48 layers of YBCO at liquid nitrogen temperature of 77.3 K is shown in Fig. 5. The places of generation of vortices and antivortices provide just a weak modulation of the density of a large amount of flux trapped in the film.

Still, in some cases, centers of the generation of vortices and antivortices can play an important role in the dynamics of flux flow in superconductors. This possibility is linked to formation of thermomagnetic instabilities [8–10]. In fact, mechanism of the excitation of vortices and antivortices was suggested following observation of thermomagnetic instabilities leading to formation of resistive domains in superconducting films with large transport current [6]. In that case, flux and antiflux formed on artificial point-like defects inside the film were accelerating in opposite directions due to the Lorentz force from the transport current.

Thermomagnetic instabilities are also frequent in thin films in external magnetic field [8–10], but in this case, one could expect instabilities leading to abrupt penetration of magnetic flux of the same direction as an applied magnetic field. Typically, it happens like this, but not always. Figure 6 shows unusual cases when pairs of dendritic flux avalanches with magnetic flux of opposite directions are coming from the same points in the sample (marked by blue arrows), which are the points of the excitation of vortex-antivortex pairs. The dendrites are of different color (dark and

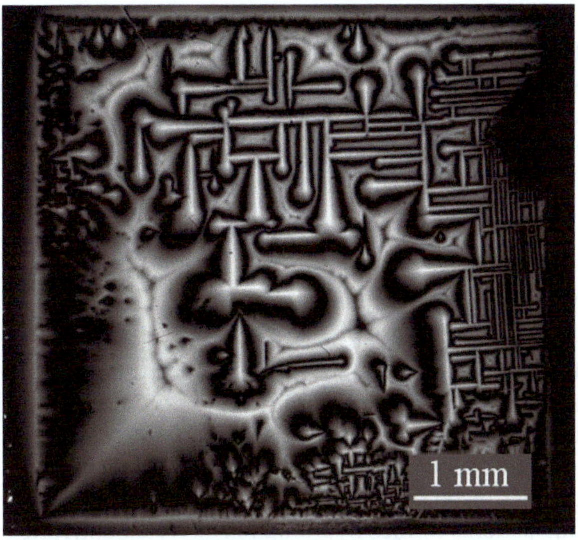

Fig. 5 MOI image of the whole sample, top-left part of which is shown in Fig. 1. The image was recorded at 77.3 K after the application and removal of magnetic field of 85 mT

white), and the imaging was, again, done with a slightly uncrossed polarizer and analyzer to distinguish between positive and negative magnetic field.

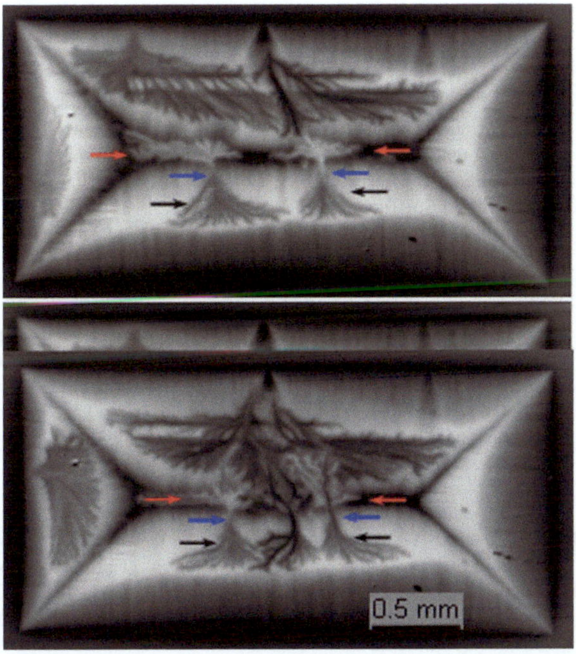

Fig. 6 Dendritic flux avalanches with magnetic flux of opposite direction that are coming from the same points marked by blue arrows. Red arrows point to avalanches with positive and the black arrows point to avalanches with negative magnetic flux

The images in Fig. 6 are related to a study of ray-optics effects in magnetic flux flow in superconductors [14]. The fragments of normal stripes affecting flux flow are seen in the upper parts of the plots. The thermomagnetic avalanches coming from the centers of the excitation of vortex-antivortex pairs may disturb flux flow under the investigation, and could be harmful for any electronic devices that may be patterned in vicinity of these centers.

4 Conclusions

Direct imaging of positive and negative trapped magnetic flux, as well as dendritic avalanches associated with excitation of vortex-antivortex pairs on internal defects in superconducting films, is reported. The local suppression of superconductivity by strong magnetic field leading to stripy features of trapped magnetic flux with balanced amount of flux and antiflux was imaged. The difference in sensitivity of different flux patterns formed by excitation of vortex-antivortex pairs to the external magnetic field is revealed. The importance of magneto-optical study allowing monitoring of local distribution of magnetic flux in superconductors is emphasized.

References

1. Finkler A, Segev Y, Myasoedov Y et al (2010) Self-aligned nanoscale SQUID on a tip. Nano Lett 10(3):1046–1049
2. Zhang P, Yaji K, Hashimoto T et al (2018) Observation of topological superconductivity on the surface of an iron-based superconductor. Science 360(6385):182–186. https://doi.org/10.1126/science.aan4596
3. Samoilenkov S, Molodyk A, Lee S et al (2015) Customised 2G HTS wire for applications. Supercond Sci Technol 29(2):024001
4. Crisan A, Dang VS, Mikheenko P (2017) Nano-engineered pinning centres in YBCO superconducting films. Phys C 553:118–132
5. Mikheenko PN, Kuzovlev YE (1993) Inductance measurements of HTSC films with high critical currents. Phys C 204(3–4):229–236
6. Ivanchenko YM, Mikheenko PN (1983) New mechanism of penetration of vortices into current-saturated superconducting films. Sov Phys JETP 58(6):1228–1234
7. Ge JY, Gladilin VN, Tempere J et al (2017) Controlled generation of quantized vortex–antivortex pairs in a superconducting condensate. Nano Lett 17(8):5003–5007
8. Johansen TH, Shantsev DV (eds) (2004) Magneto-optical imaging. Kluwer Academic Publishers, Dordrecht
9. Jooss C, Albrecht J, Kuhn H et al (2002) Magneto-optical studies of current distributions in high-Tc superconductors. Rep Prog Phys 65(5):651–788
10. Mikheenko P, Yurchenko VV, Cardwell DA et al (2013) Magneto-optical imaging of superconductors for liquid hydrogen applications. J Supercond Novel Magn 26(5):1499–1502
11. Baziljevich M, Johansen TH, Bratsberg H et al (1996) Magneto-optic observation of anomalous Meissner current flow in superconducting thin films with slits. Appl Phys Lett 69(23):3590–3592
12. Crisan A, Dang VS, Mikheenko P et al (2009) Pinning potential in thick $PrBa_2Cu_3O_y/YBa_2Cu_3O_x$ quasi-multilayers. Phys C 470(1):55–60

13. Mollatt HJS, Qureishy TH, Crisan A et al (2018) Flux channeling in $YBa_2Cu_3O_7$ superlattices. In: Proceedings of international conference nanomaterials: applications and properties NAP, Sumy State University 3:03TFNMC09, 9–14 Sept 2018
14. Mikheenko P, Johansen TH, Chaudhuri S et al (2015) Ray optics in flux avalanche propagation in superconducting films. Phys Rev B 91(6):060507(R)

Channeling of Magnetic Flux in $YBa_2Cu_3O_{7-\delta}$ Superlattices

H. J. Mollatt, T. Qureishy, A. Crisan, V. S. Dang and P. Mikheenko

Abstract We report an unusual effect of channeled magnetic flux motion in $YBa_2Cu_3O_{7-\delta}/PrBa_2Cu_3O_{7-\delta}$ superlattices grown by pulsed laser deposition. Magneto-optical imaging reveals that flux moves along a set of parallel and perpendicular lines, while optical microscopy does not show any features on the surface that may cause this effect. In contrast, scanning electron microscopy registers submicron fractures in the superlattices, corresponding to the flux lines, but the magnetic flux channels are much wider than the width of these fractures. To further clarify the origin of flux channels, electrical transport measurements on the superlattices have been performed. Their current-voltage characteristics reveal the presence of distinctive branches related to the flux motion along the selective channels, following which magnetic flux can cross the sample in a shortest and least resistive way. The application of very large current overheated the superlattice along these channels evaporating superconducting material and exposing wider than in the superconductor fractures in the substrate. It is concluded that motion of flux in the channels is controlled not only by the presence of nano-fractures in $YBa_2Cu_3O_{7-\delta}/PrBa_2Cu_3O_{7-\delta}$, but also stresses developed in the superconducting material appearing due to the fracturing of the substrate.

Keywords Superconductor · Superlattice · Magnetic flux · Channeling

H. J. Mollatt · T. Qureishy · P. Mikheenko (✉)
Department of Physics, University of Oslo, P.O.Box 1048, Blindern, 0316 Oslo, Norway
e-mail: pavlo.mikheenko@fys.uio.no

A. Crisan
National Institute for Materials Physics Bucharest, 405A Atomistilor Str., 077125 Magurele, Romania

V. S. Dang
Nano and Energy Center, VNU Hanoi University of Science, 334 Nguyen Trai, Thanh Xuan, Hanoi, Vietnam

© Springer Nature Singapore Pte Ltd. 2019

A. D. Pogrebnjak and V. Novosad (eds.), *Advances in Thin Films, Nanostructured Materials, and Coatings*, Lecture Notes in Mechanical Engineering,
https://doi.org/10.1007/978-981-13-6133-3_28

1 Introduction

For practical applications of high-temperature superconductors [1], one of them is $YBa_2Cu_3O_{7-\delta}$ (YBCO), highest possible in-plane critical current density in high magnetic fields should be achieved [2]. This requires the introduction of dense nano-arrays of pinning centers in the material. Related modern techniques for creating these arrays are described in [3]. An efficient way of forming nano-columns that are strong pinning centers for vortices in superconducting films, is to start with deposition of the array of nanoparticles of suitable material on the substrate before the deposition of the superconducting film. In particular, gold or silver nanoparticles appeared to be very effective for the growth of the nanocolumns of YBCO [4]. Experiments with other, more complicated nanoparticles, proved to be successful as well. In particular, nanoparticles of non-superconducting $PrBa_2Cu_3O_{7-\delta}$ (PBCO) with crystal-lattice parameters that are very close to those in $YBa_2Cu_3O_{7-\delta}$, were found to improve critical current density in the latter material [5]. This effect, however, becomes weaker with larger thickness of YBCO. To overcome this problem, the idea of growing $YBa_2Cu_3O_{7-\delta}$/$PrBa_2Cu_3O_{7-\delta}$ (YBCO/PBCO) superlattices was suggested with typical thickness of YBCO from tens to hundreds of nanometers and sub-nanometer or nanometer size of PBCO nanoparticles. Such approach allows to grow thick films with good superconducting properties [5, 6].

Some YBCO/PBCO superlattices, however, showed low global critical current density, as determined by magnetization measurements. To clarify what is specific about these films, magneto-optical imaging (MOI) was performed on several of them complemented by transport measurements. The results of this investigation are described in this paper.

2 Experimental

The YBCO/PBCO superlattices were grown epitaxially by pulsed laser deposition (PLD) on $SrTiO_3$ (STO) substrates. An excimer KrF 248 nm laser with pulse duration of 30 ns was used for the growth. The repetition rate of the laser was 4 Hz, and the distance between the target and substrate, which was kept at 780 °C, was 5.5 cm. The thickness of YBCO and PBCO layers was defined, after the calibration, by the number of laser pulses. Typically 1000 laser pulses results in a thickness of about 250 nm. The following superlattices have been prepared with general formula (PrBCO**n**-YBCO**m**) × **l**, where **n** and **m** are numbers of pulses on PBCO and YBCO targets, respectively, and **l** is number of repetitions of these sequences. Sample A: (PrBCO3-YBCO250) × 32; B: (PrBCO3-YBCO250) × 48 and C: (PrBCO15-YBCO3000) × 4. The superlattices A and B were of about 2 μm and C of about 3 μm thick. In spite of different total thickness, different number of layers and different thickness of individual layers, the superlattices show a common property—well expressed channeling of magnetic flux seen by MOI.

MOI is a technique allowing to visualize distribution of magnetic field in a sample using Faraday rotation [7, 8]. As indicator, Bi-substituted iron garnet films have been used [8, 9].

Electrical measurements were carried out by a four-contact technique on a rig with power supply, two digital voltmeters and a temperature sensor, all controlled by a MATLAB program. The samples have been cooled by liquid nitrogen or liquid helium. Scanning electron microscopy (SEM) was performed on FEI Quanta 200 FEG-ESEM.

3 Results and Discussion

Figures 1, 2 and 3 show MOI images of superlattices A to C, respectively. The common feature of the figures is channeled penetration of magnetic flux into the interior of the samples. All three figures contain network of the lines perpendicular to each other. The lines are not directed along the edges of the substrate. The thickness of the lines is similar in all three images, but their density and distribution are different. There are areas in all samples that are free from the lines.

Magnetic flux channeling is quite unusual effect. It cannot be predicted or suggested from magnetization or electrical transport measurements that reflect integral properties of the sample. It emphasizes importance of magneto-optical imaging that gives detailed local information about the sample. The origin of flux lines is not clear, especially taking into account that optical microscopy shows no any line features matching those in Figs. 1, 2 and 3.

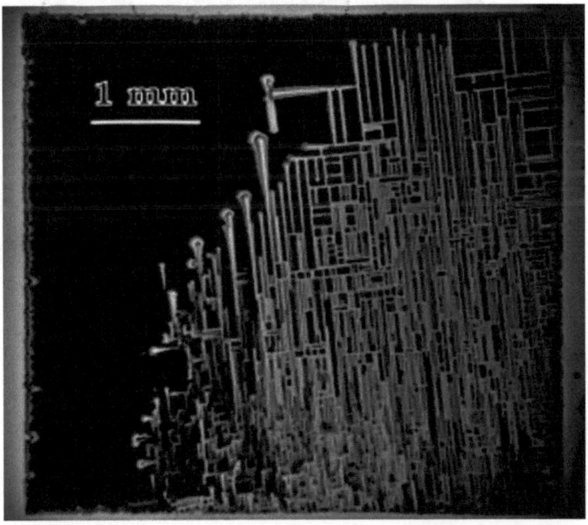

Fig. 1 Magnetic flux pattern in superlattice A composed of 32 YBCO layers. The temperature of the sample is 3.7 K. A magnetic field of 60 mT was applied after cooling sample in zero magnetic field

Fig. 2 Magnetic flux pattern in superlattice B composed of 48 YBCO layers. The temperature of the sample at the record is 3.7 K. A magnetic field of 60 mT was applied after cooling sample in zero magnetic field

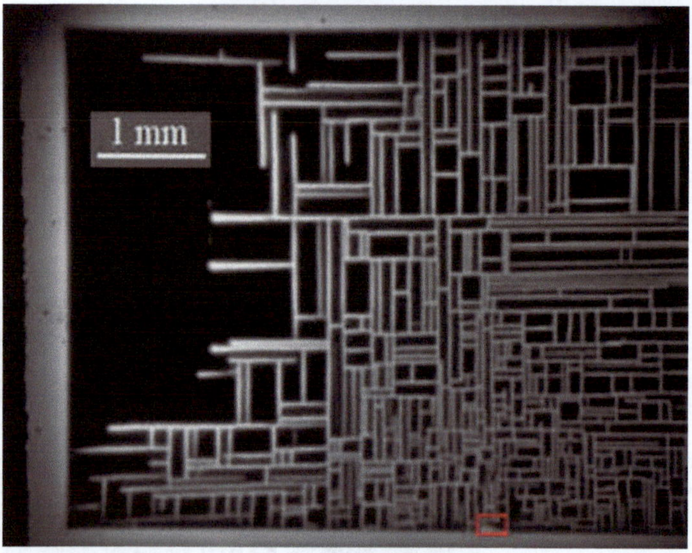

Fig. 3 Magnetic flux penetration into superlattice C composed of 4 layers of YBCO. The temperature of the record is 20 K. A magnetic field of 34 mT was applied after cooling sample in zero magnetic field

To clarify nature of lines, SEM was performed on a rectangular area outlined by red line in Fig. 3. The result, together with overlapped MOI image (dark green), is shown in Fig. 4. One can clearly see that in the middle of flux-flow lines, there are sub-micron fractures in YBCO/PBCO, and the magnetic flux penetrates along these fractures. The width of the brightest part of the lines, from which magnetic flux further penetrates into rectangular areas in critical-state fashion [10] is, however, much bigger than width of fractures.

At the increase of magnetic field, magnetic flux starts moving in the sample along the fractures that are closest to the edges of the film, gradually filling bigger areas of superlattice. Some areas, however, remain flux-free. Electrical field and corresponding voltage appears during the change of magnetic configuration in the sample, but when the distribution of magnetic flux is static, as in Figs. 1, 2, 3 and 4, voltage is absent.

The situation is different when transport current is flowing through the sample in over-critical state. It induces permanent flux flow across the sample. Figure 5 shows connection of current (red) and potential (green) leads to sample C used to record its current-voltage characteristics. When current is passed through the sample, magnetic flux moves along bright lines of easy flow revealed by MOI in Fig. 5. It is important to note that all bright horizontal lines block vertical flux flow, as it cannot cross dark areas of strong superconductivity in-between. The 'bottleneck' in the current sample is the line close to four red arrows in Fig. 5. Only four vertical segments marked by these arrows are available for flux flow. Correspondingly, one

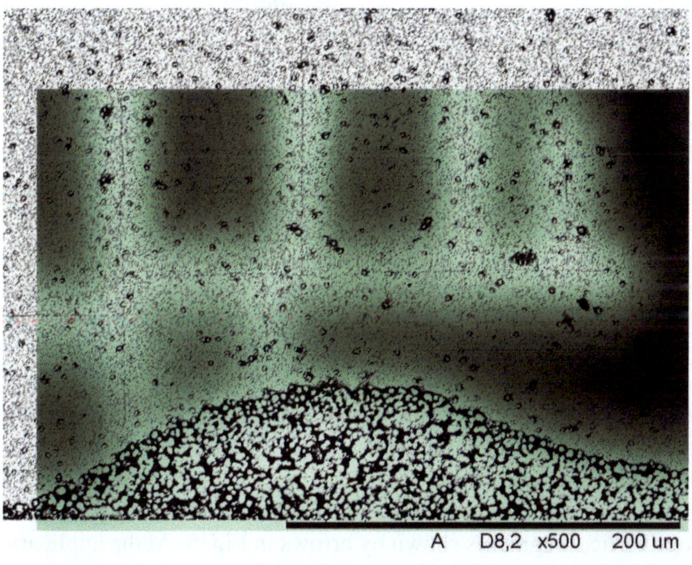

Fig. 4 Scanning electron microscopy image of the rectangular area outlined by red line in Fig. 3 with overlapped MOI image (green). Sub-micron fractures are seen along the lines of the accumulation of magnetic flux

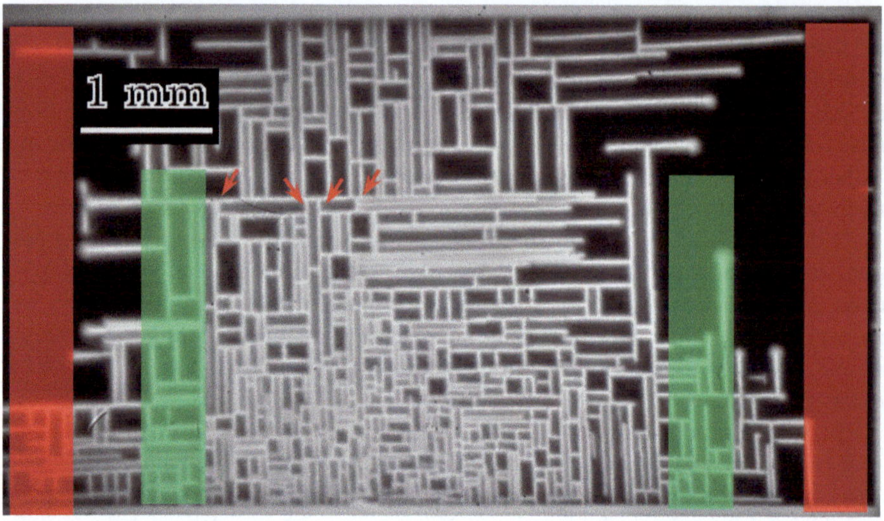

Fig. 5 Position of current (red) and potential (green) leads at the record of current-voltage characteristics of the superlattice C

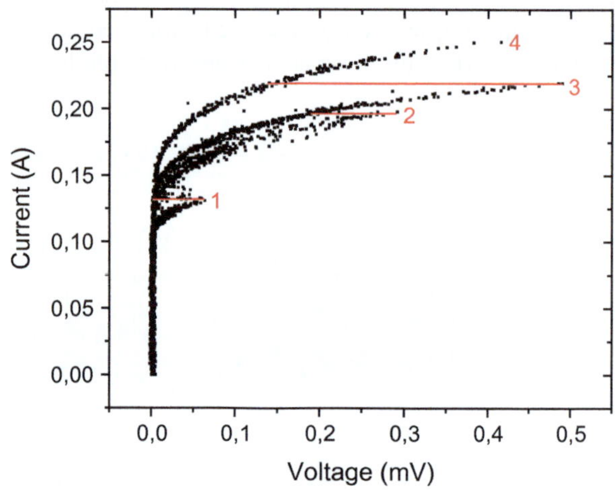

Fig. 6 Current-voltage characteristic of the superlattice C with current and potential leads as in Fig. 5. Four branches of characteristic corresponding to vertical segments marked by red arrows in Fig. 5 are seen in the plot. Red lines show how transitions between different states take place

could expect four independent branches on current-voltage characteristics. These branches indeed are registered and shown in Fig. 6. Red lines show how transitions between these branches take place.

During flux flow, dissipation of energy takes place mainly along few vertical lines linked to four vertical segments shown by arrows in Fig. 5. At the application of high current, dissipation can be so high that it can evaporate superconducting material. This is exactly what happened with superlattice C.

Figure 7 shows MOI image of the part of the sample after application of high current that burned the superlattice. This image was obtained by cooling sample in magnetic field of 17 mT to 20 K and then reducing magnetic field to 13.5 mT. Dark lines in the image correspond to areas of evaporated superconducting material. At least three marked vertical segments to the right in Fig. 5 are fully burned during application of high current.

Figure 8 shows with a higher than in Fig. 7 resolution conventional optical image of a burned channel of superlattice C. It is clearly seen that below evaporated YBCO/PBCO, there is fracture in substrate marked by red arrow in the plot.

All segments of burned lines show presence of fractures in the substrate. It means that these are fractures in the substrate, developed during the deposition, that induce channeling of magnetic flux in the superlattices. The fractures in cubic STO substrate developed along crystallographic directions. The different from 0° to 90° slope of lines in Figs. 1 and 2 with respect to the edges simply reflects the fact that substrate were cut not exactly along these directions.

In its turn, fractures in substrate induce nano-fractures in PBCO/YBCO and create stresses that weaken superconductivity in rather wide channels that are revealed

Fig. 7 MOI image of the part of the superlattice C after application of high current that burned the sample. This image is obtained by cooling sample in magnetic field of 17 mT to 20 K and then reducing magnetic field to 13.5 mT. Dark lines correspond to areas of evaporated superconducting material

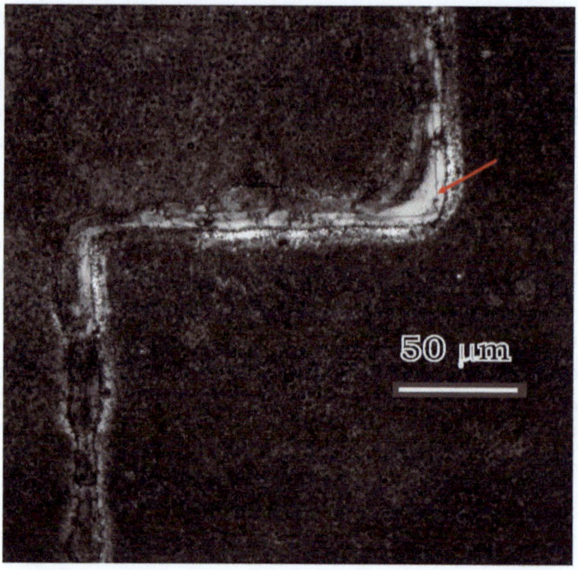

Fig. 8 Conventional optical image of a burned channel of the superlattice C

by MOI. The rectangular array of flux channels is not only a remarkable phenomenon. Being properly controlled, it can find use in practical applications combining strong superconductivity in rectangular 'windows' and weak, possibly Josephson-like behaviour in nano-fractures in-between.

4 Conclusions

Channeling of magnetic flux along a network of mutually perpendicular lines in $YBa_2Cu_3O_{7-\delta}/PrBa_2Cu_3O_{7-\delta}$ superlattices is reported. A range of experiments have been performed to clarify the nature of this effect. It is found that channels appear due to fracturing of substrate in the process of pulsed laser deposition and the corresponding stress and nano-fracturing induced in the superconductor. A combination of magneto-optical imaging and electrical transport measurements allows to follow details of flux motion along the network of weak superconducting channels.

References

1. Marchionini BG, Yamada Y, Martini L et al (2017) High-temperature superconductivity: a roadmap for electric power sector applications, 2015–2030. IEEE Trans Appl Supercond 27(4):0500907

2. Nault RM (ed) (2006) Basic research needs for superconductivity. Report on the basic energy sciences workshop on superconductivity, Argonne National Laboratory, Arlington, 8–11 May 2006
3. Crisan A, Dang VS, Mikheenko P (2017) Nano-engineered pinning centres in YBCO superconducting films. Phys C 553:118–132
4. Mikheenko P, Sarkar A, Dang VS et al (2009) c-Axis correlated extended defects and critical current in $YBa_2Cu_3O_x$ films grown on Au and Ag-nano dot decorated substrates. Phys C 469(14):798–804
5. Crisan A, Dang VS, Mikheenko P et al (2009) Pinning potential in thick $PrBa_2Cu_3O_y/YBa_2Cu_3O_x$ quasi-multilayers. Phys C 470(1):55–60
6. Dang VS, Mikheenko P, Sarkar A et al (2010) Critical current density in $Ag/YBa_2Cu_3O_x$ and $PrBa_2Cu_3O_y/YBa_2Cu_3O_x$ multilayers. JPCS 234(1):012010
7. Johansen TH, Shantsev DV (eds) (2004) Magneto-optical imaging. Kluwer Academic Publishers, Dordrecht
8. Jooss C, Albrecht J, Kuhn H et al (2002) Magneto-optical studies of current distributions in high-Tc superconductors. Rep Prog Phys 65(5):651–788
9. Mikheenko P, Yurchenko VV, Cardwell DA et al (2013) Magneto-optical imaging of superconductors for liquid hydrogen applications. J Supercond Novel Magn 26(5):1499–1502
10. Mikheenko PN, Kuzovlev YE (1993) Inductance measurements of HTSC films with high critical currents. Phys C 204(3–4):229–236

Effect of Temperature on Phase Formation in Thin Bilayer Ni/GaAs Films

S. V. Dukarov, S. I. Petrushenko, V. V. Miroshnychenko, O. O. Nevgasimov and V. N. Sukhov

Abstract The results of a study of the effect of sample condensation conditions and annealing temperature on the phase composition of bilayer Ni/GaAs films are presented. A layer of GaAs was deposited at room temperature and a layer of nickel at different substrate temperatures. The temperature intervals for the existence of a low-temperature amorphous phase and an intermediate phase of the ternary Ni_3GaAs system were established by means of electron diffraction studies. The Ni_3GaAs phase decomposes into the phases of the binary systems NiAs and γ-Ni_3Ga_2 with increasing temperature. Annealing of the films or an increase in the temperature of the substrate during their condensation causes a decrease in the crystallization and phase decomposition temperatures.

Keywords Thin films · Metal-semiconductor interface · Gallium arsenide

1 Introduction

The processes occurring at the metal-semiconductor boundary attract the constant attention of researchers in connection with the wide and diverse use of such interfaces in modern electronics. Knowing the specifics of the interaction of components during the creation of contacts and in the process of their working under conditions of temperature influences is necessary to ensure the stability of the operation of the end devices. By now, the interaction of metals with traditional semiconductors, silicon, and germanium, has been studied in sufficient detail. Much fewer works are devoted to the reactions between metals and semiconductor compounds. At the same time, such structures are also very promising. In particular, thin-film systems of nickel and gallium arsenide form structures with a Schottky barrier diode and can be the basis of highly efficient photoconverters of visible and ultraviolet radiation [1, 2].

S. V. Dukarov (✉) · S. I. Petrushenko · V. V. Miroshnychenko · O. O. Nevgasimov · V. N. Sukhov
V. N. Karazin Kharkiv National University, Kharkiv, Ukraine
e-mail: dsv@univer.kharkov.ua

© Springer Nature Singapore Pte Ltd. 2019

A. D. Pogrebnjak and V. Novosad (eds.), *Advances in Thin Films, Nanostructured Materials, and Coatings*, Lecture Notes in Mechanical Engineering,
https://doi.org/10.1007/978-981-13-6133-3_29

In addition, hybrid structures based on a ferromagnetic metal and a semiconductor material have unique magnetic properties and can be promising as elements of spin electronics [3, 4].

In [5] lead wetting of thin nickel films of different thickness, deposited on single crystals of gallium arsenide, was studied. It was found that the contact angle (θ) in this system varies within the values corresponding to wetting a clean GaAs substrate ($\theta \approx 120°$ at nickel film thickness $t \to 0$) and wetting nickel in bulk ($\theta \approx 20°$ at $t > 18$ nm). In contrast to previously studied analogous contact systems [6], a plateau ($\theta \approx 75°$) is observed for the $\theta(t)$ dependence in the thickness range $6 < t < 14$ nm. This region corresponds to the wetting of the intermediate phase, which is formed due to the chemical interaction of nickel with gallium arsenide.

Electron-diffraction studies of the interaction between Ni and GaAs films as a function of the ratio of the thicknesses of the layers [5] have shown that in the given system at a temperature of 400 °C several phases, known for the corresponding binary Ni-Ga and Ni-As systems, are possible. The complexity of the interpretation of the results is due to the fact that Ni-Ga and Ni-As binary systems are characterized by a large number of phases of different symmetry, namely: cubic: $NiGa_4$, Ni_3Ga_4, α'-Ni_3Ga_4, GaAs; hexagonal: NiAs, Ni_5As_2, β'–Ni_2Ga_3,γ-Ni_3Ga_2, γ'-Ni_3Ga_2; tetragonal: $Ni_{11}As_5$ and rhombic α'-$NiAs_2$. A significant part of these phases has very close values of the interplanar spacing and cannot always be reliably identified from the data of diffraction studies. In the literature, there is practically no data on the structure of the phases of the ternary system Ni-Ga-As. In addition, the interaction of components in the film state can differ from the bulk due to the size effects of diffusion [7] and solubility [8]. To clarify the mechanism of formation of compounds on the boundary of nickel with gallium arsenide, it seems expedient to trace the change in the phase composition of Ni/GaAs films as a function of their condensation temperature and subsequent annealing.

2 Experimental Approach

Samples for electron diffraction and electron microscopy studies were prepared by evaporating and condensing Ni and GaAs in a vacuum of 10^{-6} Torr. A method of variable composition and variable state was used to obtain a set of samples [6]. In this method, gradients of independent variables—temperature, concentration or thickness of the component layers—are created on the substrate in mutually perpendicular directions (or in the simplest case in one direction). This makes it possible to obtain in one experiment and under identical conditions a series of samples differing only in these parameters.

Moreover, the temperature gradient on the substrate can be created both before and after the condensation of the films. The first case is convenient for studies of metastable states, for example, supercooling during crystallization [9], or regularities of a condensed films formation [10]. In the second case, the film degradation processes [11, 12] and the size effects associated with melting are studied [13]. An extended polished stainless steel plate was used as the substrate, and a thin layer of NaCl was deposited immediately before condensing GaAs. To obtain GaAs films, the flash evaporation method was used [14], that is, gallium arsenide was supplied to the incandescent evaporator in the form of separate powder particles. This made it possible to obtain films that did not differ in composition from the starting material. Next, a nickel film of the required thickness was deposited on the substrate without breaking the vacuum by thermal evaporation from a tungsten helix. The films mass thicknesses were controlled during the condensation by a quartz resonator and amounted to 20–50 nm.

Four series of samples were obtained under different conditions of films preparation. The ratio of the components concentrations in films X_{Ni}: X_{Ga}: X_{As} was 3: 2: 2 (42.8, 28.6, 28.6 at.%, series A) and 1: 1: 1 (33.3, 33.3, 33.3 at.%, series B, C, D). Condensation of NaCl and GaAs was carried out in all cases at room temperature of the substrate. The gallium arsenide films condensed under such conditions are amorphous. In series A and B, before the deposition of nickel along the substrate, a temperature gradient was created within the limits of $80 < T < 450$ °C. The substrate was cooled immediately after the completion of the Ni condensation. In series C and D, the nickel film was deposited at room temperature. Then a temperature gradient was established along the substrate, after what the C-series samples were cooled immediately, and the D-series samples were additionally annealed within one hour. After being cooled under vacuum to the room temperature, the samples were taken from the vacuum chamber, separated from the substrate by dissolving the NaCl layer in distilled water, and examined in an SELMI EMV-100BR transmission electron microscope to determine the microstructure and phase composition. Electron diffraction patterns from Ni/GaAs films were obtained in the general diffraction mode. The interpretation of the electron diffraction patterns was made by comparing experimental values of the interplanar distances with the values calculated for the known phases of the Ni-Ga and Ni-As binary systems [15–17]. It was taken into account that phases formed in the ternary system can have different parameters of the crystal lattice and give reflections that are forbidden for the corresponding phases of binary systems.

3 Results and Discussion

As a result of the studies carried out, it has been established that for all series of samples in the researched temperature range four temperature intervals are distinguishable. They differ significantly both in the patterns of electron diffraction and the microstructure of the samples. Figure 1 shows electron microscopic images and their corresponding electron diffraction patterns, characteristic for different temperature intervals. Data on the interplanar distances for the phases formed during the interaction are given in Tables 1, 2, 3 and 4. The results of the phase composition interpretation for all series of samples are illustrated by the diagrams in Fig. 2.

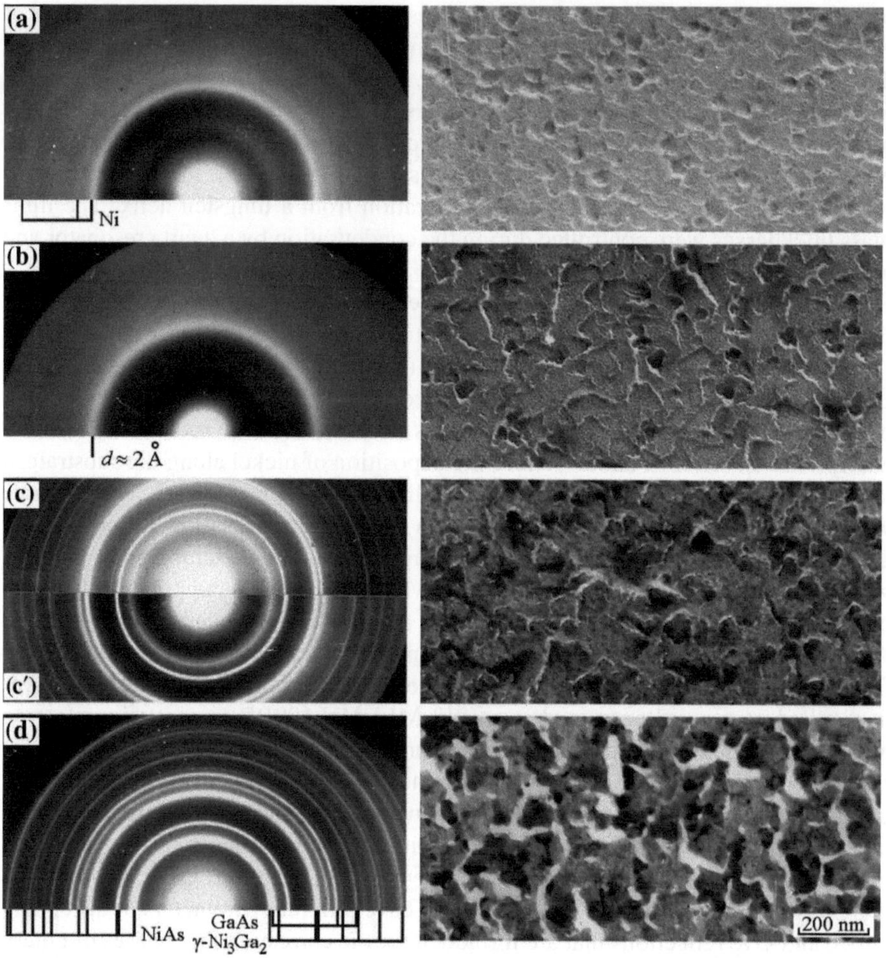

Fig. 1 TEM images and electron diffraction patterns of bilayer Ni/GaAs films (series C). Temperatures to which the films were heated are: **a** 80 °C, **b** 150 °C, **c** 220 °C, **c** 300 °C and **d** 350 °C

In the first temperature range ($T < 100$ °C), the haloes from amorphous gallium arsenide and fuzzy lines belonging to the fcc lattice with a period of 3.51 Å are observed on electron diffraction patterns (for a nickel the lattice parameter is $a = 3.516$ Å).

As it can be seen from the electron microscopic images, Ni/GaAs films in this and the following temperature range are fairly homogeneous and do not contain large crystallites. The presence on the electron diffraction patterns of reflexes only from the initial substances gives ground to conclude that at low temperatures the amorphous phase of GaAs practically does not interact with the polycrystalline layer of nickel.

Table 1 The results of the interpretation of the electron diffraction patterns of the Ni/GaAs film system (series A)

80–115 °C	120–210 °C	220–330 °C		NiAs (hkl)	340–450 °C		γ-Ni_3Ga_2 (hkl)	GaAs d_{tabl}
					3.45		(100)	
halo	halo	3.2 + halo		3.22	3.26			3.26
						2.80	(101)	2.82
		2.76	2.66	(101)	2.67	2.66		
		2.50		(002)		2.51	(002)	
					2.02	2.02	(102)	1.999
1.97	halo	1.99	1.96	(102)	1.96	1.96	(110)	
		1.90	1.81	(110)	1.81	1.81		
1.75					1.69	1.70	(200)	1.70
		1.56		(201)				
		1.48	1.48	(103)	1.48	1.48		
					1.42	1.41	(202)	1.41
		1.37	1.33	(202)	1.33	1.33		
					1.28	1.29		1.29
1.24		1.24	1.25	(004)	1.25	1.25	(121)	
		1.17	1.16	(104)	1.16			1.154
		1.17	1.45	(203)	1.14	1.145	(122)	
		1.12	1.07	(122)	1.07	1.07	(114)	1.088
1.06		1.04	1.03	(114)	1.03	1.03		

With increasing temperature, the nickel lines gradually disappear and in the second temperature range (100 < T < 200 °C) the result of the interaction in the system is an amorphous phase characterized by a strong halo (with $d \approx 2$ Å), whose position on the electron diffraction patterns coincides with the position of the most intense lines of nickel and a number of compounds of the Ni-Ga and Ni-As systems (NiAs, γ-Ni_3Ga_2, $NiGa_4$, Ni_5As_2).

Crystallization of the amorphous phase takes place at a temperature of 205–220 °C or $T \approx 175$ °C for samples subjected to hourly annealing. In the third temperature range (200 < T < 300 °C) the pattern of electron diffraction does not qualitatively change. Along with the crystalline phases, a halo of the amorphous phase (GaAs) is also observed.

For all series of samples, as can be seen from Tables 1, 2, 3 and 4, the interplanar distances of some lines remain practically constant (2.50, 1.48, 1.25, 1.16, 1.04 Å), while for other lines they vary significantly with temperature. The same groups of lines can be separated in clarity: lines with constant value d turn out to be clear, and lines with varying interplanar distances are blurred with increasing temperature.

Table 2 The results of the interpretation of the electron diffraction patterns of the Ni/GaAs film system (series B)

80–100 °C	110–195 °C	205–300 °C		NiAs (hkl)	310–420 °C		γ-Ni_3Ga_2 (hkl)	GaAs d_{tabl}
halo	halo	3.2	halo		3.2	3.25	(100)	3.26
3.2		3.25		(100)				3.21
		2.74	2.66	(101)	2.65	2.62		
		2.49		(002)				
1.99	halo	1.99	1.96	(102)	1.97	2.00	(110)	
		1.89	1.79	(110)	1.79	1.79		
1.76						1.70	(200)	1.70
		1.56		(201)		1.54		
		1.48	1.47	(103)	1.48	1.47		
					1.42	1.41	(203)	1.41
		1.37	1.32	(202)	1.32	1.32		
1.25		1.24	1.24	(004)	1.24	1.25	(121)	
		1.16	1.15	(104)	1.15	1.15		1.154
				(203)		1.14	(122)	
		1.11	1.06	(122)	1.06	1.06	(114)	
1.06		1.04	1.03	(114)	1.025	1.025		

At the same time, if we follow the variation with the temperature of the interplanar distances for all the lines, we can find that they fit into a system which corresponds to a hexagonal lattice of the NiAs type with a variable parameter *a*. The calculation results of the parameter *a* by the method of ordinary least squares over all lines with interplanar distances $d > 1$ Å are shown in Fig. 2. The parameter *c* remains approximately constant ($c \approx 4.98$ Å). The ratio of axes *c*/*a* varies from 1.31 to 1.39 (the tabulated values for the phases of γ-Ni_3Ga_2 and NiAs possessing lattices of the NiAs type are 1.245 and 1.390, respectively).

The existence of NiAs type phases with a different ratio of components and with different mass ratios of the starting materials is explained by the peculiarities of this structure. Analysis of various compounds with a nickel-arsenide structure [18] shows that a lattice of this type is conserved not only in AB compounds but also in the form of A_2B (filling internodes) and AB_2 (the presence of vacant sites). Suppose that in a lattice of the NiAs type the positions in the basal planes are occupied by nickel atoms and in layers at half-height—by Ga or As atoms. Then, taking into account the parameters *a* of the phases of γ-Ni_3Ga_2 and NiAs (3.99 and 3.62 Å, respectively), and also the covalent radii of Ga and As ($r_{Ga} = 1.39$ Å, $r_{As} = 1.21$ Å), we can state that the replacement of arsenic by gallium in the structure of NiAs leads to an increase in the lattice period. From this point of view, the phase of the NiAs type, formed at the beginning of the third temperature interval with the parameter $a = 3.82$ Å,

Table 3 The results of the interpretation of the electron diffraction patterns of the Ni/GaAs film system (series C)

90–100 °C	110–215 °C	220–310 °C		NiAs (hkl)	320–450 °C		γ-Ni_3Ga_2 (hkl)	GaAs d_{tabl}
					3.42	3.45	(100)	
halo	halo	3.2 + halo			3.20	3.23		3.26
3.2		3.28	3.18	(100)				
					2.80		(101)	2.82
		2.73	2.67	(101)	2.65	2.66		
		2.50	2.49	(002)		2.59	(002)	
2.03	halo	1.98	1.96	(102)	1.97	1.99	(110)	
		1.90	1.81	(110)	1.80	1.80		
1.80			1.68		1.68	1.69	(200)	1.70
		1.62		(200)	1.62			1.63
		1.56	1.51	(201)				
		1.48	1.47	(103)	1.47	1.48		
					1.41	1.41	(202)	1.41
		1.37	1.33	(202)	1.34	1.33		
					1.29	1.29	(120)	1.29
1.26		1.25	1.25	(004)	1.25	1.26	(121)	
		1.20	1.21					
		1.16	1.16	(104)	1.17			
					1.14	1.145	(122)	1.154
		1.11	1.095	(122)	1.08	1.07		1.088
		1.095	1.08	(300)	1.06	1.065	(114)	
1.06		1.04	1.03	(114)	1.03	1.03		

occupying an intermediate position between the values of a for NiAs and γ-Ni_3Ga_2, is the phase of the ternary system. In this phase, as follows from the dependences $a(T)$, as the temperature increases, the proportion of gallium decreases due to the release of the phase of γ-Ni_3Ga_2. It is also possible to form the $NiGa_4$ phase, the most intense lines of which coincide with the reflexes from NiAs. Assuming that the parameter a of the ternary phase varies linearly with the concentration of Ga and As in the range from $a = 3.62$ Å (phase NiAs, $X_{As} = 50$ at.%, $X_{Ga} = 0\%$) to $a = 3.99$ Å (phase γ-Ni_3Ga_2, $X_{As} = 0\%$, $X_{Ga} = 36$ at.% [15–17]), the composition of the formed intermediate compounds can be estimated. Figure 3 shows the temperature dependences of gallium and arsenic concentrations calculated within the framework of this assumption for the B series. It can be seen that in the case when nickel condensation was carried out on a substrate with a temperature gradient (series A and B), a Ni_3GaAs phase was initially formed with approximately equal gallium

Table 4 The results of the interpretation of the electron diffraction patterns of the Ni/GaAs film system (series D)

130–170 °C	175–250 °C		NiAs (hkl)	260 °C	GaAs $d_{tabl.}$	270–410 °C		γ-Ni_3Ga_2 (hkl)
						3.42	3.47	(100)
halo	3.2 + halo			3.26	3.26	3.26	3.25	
	3.25	3.18	(100)	3.17		3.13	3.10	
					2.82	2.80	2.80	
	2.73	2.68	(101)	2.67		2.67	2.65	
	2.50	2.49	(002)	2.50				
				1.98	1.999	1.99	2.00	(102)
halo	1.99	1.96	(102)	1.96	1.95	1.95	110	
	1.90	1.84	(110)	1.82		1.82	1.80	
				1.68	1.70	1.69	1.69	(200)
		1.61	(200)	1.60	1.633	1.62	1.63	
							1.56	(103)
	1.56	1.52	(201)	1.51		1.50	1.49	
	1.48	1.47	(103)	1.47		1.47	1.48	
				1.41	1.41	1.40	1.41	(202)
	1.37	1.35	(202)	1.34		1.33	1.33	
				1.29	1.29	1.29	1.30	
	1.25	1.25	(004)	1.25		1.25	1.26	(121)
	1.19	1.16	(121)	1.16	1.154	1.16	1.16	
	1.16		(104)	1.14		1.145	1.15	
	1.11	1.08	(122)	1.08	1.088	1.08	1.085	
	1.08		(300)			1.065	1.07	(114)
	1.04	1.03	(114)	1.03		1.03	1.03	

and arsenic content ($X_{As} \approx X_{Ga} \approx 21$ at.%). As the temperature was raised, a rapid decrease in the gallium concentration and the appearance of a phase of γ-Ni_3Ga_2 containing up to 10 at.% As were observed. Note that series A and B are almost identical in composition and temperature of phase formation.

When nickel is condensed on a cold substrate, followed by heating (series C) and annealing (series D), a phase enriched in arsenic appears, and the concentration changes more smoothly. As can be seen from the comparison of Fig. 2c, d, annealing of the films leads to a shift in the temperature dependence of the lattice parameter of the triple phase to lower temperatures. This indicates that the phase composition of the unannealed films is not in equilibrium. The blurring of certain lines of the triple phase, which is more pronounced for the A, B and C series, is a consequence of the incompleteness of the phase decay.

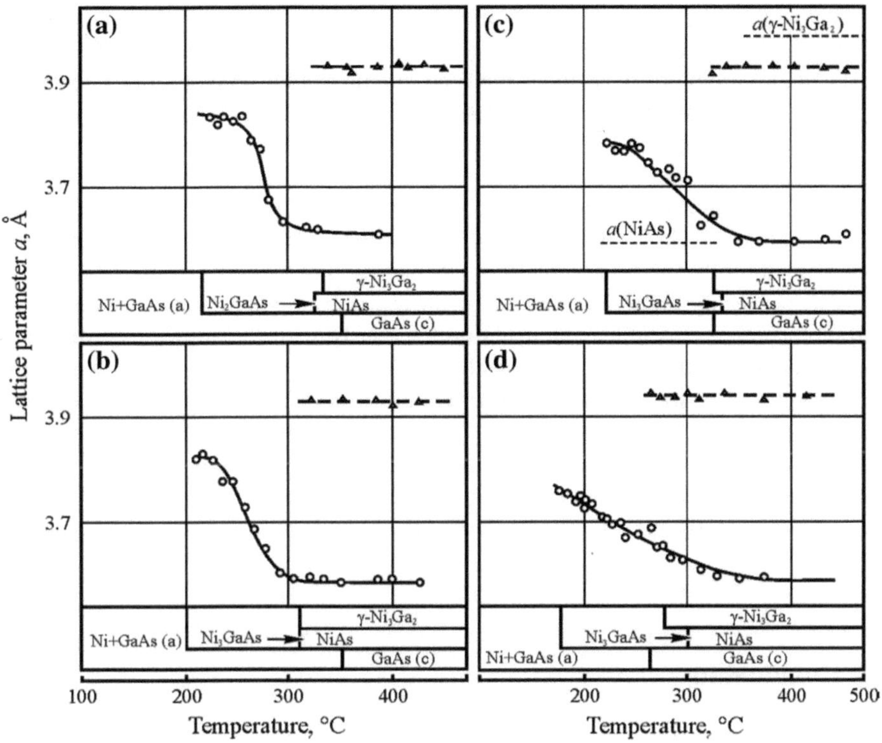

Fig. 2 Phase composition of bilayer Ni/GaAs films (A-D series) and the temperature variation of the parameter a of Ni_3GaAs (empty circle) and γ-Ni_3Ga_2 (empty triangle) phases

Fig. 3 Estimation of the concentration temperature dependence of Ga (empty circle, empty triangle) and As (filled circle, filled triangle) in Ni_3GaAs (empty circle, filled circle) and γ-Ni_3Ga_2 (empty triangle, filled triangle) phases for Ni/GaAs films condensed on a substrate with a temperature gradient (series B)

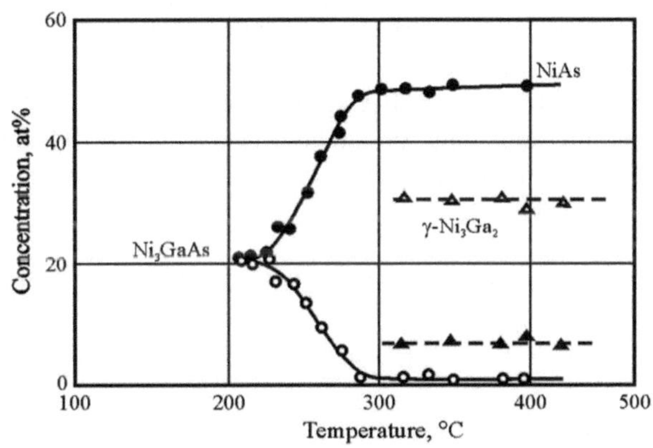

This conclusion is directly confirmed by the following. The lines whose interplanar distances are determined mainly by the variable parameter *a* ((100), (110), (201), (121)) are blurred. At the same time, the lines whose position is determined mainly by the constant parameter *c* ((002), (103), (004), (104), (114)) remain clear throughout the temperature range. We note that similar results were obtained when studying the solid-state reaction of a thin Ni film with a single-crystal GaAs substrate using electron microscopy and in situ X-ray diffraction [19, 20]. The formation of the crystalline Ni_3GaAs phase at 200 °C was observed, as well as its decay with increasing temperature under conditions of excess GaAs.

With increasing temperature, that is, in the fourth temperature range, the unreacted gallium arsenide crystallizes almost simultaneously with the appearance of the γ-Ni_3Ga_2 phase. In this case, the annealing of the films substantially decreases the temperature of crystallization of GaAs. On electron diffraction patterns from pure gallium arsenide films prepared under conditions corresponding to the D series, GaAs lines are detected at higher temperatures ($\geq$350 °C). This indicates that for gallium arsenide, just as for germanium [21, 22], metal-induced crystallization takes place. Consequently, nickel is the initiator of the crystallization of amorphous gallium arsenide.

4 Conclusions

The phase composition of two-layer Ni/GaAs films prepared at different temperatures was studied by electron diffraction. It was established that the interaction in the system begins with the formation of an amorphous phase at a temperature $T \approx$ 100 °C. Crystallization of the amorphous phase takes place at a temperature $T \approx$ 200 °C with the formation of a ternary phase of Ni_3GaAs, which has a hexagonal structure of the NiAs type. As the temperature is raised, the Ni_3GaAs phase decays into NiAs and the γ-Ni_3Ga_2 phase, containing up to 10 at.% As. It is also shown that the films annealing for one hour leads to a 50–70 K decrease in the temperatures of phases formation and decay, as well as the crystallization temperature of amorphous gallium arsenide.

Acknowledgements This study was supported by the Ministry of Education and Science of Ukraine.

References

1. Melebaev D, Melebaeva GD, Rud VY et al (2009) Photosensitivity of the Ni-n-GaAs Schottky barriers. Semiconductors 43(1):29–32. https://doi.org/10.1134/S1063782609010072
2. Yildirim N, Korkut H, Türüt A (2009) Temperature-dependent Schottky barrier inhomogeneity of Ni/n-GaAs diodes. EPJ AP 45(1): 10302-1–10302-7. https://doi.org/10.1051/epjap:2008191

3. Filippov DA, Firsova TO, Laletin VM et al (2017) The magnetoelectric effect in nickel–GaAs–nickel structures. Tech Phys Lett +43(3): 313–315. https://doi.org/10.1134/S106378501703018X
4. Soda T, Minakawa S, Futamoto M et al (2018) Structure characterization of Fe Co, and Ni thin films epitaxially grown on GaAs (111) substrate. JMSJ 42(2):30–36. https://doi.org/10.3379/msjmag.1803R005
5. Dukarov SV, Pojda VP, Churilov IG (2017) Wetting of nickel films of variable thickness by island lead condensates. In: Proceedings of the 2017 IEEE 7th international conference on nanomaterials: applications & properties (NAP-2017) Part 1, Zatoka, Ukraine, 10–15 Sept, 2017, p 01PCSI08. https://doi.org/10.1109/NAP.2017.8190145
6. Gladkikh NT (ed), Dukarov SV, Kryshal AP, Larin VI, Sukhov VN, Bogatyrenko SI (2004) Poverkhnostnye yavleniya i fazovye prevrashcheniya v kondensirovannykh plenkakh (Surface phenomena and phase transformations in condensed films). V. N. Karazin Kharkiv National University, Kharkiv. (in Russian)
7. Kryshal AP, Bogatyrenko SI, Sukhov RV et al (2014) The kinetics of the formation of a solid solution in an Ag–Pd polycrystalline film system. Appl Phys A 116(4):1891–1896. https://doi.org/10.1007/s00339-014-8349-8
8. Dukarov SV, Petrushenko SI, Sukhov VN et al (2016) In situ research on temperature dependence of the lattice parameters of fusible metals in thin Cu-Pb and Cu-Bi films. Funct Mater 23(2):218–223. http://dx.doi.org/10.15407/fm23.02.218
9. Gladkikh NT, Dukarov SV, Sukhov VN et al (2011) Condensation mechanism of AgCl and NaCl island films on a nickel substrate. Funct Mater 18(4):529–533
10. Dukarov SV, Petrushenko SI, Sukhov VN (2018) Growth of island films during vapor-liquid condensation. J Nano-Electron Phys 10(1):01023. https://doi.org/10.21272/jnep.10(1).01023
11. Petrushenko SI, Dukarov SV, Sukhov VN (2016) Effect of lead on the thermal dispersion of continuous polycrystalline copper films. Vacuum 142:29–36. https://doi.org/10.1016/j.vacuum.2017.04.037
12. Petrushenko SI, Dukarov SV, Sukhov VN (2016) Growth of through pores and thermal dispersion of continuous polycrystalline films of copper. Metallofizika i Noveishie Tekhnologii 38(10):1351–1366. (In Ukrainian). https://doi.org/10.15407/mfint.38.10.1351
13. Petrushenko SI, Dukarov SV, Sukhov VN et al (2015) Inner size effect in the polycrystalline metal films of fusible metals. J Nano-Electron Phys 7(2):02033
14. Hollan L, Hallais JP, Brice JC (1980) The preparation of gallium arsenide. In: Kaldis E (ed) Current topics in materials science, vol 5. Elsevier North-Holland, Inc., pp 1–217. https://doi.org/10.1002/bbpc.19810850426
15. Okamoto H (2000) Desk handbook. Phase diagrams for binary alloys. ASM International, Ohio
16. Vol AE (1962) Stroenie i svojstva dvojnykh metallicheskikh sistem (The structure and properties of binary metal systems). GIFML, Moscow [In Russian]
17. Lyakishev NP (2001) Phase diagrams of binary metal systems. Mashinostroenie, Moscow
18. Schubert K (1964) Kristallstrukturen zweikomponentiger Phasen. Springer-Verlag, Berlin
19. Sands T, Keramidas VG, Washburn J et al (1986) Structure and composition of NixGaAs. Appl Phys Lett 48(6):402–404. https://doi.org/10.1063/1.96511
20. Rabhi S, Perrin-Pellegrino C, Zhiou S et al (2017) Phase formation between Ni thin films and GaAs substrate. Scripta Mater 141:28–31. https://doi.org/10.1016/j.scriptamat.2017.07.011
21. Kryshtal AP, Minenkov AA, Ferreira PJ (2017) Interfacial kinetics in nanosized Au/Ge films: an in situ TEM study. Appl Surf Sci 409:343–349. https://doi.org/10.1016/j.apsusc.2017.03.037
22. Kryshtal AP, Minenkov AA, Bogatyrenko SI (2018) Effect of size on phase transformation temperatures in Ge/Bi/Ge films. J Alloy Compd 756:50–56. https://doi.org/10.1016/j.jallcom.2018.04.335

Detection of Unreliable Superluminescent Diode Chips Using Gamma-Irradiation

P. B. Lagov and V. M. Maslovsky

Abstract The influence of ^{60}Co-irradiation accompanied by low-temperature treatment on AlGaAs/GaAs chips of 835 ± 25 nm superluminescent diodes (SLD) is addressed. It is shown that irradiation of potentially unreliable chips with latent defects in active 20-nm layer increases optical power degradation rate during the subsequent burn-in test. The possible cause of enhanced degradation of potentially unreliable chips during a long-term operation or gamma-ray irradiation is the presence of local defects, which can be rearranged into larger clusters commensurate with the active layer. The decrease in the degradation rate during the burn-in test for irradiated reliable chips probably caused by the mechanical stresses relaxation and its homogenization. A method for rejecting unreliable chips using gamma irradiation processing is proposed.

Keywords Irradiation · Chip · Heterostructures · Superluminescent diode · Optical power · Degradation · Burn-in test

1 Introduction

SLD can be used as a light source for different applications, such as in gyroscopes, imaging of organics, low-coherence interferometry, optical reflectometry, etc. These techniques can be applied in medicine, telecommunication and navigation systems. On the one hand, it is possible to obtain SLDs with a broad emission spectrum comparable with that for light-emitting diodes (LED). On the other hand, SLDs have high output optical power and spatial brightness of emitted light comparable with those for laser diodes. This makes SLDs attractive for the application in fiber-optic communication lines because of the higher efficiency of the light input into fiber

P. B. Lagov (✉)
National University of Science and Technology "MISiS", Moscow, Russia
e-mail: lagov2000@mail.ru

V. M. Maslovsky
Moscow Institute of Physics and Technology, Dolgoprudny, Moscow Region, Russia

© Springer Nature Singapore Pte Ltd. 2019

A. D. Pogrebnjak and V. Novosad (eds.), *Advances in Thin Films, Nanostructured Materials, and Coatings*, Lecture Notes in Mechanical Engineering,
https://doi.org/10.1007/978-981-13-6133-3_30

compared to that for LED. Defect formation in LEDs now is actively investigated [1–4] using electron irradiation equipment [5, 6]. To detect unreliable chips during manufacturing, gamma irradiation should be used. In this case, the displacement of atoms is minimal, in contrast to lifetime killing techniques based on an electron or proton irradiation [6, 7]. That is why gamma-irradiation processing eliminates the damage of good chips without hidden defects in the active layer.

Reliability is one of the most important requirements specified for SLDs. For instance, SLDs intended for the application in sensors of fiber-optic gyroscopes are required to have the lifetime not less than ten years. To evaluate the quality of fabricated SLDs the manufacturer (or customer) provides an acceptance testing, the first step of which is a qualification of SLDs by measuring their characteristics (current-voltage characteristics, light output versus current, light output spectra, etc.) in accordance with the device specification. Then for all the devices, which successfully passed the qualification, the burn-in test is performed during 168 h at a specified temperature and forward current. During this testing, the light output is measured every 24 h, and typical criteria for the rejection of potentially unreliable devices are following: the decrease in the light output of more than 0.15% for the last 24 h, more than 0.3% for the last 48 h and more than 0.6% for the last 72 h. The devices, which show the fall in light output of more than 10% after 168-h burn-in test are also rejected.

For the devices passed the burn-in test the accelerated life testing is performed (typically 5000 h). This testing is carried out for a limited sample size at a fixed temperature and forward current. Typically, the criterion of failure for such a test is the fall in light output of more than 50% in comparison with the initial value. The life testing is time-consuming, and it may be quite expensive. The purpose of this paper is to investigate the impact of preliminary irradiation by gamma rays on the degradation of SLDs during a burn-in testing and to show that such approach can be used for the accelerated rejection of unreliable SLDs containing latent defects.

2 Experiment and Results

2.1 Samples, Testing and Irradiation Conditions

SLDs were fabricated by MOCVD on a GaAs substrate. The peak emission wavelength 810–860 nm, the average light output 10 mW. Figure 1 schematically shows the SLD-structure, and Table 1 presents the characteristics of layers.

Eight samples SLDs was investigated and irradiated by ^{60}Co with the dose of 24 Mrad (GaAs). Before the irradiation, the burn-in test during 240 h at a 25 °C and forward current of 140 mA was carried out. After the irradiation, the burn-in testing was continued for all studied devices during 2148 h under the same conditions. Figure 2 shows the change in light output of studied devices during the preliminary burn-in testing. This testing does not reveal any device from the studied sample that

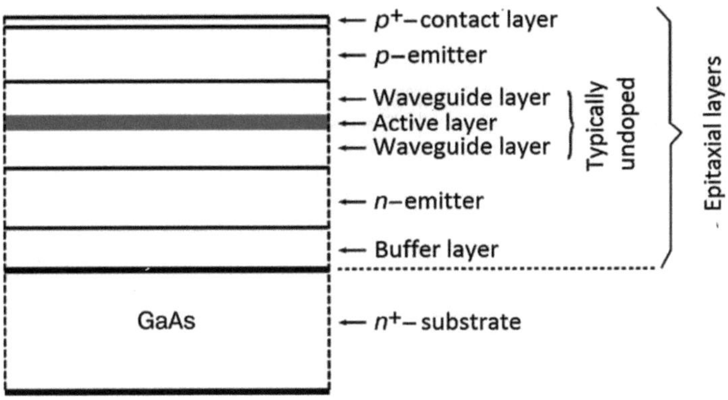

Fig. 1 Schematic cross-section of SLD AlGaAs/GaAs heterostruture

Table 1 Characteristics of the AlGaAs/GaAs SLD

Layer	Heterostructure characteristics			
	Material	Doping (cm^{-3})	Thickness (nm)	Bandgap (eV)
p+-contact	GaAs:Zn	$2 \cdot 10^{19}$	400	–
p-emitter	$Al_{0.52}Ga_{0.48}As$:Zn	7×10^{17}	1400	2.07
Waveguide	$Al_{0.41}Ga_{0.59}As$	–	120	1.94
Active layer	GaAs	–	20	1.42
Waveguide	$Al_{0.41}Ga_{0.59}As$	–	120	1.94
n-emitter	$Al_{0.52}Ga_{0.48}As$:Si	8×10^{17}	1500	2.07
n+-buffer	GaAs:Si	1×10^{18}	500	–
n+substrate	GaAs:Si	3×10^{18}	350	–

showed an anomalously high degradation rate and thus could be considered as a potentially unreliable. The maximum degradation rate did not exceed the value of 8.8% per 1000 h.

2.2 *Irradiation and Burn-In Test Experiments*

The irradiation by gamma rays did not lead to any marked change in basic spectral characteristics of studied SLDs, such as the peak emission wavelength, spectral radiation half bandwidth, and spectral modulation coefficient. However, it led to the change in the light output degradation rate during the burn-in testing after the irradiation and Fig. 3 illustrates it.

While the light output degradation rate for most of the tested devices decreased by approximately 1.5–3 times compared with the initial value, for two devices (number 1 and 6), which were classified as potentially unreliable, this rate increased after the

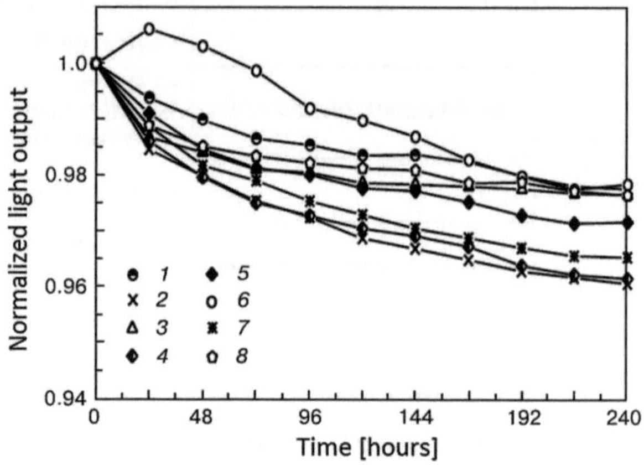

Fig. 2 Change in SLD light output during usual burn-in test (25 °C, 140 mA)

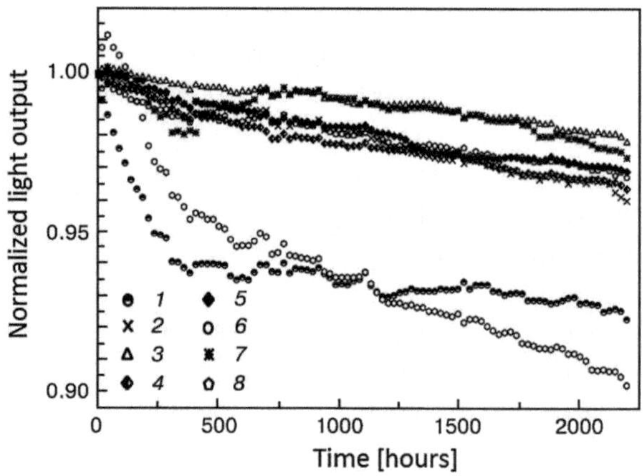

Fig. 3 Change in SLDs light output during the burn-in testing (25 °C, 140 mA) after the gamma irradiation 24 Mrad(GaAs)

gamma irradiation by more than five times. Table 2 presents the comparison of light output degradation rate for studied SLDs before and after the gamma irradiation.

Table 3 sums up the changes in the light output during the whole experiment. This table presents the values of light output for each tested device, which were measured before the gamma irradiation (Φ_0), immediately after the gamma irradiation (Φ_γ), and during the post-irradiation burn-in test in 5 min after the beginning (Φ_5) and at the end of the test (Φ_{2148}). The decrease in the light output during the 2148-h burn-in testing $\Delta\Phi_{2148}$ was calculated by the equation

$$\Delta\Phi_{2148} = (\Phi_{2148}/\Phi_5 - 1)100\% \quad (1)$$

Table 2 Light output degradation rate for studied SLD before and after the gamma irradiation 24 Mrad(GaAs)

Sample #	Light output degradation rate (%/10^3 h)	
	Before the irradiation	After the irradiation
1	5.5	18.8
2	8.8	2.4
3	2.7	1.3
4	8.4	2.8
5	6.3	1.9
6	1.2	22.9
7	8.2	6.3
8	4.1	2.8

Table 3 Change in the light output of studied SLD

Sample #	Light output (mW)				
	Φ_0	Φ_γ	Φ_5	Φ_{2148}	$\Delta\Phi_{2148}$
1	4.39	3.86	4.37	4.07	– 7.02
2	4.30	3.91	4.30	4.14	– 3.84
3	5.08	4.70	4.98	4.98	– 0.16
4	4.19	3.82	4.22	4.05	– 4.11
5	4.76	4.27	4.75	4.63	– 2.67
6	3.80	3.42	3.82	3.45	– 9.63
7	5.06	4.63	5.05	4.93	– 2.34
8	4.92	4.46	4.90	4.77	– 2.61

2.3 *Active Region Investigation*

To investigate physical processes, which cause the degradation of SLDs during the burn-in testing and gamma irradiation, profiles of effective density of charged defect centers in the active region of tested structures were studied, and scanning electron microscopy (SEM) images of active regions of tested structures were obtained. This investigation was carried out with the additional four samples (SLD numbers from 9 to 12). The burn-in testing was performed for all these chips, and two of them (#10 and #12) were preliminary gamma irradiated. Figure 4 presents the results of this testing.

The profiles of the effective density of charged defect centers (CDC) were obtained with the method of dynamic barrier capacity. Figure 5 presents the results obtained for the SLDs number 9–11. For device #12, the profile was not measured because of the high leakage. The figure shows that the CDC density in the active region of the preliminary irradiated reliable SLD #10 is less than that for the unirradiated reliable device #9, while for the potentially unreliable device #11, which was rejected during the burn-in testi, the density of CDC in the active region is substantially higher than that for reliable structures (both preliminary irradiated and unirradiated).

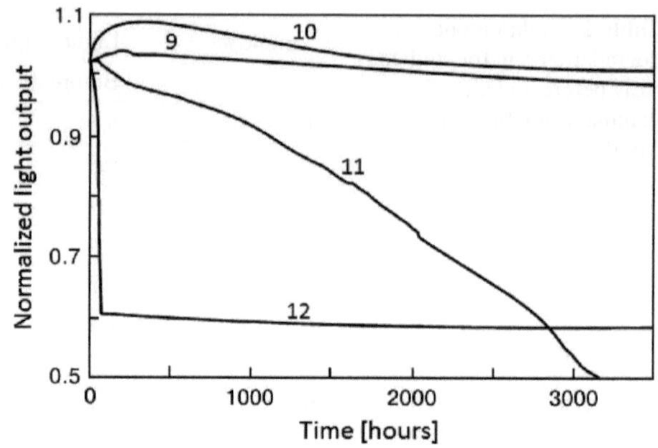

Fig. 4 Change in light output during the burn-in testing (25 °C, 140 mA) for SLDs additional samples: 9—unirradiated reliable SLD; 10—irradiated reliable SLD; 11—unirradiated SLD rejected during burn-in testing; 12—irradiated SLD rejected during burn-in testing

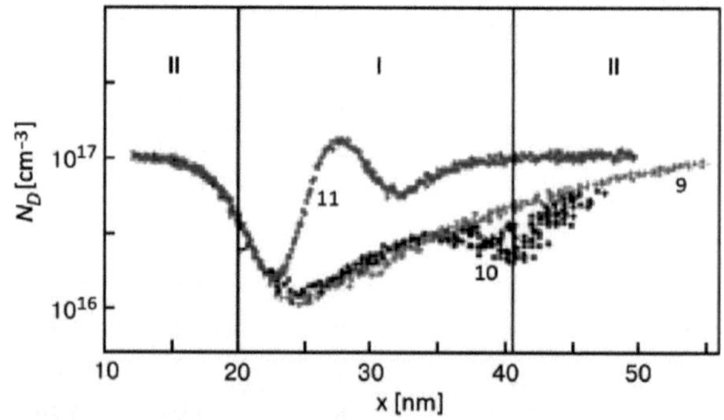

Fig. 5 Profiles of effective density of charged defect centers in the SLD heterostructure: I—active layer (GaAs); II—waveguides ($Al_{0.41}Ga_{0.59}As$)

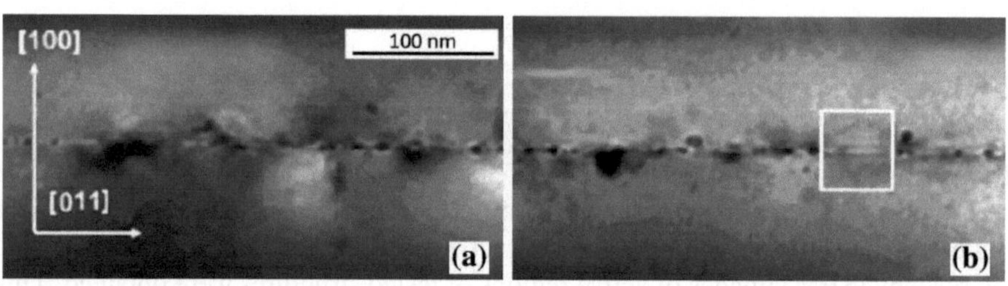

Fig. 6 Unreliable SLDs SEM images of the active region: #11 (**a**), #12 (**b**)

For the chips, which showed the highest light output degradation rate during the burn-in test (#11 and #12), the SEM images of active region were obtained using a Hitachi SCM-800 electron microscope. Figure 6 presents these results.

Figure 6 demonstrates that unreliable devices, which were rejected by burn-in testing either before or after the gamma irradiation, contain a considerable number of defects in their active regions. In SEM images presented in Fig. 6, these defects look like black spots of various sizes. The variety of defect sizes suggests that local defects in the active region can rearrange into the clusters of defects of various size, and such a process can take place either during a burn-in testing or during the irradiation by gamma rays. For reliable (defect-free) chips the active layer looks like the region marked by the white rectangle in Fig. 6b.

3 Discussion

It is known that the borders of layers in heteroepitaxial structures have their own relief and contain the major part of defects created during growing the structure. These defects cause the creation of mechanical stress. The gamma irradiation of heterostructures produces point defects, which can take part in different processes leading to the evolution of already existing technological defects or their compensation (annihilation). Existing experimental data show that $A^{III}B^{V}$ semiconductors are characterized by higher relaxation of mechanical stress during the irradiation compared to Si and Ge [8] that can be caused by the substantial ionicity of chemical bonds in compound semiconductors. Therefore, for such materials, the radiation-induced defect migration will be more visible than that for simple semiconductors.

According to this and taking into account the results presented in Figs. 5 and 6 it is possible to suggest that the cause of enhanced degradation of potentially unreliable structures is the presence of local defects, which can be rearranged into clusters of the size comparable with the thickness of active region during long-term operation or gamma-ray irradiation. The decrease in the degradation rate during burn-in test for reliable gamma irradiated structures can be attributed to the relaxation of mechanical stresses and improving the homogeneity of layers in heteroepitaxial structures which is in a good agreement with the existing data on the radiation-induced defect formation in semiconductor structures [8–12].

The fact that gamma irradiation leads to the increase in the light output degradation rate for unreliable SLDs and the decrease in that for reliable (defect-free) SLDs suggest that preliminary gamma irradiation can be used for rejection of potentially unreliable structures. Such a method may consist in the irradiation of all the lot of SLDs by gamma rays followed by the standard burn-in testing. Since the gamma irradiation causes the increase in the density of defects in the active region of SLD and rearranging them to the clusters of defects, the implementation of the preliminary gamma irradiation to the acceptance testing will raise the probability of rejection of potentially unreliable devices at the stage of burn-in testing.

4 Conclusions

The results presented in the paper show that the preliminary gamma irradiation can be successfully applied for the rejection of potentially unreliable superluminescent diodes. This irradiation does not cause any considerable change in basic spectral characteristics of SLDs, such as the peak emission wavelength, spectral radiation half bandwidth, and spectral modulation coefficient. However, it leads to the increase by more than five times in the light output degradation rate of potentially unreliable structures during the consecutive burn-in testing. On the other hand, in the case of reliable structures, gamma irradiation reduces such degradation by about 1.5–3 times. It suggests that the implementation of preliminary gamma irradiation to the acceptance test procedure will raise the probability of revealing and rejection of potentially unreliable devices at the stage of burn-in testing.

The possible cause of enhanced degradation of potentially unreliable structures is the presence of local defects, which can be rearranged into clusters of the size comparable with the thickness of active region during long-term operation (burn-in testing) or gamma-ray irradiation. The decrease in the degradation rate during burn-in test for reliable structures preliminary irradiated by gamma rays can be attributed to the relaxation of mechanical stress and the increase of homogeneity of layers in heteroepitaxial structures.

It should be noted that successful implementation of gamma irradiation to the acceptance test procedure requires the preliminary determination the optimal conditions for the irradiation (the value of total dose, electrical and temperature conditions during the irradiation) and for the consecutive burn-in testing (temperature, forward current, duration, criteria for the rejection of potentially unreliable devices). Such optimal conditions should be found for each type of studied SLDs, and it may require additional experiments.

References

1. Polyakov AY, Shmidt NM, Smirnov NB et al (2018) Defect states induced in GaN-based green light emitting diodes by electron irradiation. ECS J Solid State Sci Technol 7(6):323–328
2. Lee I-H, Polyakov AY, Smirnov NB et al (2017) Point defects controlling non-radiative recombination in GaN blue light emitting diodes: insights from radiation damage experiments. J Appl Phys 122:115704
3. Lee I-H, Polyakov AY, Smirnov NB et al (2017) Electron irradiation of near-UV GaN/InGaN light emitting diodes. Phys Status Solidi A 214(10):1700372
4. Lee I-H, Polyakov AY, Smirnov NB et al (2017) Deep electron and hole traps in electron-irradiated green GaN/InGaN light emitting diodes. ECS J Solid State Sci Technol 6(10):Q127–Q131
5. Pavlov YS, Lagov PB (2015) Magnetic buncher accelerator for radiation hardness research and pulse detector characterization. In: Proceedings of the european conference on radiation and its effects on components and systems, RADECS 2015, pp 336–338, 7365629, Dec 2015
6. Pavlov YS, Surma AM, Lagov PB et al (2016) Accelerator-based electron beam technologies for modification of bipolar semiconductor devices. JPCS 747(1):012085
7. Lagov PB, Drenin AS, Zinovjev MA (2017) Proton-irradiation technology for high-frequency high-current silicon welding diode manufacturing. JPCS 830(1):012152
8. Clayes C, Simoen C (2002) Radiation effects in advanced semiconductor materials and devices. Springer, Berlin, Heidelberg, pp 281–330
9. Pogrebnyak AD, Shpak AP, Azarenkov NA et al (2009) Structures and properties of hard and superhard nanocomposite coatings. Phys-Usp 52:29–54
10. Pogrebnjak AD, Bagdasaryan AA, Yakushchenko IV (2014) The structure and properties of high-entropy alloys and nitride coatings based on them. Russ Chem Rev 83:1027–1061

11. Pogrebnjak AD, Beresnev VM, Smyrnova KV et al (2018) The influence of nitrogen pressure on the fabrication of the two-phase superhard nanocomposite (TiZrNbAlYCr)N coatings. Mater Lett 211:316–318
12. Smyrnova KV, Pogrebnjak AD, Beresnev VM et al (2018) Microstructure and physical-mechanical properties of (TiAlSiY)N nanostructured coatings under different energy conditions. Met Mater Int 24(5):1024–1035

Structural Features and Properties of Biocompatible Ti-Based Alloys with β-Stabilizing Elements

K. V. Smyrnova, Alexander D. Pogrebnjak and L. G. Kassenova

Abstract Nowadays titanium-based alloys are widely used in various spheres, such as metallurgy, aerospace, aircraft industries, and medicine. Recent research has demonstrated that proper heat treatment and the addition of elements, stabilizing β phase, the properties of commercially pure (CP) Ti can be successfully modified application as an implant material. The paper discusses the titanium-based alloys containing Ta, Nb, and Zr, their preparation techniques, and the influence of these elements on their microstructure, biocompatibility, mechanical properties, and corrosion resistance. Furthermore, this article examines current information about recent achievements in designing β-type titanium alloys.

Keywords Biocompatibility · Titanium alloys · Mechanical properties · Superelasticity

1 Introduction

Lately, materials engineering related to the design of new biomaterials have become increasingly popular. This type of materials are widely used in medical practice, for instance, they can be utilized to replace the damaged tissues of the human body in orthopedics—staples, stomatology—orthodontic implants, traumatology—artificial joints and bones [1]. Biomaterials have to meet high requirements to be successfully applied: they must contain non-toxic constituents, possess superelasticity, shape memory effect, long fatigue life, high biocompatibility, strength, plasticity, wear and corrosion resistance in combination with a low modulus of elasticity [2]. Stainless steel, cobalt-based alloys, ceramics, polymers, commercially pure (CP)

K. V. Smyrnova (✉) · A. D. Pogrebnjak
Sumy State University, Sumy, Ukraine
e-mail: kateryna.v.smyrnova@gmail.com

L. G. Kassenova
The Kazakh University of Economics, Finance and International Trade, Astana, Republic of Kazakhstan

© Springer Nature Singapore Pte Ltd. 2019
A. D. Pogrebnjak and V. Novosad (eds.), *Advances in Thin Films, Nanostructured Materials, and Coatings*, Lecture Notes in Mechanical Engineering,
https://doi.org/10.1007/978-981-13-6133-3_31

titanium and its alloys are among the most ubiquitous materials. All of these types of materials have advantages and disadvantages, but the most popular are titanium-based alloys. They demonstrate chemical inertness in the internal environment of the human body—high biocompatibility and allow to obtain values of Young's modulus close to the modulus of elasticity of the human bone tissue (4–30 GPa) to avoid bone resorption and, as a result, loosening of the implant [3].

Recent research has explored the occurrence of several different phases in biocompatible titanium-based alloys. The following phases can be found in within this alloys: α—hexagonal close-packed structure (hcp), α′—hexagonal structure, α″—base-centered orthorhombic structure, β—body-centered cubic structure (bcc), and ω—hexagonal structure. In addition, every phase has different Young's modulus, for example, the β-phase has the lowest and ω-phase has the highest [4]. Thus, the β crystalline structure is the most desirable to be formed in the biocompatible Ti-based alloys. Mostly, titanium alloys depending on the structure are classified into α, α + β, β alloys.

Over the last decades, CP Ti the Ti–6Al–4V alloy has been widely used for biomedical applications [5]. Despite its popularity, this material releases Al and V ions that after the accumulation of a certain concentration can cause different allergic reactions and neurological problems. Furthermore, the elastic modulus of the Ti–6Al–4V alloy is approximately 110 GPa that results in stress-shielding effect—decline in bone density caused by the removal of the normal load to the bone, which occurs when metal implants are used to restore dysfunctional tissue [6, 7]. Therefore, the new low modulus titanium alloys with a composition containing only non-toxic elements have to be designed to enable the high performance. Prior research has shown that characteristics of titanium can be significantly improved by its alloying with Ta, Nb, and Zr. First, these elements do not induce an adverse reaction in the human body. Second, they foster the formation of β-phase in titanium alloys that helps to enhance mechanically and wear properties [8, 9]. Finally, they exhibit excellent biocompatibility. Niobium has been found to reduce Young's modulus and improve the corrosion resistance of titanium; tantalum is well-known for its high mechanical properties [10]. Besides these two elements, there is another one that is frequently added to the alloys for biomedical applications—zirconium [11]. Together with Sn, it belongs to the group of neutral elements, which do not influence the phase transformations of titanium but form solid solutions [12]. Zirconium excludes the process of corrosion in α-Ti and eliminates the formation of unwanted athermal ω phase [13, 14]. However, although Zn is not among the list of β-stabilizers, it demonstrates a similar impact in Ti–Nb–Zr alloys [15]. Thus, titanium alloys containing these elements must demonstrate appropriate for biomedical applications characteristics.

The paper presents an overview of the preparation techniques of biomedical titanium alloys with β-stabilizing elements, such as tantalum, niobium and zirconium, their microstructure, mechanical properties, biocompatibility, the shortcomings and prospects for further development are undertaken.

2 Preparation Techniques and Structural Features

The choice of a particular preparation method significantly affects the properties of the produced material. Regarding titanium alloys, such as Ti–Ta, Ti–Ta–Nb, Ti–Nb–Zr, Ti–Ta–Nb–Zr, the most commonly used technique for their manufacturing is an arc-melting [16–18]. Plasma arc melting is one of the most used methods for preparing alloys from refractory metals due to the intense controllable energy source—electric arc discharge [19–24]. The properties of CP Ti, Nb, Ta, and Zr are given in Table 1. For instance, Cremasco et al. [25] investigated (CP) Ti, Ti–35Nb, Ti–35Nb–7.5Ta, and Ti–25Nb–15Zr alloys, which were prepared via arc-melting method under Ar atmosphere, then homogenized at 1000 °C, subjected to plastic deformation, quenched by water, and milled to the final dimension. The X-ray diffraction analysis revealed that CP Ti consisted of only martensite hexagonal α'-phase, Ti–35Nb alloy had an orthorhombic α'' crystalline structure in a β matrix, but Ti–35Nb–7.5Ta and Ti–25Nb–15Zr alloys exhibited a pure β-phase. It is known that titanium has several allotropic forms: at the temperature above 883 °C only bcc structure can be found, however, when the temperature decreases, it transforms the hcp lattice (α phase). Hence, in order to prepare near-β or β titanium alloys, we need to modify them with proper elements that reduce the temperature of $\beta \rightarrow \alpha$ transformation and stabilize bcc crystalline structure.

When only bcc phase has to be formed in the alloys the reaction time should be taken into consideration [26]. To eliminate the $\beta \rightarrow \alpha$ allotropic transformations, the cooling rate after the melting process must be fast enough. In Ref. [27] in order to prepare fully recrystallized β–Ti–25Ta–25Nb (mass%) alloy with no inclusions of α, α', α'' or ω phases, the melting process was followed by annealing at 950 °C, water quenching, then 80% cold rolling, solution treating at 850 °C, and again water quenching. The resulting structure of Ti–25Ta–25Nb alloy compared to the CP Ti is shown in X-ray diffraction patterns in Fig. 1. The preparation procedure allowed Bertrand et al. [27] to obtain the Ti-25Ta-25N alloy with bcc structure and lattice parameter $a = 0.3294$ nm, which was close to that of pure Ti, Ta, and Nb.

Hagihara and Nakano in the paper [28] proposed a "single-crystal" quaternary alloys—Ti–25Nb–10Ta–5Zr, Ti–29Nb–13Ta–4.6Zr, and Ti–35Nb–10Ta–5Zr (mass%). These alloys were prepared by the same melting technique, which was mentioned above, however, after that single crystal were grown by the floating-zone method in Ar atmosphere, then they were cooled, solution treated at 690 °C, and water

Table 1 Properties of Ti and some refractory metals

Element	Properties	
	Melting point (°C)	Density (kg/m^3)
Titanium	1668	4500
Niobium	2477	8570
Tantalum	3017	16650
Zirconium	1855	6570

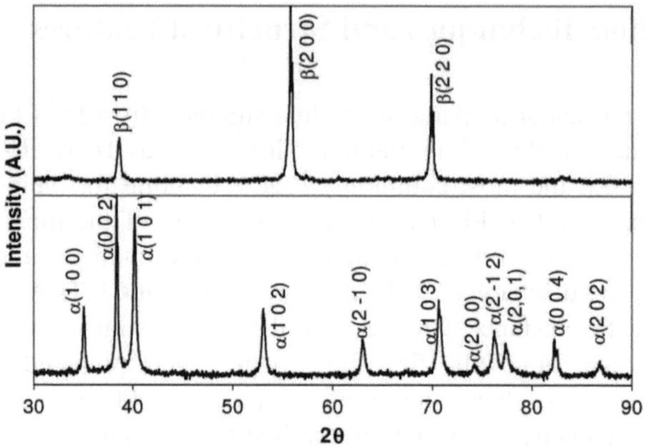

Fig. 1 X-ray diffraction patterns of **a** Ti–25Ta–25Nb alloy and **b** CP Ti [27]

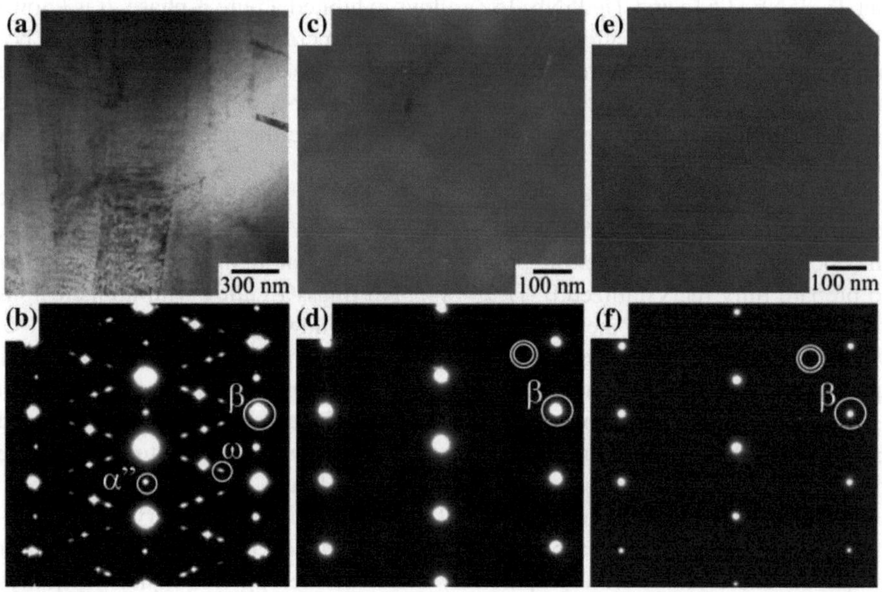

Fig. 2 TEM images of microstructure (**a**, **c**, **e**) and selected-area electron diffraction patterns (**b**, **d**, **f**) along [113] for Ti–25Nb–10Ta–5Zr, Ti–29Nb–13Ta–4.6Zr, and Ti-35Nb–10Ta–5Zr alloys [28]

quenched to obtain a single bcc structure. Figure 2 demonstrates TEM and SAED images for these alloys. It can be seen that the certain proportion of β-stabilizing elements makes it possible to prepare samples with bcc lattice.

Titanium alloys are studied not only experimentally but also theoretically. Karre et al. [29] employed the first principles calculations to discover the minimum content of Nb and Zr in binary (Ti–Nb) and ternary (Ti–Nb–Zr) alloys, which is sufficient to stabilize the bcc structure. It has been found that stable β phase in Ti–Nb alloy can

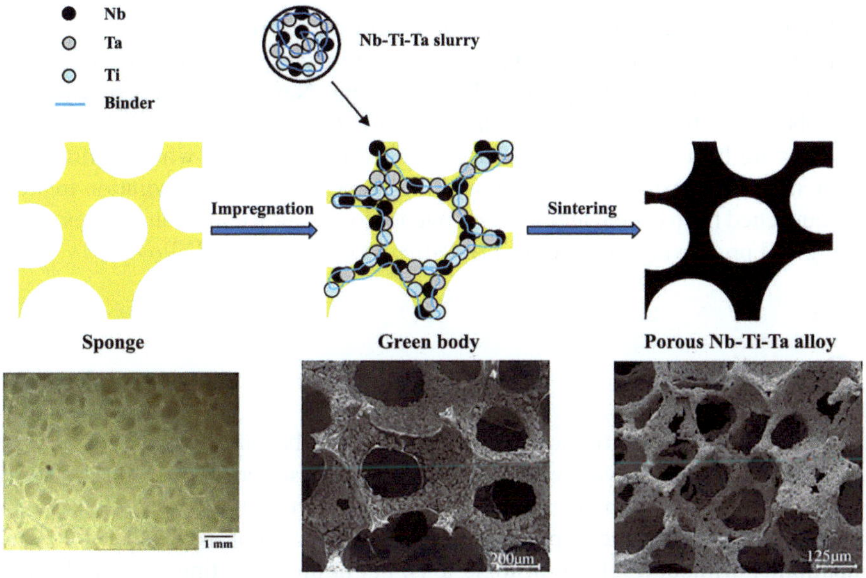

Fig. 3 Scheme of the pore formation process in Nb–Ti–Ta alloy via sponge impregnation method [31]

be obtained when the critical concentration of Nb is approximately 22% that is in a good agreement with experimental studies (~25%) and addition of 6.25 at.% of Zr enhances the stability of bcc structure. Therefore, the usage of theoretical calculations makes a process of designing new alloys more effective; it is possible to identify the most suitable ration of the content of constituent elements.

However, as was pointed out by Liu et al. [30, 31], the titanium alloys with porous structure can demonstrate high-performance characteristics. In Ref. [30] Nb–25Ti–xTa (x = 10, 15, 20, 25, 35 at.%) alloys were prepared from the powders, then they were blended, compacted under uniaxial pressure to form a green compact by a hydraulic press, sintered at vacuum furnace at 1600–1800 °C, and finally cooled to room temperature. This process allowed to obtain β-type alloys with a porous structure. As it is known, the size of pores and the level of porosity can effectively influence the properties of metallic alloys as pores are suitable for the growth of new bone tissue. Porous materials can also be successfully produced by freeze casting method [32] and Selective laser melting with rapid prototyping [33]. In Ref. [31] Liu et al. employed a commercial polyurethane sponge, where mixed metal powders with polyvinyl alcohol aqueous solution as a binder were placed with the following sintering to fabricate the porous Nb–Ti–Ta alloy. In Fig. 3 the process of producing porous structure by sponge impregnation technique is schematically depicted. Titanium-based alloys with pores demonstrate bcc structure with high properties.

Taking into account the examples mentioned above it follows that titanium alloys with two or more non-toxic biocompatible refractory metals have been widely produced and investigated. There are some successful studies where methods providing formation of the titanium alloys with porous structure were employed. However, the most commonly used technique is still the melting. Nevertheless, every scientific group obtains Ti-based alloys with different structure and properties. The reason is a following heat treatment procedures applied to the alloys, which define their characteristics; for example, after the melting process they can be solution treated, water quenched from β-field temperature, aged at low or high temperature. Therefore, alloying and processing fully define properties of titanium-based alloys.

3 Properties

Titanium is highly amenable to change its properties because of the alloying elements [34–36]. Thus, a wide range of different Ti-based alloys can be produced and depending on their microstructure; the different properties can be realized. To be successfully applied the implant material has to show not only good biological compatibility without a release of toxic ions, but at the same time it must demonstrate proper mechanical, corrosion, and wear behavior for long-term performance in an aggressive internal environment of a human body. The essential properties that materials for biomedical applications must possess are high strength, corrosion and wear resistance, biocompatibility, superelasticity, and low Young's modulus.

There are several ways to modify the properties of titanium alloys [12, 37, 38]. The first, alloying of Ti with proper elements, such as α-stabilizers (Al, O, etc.), β-stabilizers (Nb, V, Fe, Cr, etc.), and neutral ones (Zr and Sn). Using this method, it is possible to determine a structure and physical-mechanical properties, for instance, corrosion resistance, elastic modulus, strength. Another way to change the characteristics of the alloy is its processing, which makes it possible to control the microstructure, for example, hardening, solid solution treatment, aging, rapid solidification, etc.

3.1 Mechanical Behavior

Implant material has to demonstrate mechanical properties that are close to those of the human bone. One of the main reasons of an implant loosening is a mismatch in elastic modulus between bone and implant material that may result in the stress shielding effect. Titanium alloys initially show much higher Young's modulus, so that is why some modifications have to be done. Combining proper alloy composition and heat treatment, physical-mechanical properties of CP titanium can be significantly improved. Some achievements in designing Ti-based alloys with β-phase structure and low Young's modulus are shown in Table 2, where elastic modulus of different human bones are given for comparison. As it can be seen, alloying of CP titanium reduces Young's modulus to values that are closed to that of different bone tissues.

Table 2 Young's modulus of some Ti-based alloys and human bones

Sample	Young's modulus (GPa)	References
Femur (human bone)	16.7	[39]
Tibia (human bone)	23.8	[40]
Cortical human bone	17.0	[41]
CP Ti	105.0	[40]
Zr–20Nb–11Ti	29.3	[6]
Zr–12Nb–8Ti	23.1	[6]
Ti–25Ta–25Nb (as-cast)	40.1	[42]
Ti–24.2Nb–2.0Ta–5.1Zr	48.0	[43]
Nb–25Ti–10Ta	71.0	[30]
Nb–25Ti–10Ta (50% porosity)	2.08	[31]
Ti–15Mo–2.8Zr–3Al	82.0	[44]

The tensile mechanical properties of Ti–25Ta–25Nb biomedical alloy were studied by Bertrand et al. [27]. This β-type titanium alloy showed clearly a non-linear elastic behavior, and its stress-strain curves exhibited a stress plateau—an indication of a superelasticity that can be caused by strain induced martensite formation or stress. Furthermore, it also demonstrated a relatively low Young's modulus of about 55 GPa; this value is close to that of human cortical bone. Tensile properties of Ti–35.3Nb–5.7Ta–7.3Zr alloys can be enhanced by the addition of oxygen: the yield stress reached 1000 MPa, however, its elastic modulus experienced a rise from 65 to 80 GPa [45]. In order to investigate mechanical properties, compression tests are also used. For example, among Zr–12Nb–xTi ($x = 0, 2, 4, 8, 16$ at.%) alloys, studied in Ref. [6], the Zr–12Nb–8Ti exhibited an excellent combination of strength and plasticity: compressive strength was about 1940 MPa, yield strength reached 774 MPa, elastic modulus was 23.12 GPa, and the plastic strain was approximately 46%. These results indicate its high mechanical properties. Therefore, as He et al. have pointed out [46] the low Young's modulus and the high strength can be reached when titanium is alloyed with the elements stabilizing bcc phase and is properly processed.

3.2 *Biocompatibility*

Ti, Ta, Nb, and Zr belong to the loose connective vital group and demonstrate great biocompatibility and osteogenesis as that is extremely important in the case of biomedical applications [30, 47]. Thus, many studies connected with these alloys demonstrate promising results. To investigate biological compatibility, an apatite formation ability in stimulated body fluid (SBF), cell adhesion, and proliferation are evaluated. Metal biomaterials can bond with a bone using a formation of the apatite coating on the surface of the alloy within the human body fluid. To measure the toxicity of alloys, they are incubated with cultured human osteoblast-like cells. For further investigations, the animal experiments are performed to understand all processes occurred when implant material is placed into the body. Before being commercially used, alloys for biomedical applications must be thoroughly tested, as there can be numerous of reactions, such as fibrosis, activation of the clotting cascade, tumor formation, antibody production, etc. [48]. For instance, Ti–Ta–Nb–Zr alloy was used as an implant for rats in Ref. [49], it was placed in the contralateral tibial metaphysis, and results showed no sign of cytotoxicity and adverse tissue reactions, as well as low inflammatory response. Testing of other Nb–Ti–Ta porous alloy scaffolds [31] exhibited a good/excellent adhesion and proliferation ability. Moreover, this alloy was implanted on the rabbit model and results demonstrated a new bone tissue generation. Extensive research shows that titanium-based alloys with β-stabilizing elements are highly promising candidates for implant materials due to good osseointegration (the ability of implant material to connect with an adjacent bone) and biocompatibility.

3.3 *Corrosion Behavior*

Corrosion is one of the major reasons of an implant loosening due to the accumulation of debris in human tissues that can cause inflammation and cell damage. That is why materials for biomedical applications have to demonstrate high corrosion resistance in various environments as a human body has a wide range of pH level [50].

Titanium on its own is chemically reactive metal, and in the very oxidizing environment, an impact of the TiO_2 thin film on its surface diminishes. Hence, corrosion resistance is significantly reduced. The stability of an oxide film can be enhanced by means of adding elements [51–53]. Since alloys containing hcp phase tend to fracture and crack in an aggressive environment and bcc phase is more stable, titanium-based alloys with β-stabilizers should be considered as the best candidates for implant materials that work in the human body fluids.

In Ref. [27] a corrosion resistance of Ti–25Ta–25Nb biomedical alloy was determined by cyclic potentiodynamic and linear polarization measurements in neutral Ringer's solution. The results showed lower corrosion and ion release rate compared to the CP Ti. Hence, the addition of β-stabilizing elements to Ti–25Ta–25Nb alloy has a positive influence on its corrosion behavior. An extensive corrosion tests carried out for Ti–5%Ta–1.8%Nb alloys indicated the superior corrosion resistance i\n highly oxidizing solutions [54]. Hence, Ti alloys with Ta, Nb, Zr exhibit good corrosion behavior in aggressive environments and can perform long-term use as an implant material.

4 Conclusions

Ti-based alloys are successfully applied in medicine as implant materials due to their unique properties that can be easily changed with element alloying or heat processing (solid solution treatment, hardening, aging, etc.). The most interesting characteristics are observed in alloys consisted of β phase (bcc structure) as they exhibit low Young's modulus, high strength, biocompatibility, corrosion and wear resistance. Among the β-stabilizing elements that are non-toxic, tantalum and niobium demonstrate a positive impact on the titanium properties. However, not only these elements can lower the temperature of the β-phase field, but zirconium (element of the neutral group that must not influence the phase transformation of Ti) also can stabilize bcc structure. The literature reveals that authors have successfully prepared Ti-based alloys with high strength and Young's modulus that is close to that of human bone tissue (4–30 GPa). However, for mass production implant materials apart from outstanding properties must not be expansive. Concerning alloys with Ta, Nb, and Zr, the cost of their preparation is high, so further improvements must be implemented.

Acknowledgements This research was carried out under the aegis of Science Committee of the Ministry of Education and Science of the Republic of Kazakhstan in the framework of the targeted financing program "Targeted scientific and technical program of the D. Serikbayev East Kazakhstan State Technical University, focused on the development of new types of products for fabrication at the leading industrial enterprises of the East Kazakhstan region" for 2017–2019 years, and was supported by the budget program of Ministry of Education and Science of Ukraine "Multilayer and multicomponent coatings with adaptive behavior in wear and friction conditions" (No. 0118U003579).

References

1. Biesiekierski A, Wang J, Abdel-Hady Gepreel M et al (2012) A new look at biomedical Ti-based shape memory alloys. Acta Biomater 8(5):1661–1669. https://doi.org/10.1016/j.actbio.2012.01.018
2. Geetha M, Singh AK, Asokamani R et al (2009) Ti-based biomaterials, the ultimate choice for orthopedic implants—a review. Prog Mater Sci 54(3):397–425. https://doi.org/10.1016/j.pmatsci.2008.06.004
3. Liu J, Chang L, Liu H, Li Y et al (2017) Microstructure, mechanical behavior and biocompatibility of powder metallurgy Nb–Ti–Ta alloys as biomedical material. Mater Sci Eng C Mater Biol Appl 71:512–519. https://doi.org/10.1016/j.msec.2016.10.043
4. Hao YL, Yang R, Niinomi M et al (2002) Young's modulus and mechanical properties of Ti–29Nb–13Ta–4.6Zr about α'' martensite. Metall Mater Trans A 33(10):3137–3144. https://doi.org/10.1007/s11661-002-0299-7
5. Zhang Y, Xiu P, Jia Z et al (2018) Effect of vanadium released from micro-arc oxidized porous Ti6Al4V on biocompatibility in orthopedic applications. Colloids Surf B 169:366–374. https://doi.org/10.1016/j.colsurfb.2018.05.044
6. Chui P (2017) Near β-type Zr–Nb–Ti biomedical alloys with high strength and low modulus. Vacuum 143:54–58. https://doi.org/10.1016/j.vacuum.2017.05.039
7. Millis DL (2014) Responses of musculoskeletal tissues to disuse and remobilization. In: Canine rehabilitation and physical therapy, Elsevier, pp 92–153

8. Zheng XH, Sui JH, Zhang X et al (2013) Thermal stability and high-temperature shape memory effect of Ti–Ta–Zr alloy. Scr Mater 68(12):1008–1011. https://doi.org/10.1016/j.scriptamat.2013.03.008
9. Geetha M, Kamachi Mudali U, Gogia AK et al (2004) Influence of microstructure and alloying elements on corrosion behavior of Ti–13Nb–13Zr alloy. Corros Sci 46(4):877–892. https://doi.org/10.1016/S0010-938X(03)00186-0
10. Pogrebnjak AD, Rogoz VM, Bondar OV et al (2016) Structure and physicomechanical properties of NbN-based protective nanocomposite coatings: a review. Prot Met Phys Chem Surf 52(5):802–813. https://doi.org/10.1134/S2070205116050191
11. Kadyrzhanov EY, Zdorovets M, Kozlovskiy AL et al (2018) Influence of ionizing irradiation on the parameters of Zn nanotubes arrays for design of flexible electronics elements. Devices Methods Meas 9(1):66–73. https://doi.org/10.21122/2220-9506-2018-9-1-66-73
12. Peters M, Hemptenmacher J, Kumpfert J et al (2003) Structure and properties of titanium and titanium alloys. In: Leyens C, Peters M (eds) Titanium and titanium alloys: fundamentals and applications, Wiley-VCH Verlag GmbH & Co, pp 1–36
13. Hao YL, Li SJ, Sun SY et al (2006) Effect of Zr and Sn on Young's modulus and superelasticity of Ti–Nb-based alloys. Mater Sci Eng A 441(1):112–118. https://doi.org/10.1016/j.msea.2006.09.051
14. Cojocaru VD, Raducanu D, Gloriant T et al (2012) Structural observation of twinning deformation mechanism in a Ti–Ta–Nb alloy. Solid State Phenom 188:46–51. https://doi.org/10.4028/www.scientific.net/SSP.188.46
15. Nie L, Zhan Y, Hu T et al (2014) β-Type Zr–Nb–Ti biomedical materials with high plasticity and low modulus for hard tissue replacements. J Mech Behav Biomed Mater 29:1–6. https://doi.org/10.1016/j.jmbbm.2013.08.019
16. Drob SI, Vasilescu C, Drob P et al (2015) Corrosion behaviour of nitrogen-implantation Ti—Ta–Nb alloy in physiological solutions simulating real conditions from human body. JOM 67:818–829. https://doi.org/10.1007/s11837-015-1351-6
17. Raducanu D, Vasilescu E, Cojocaru VD et al (2011) Mechanical and corrosion resistance of a new nanostructured Ti–Zr–Ta–Nb alloy. J Mech Behav Biomed Mater 4:1421–1430. https://doi.org/10.1016/j.jmbbm.2011.05.012
18. Hussein AH, Gepreel MAH, Gouda MK et al (2016) Biocompatibility of new Ti–Nb–Ta base alloys. Mater Sci Eng C 61:574–578. https://doi.org/10.1016/j.msec.2015.12.071
19. Berladir KV, Budnik OA, Dyadyura KA et al (2016) Physicochemical principles of the technology of formation of polymer composite materials based on polytetrafluoroethylene—a review. High Temp Mater Process Int Q High Technol Plasma Process 20(2):157–184. https://doi.org/10.1615/HighTempMatProc.2016017875
20. Boiko O, Koltunowicz TN, Zukowski P et al (2017) The effect of sputtering atmosphere parameters on dielectric properties of the ferromagnetic alloy—ferroelectric ceramics nanocomposite (FeCoZr)x(PbZrTiO3)(100–x). Ceram Int 43(2):2511–2516. https://doi.org/10.1016/j.ceramint.2016.11.052
21. Ivashchenko VI, Veprek S, Argon AS et al (2015) First-principles quantum molecular calculations of structural and mechanical properties of TiN/SiNx heterostructures, and the achievable hardness of the nc-TiN/SiNx nanocomposites. Thin Solid Films 578:83–92. https://doi.org/10.1016/j.tsf.2015.02.013
22. Ivashchenko VI, Veprek S, Turchi PEA et al (2012) First-principles study of TiN/SiC/TiN interfaces in superhard nanocomposites. Phys Rev B 86(1):14110. https://doi.org/10.1103/PhysRevB.86.014110
23. Kasiuk V, Fedotova JA, Koltunowicz TN et al (2014) Correlation between local Fe states and magnetoresistivity in granular films containing FeCoZr nanoparticles embedded into oxygen-free dielectric matrix. J Alloy Compd 586:S432–S435. https://doi.org/10.1016/j.jallcom.2012.09.058
24. Knight R, Smith RW, Apelian D (1991) Application of plasma arc melting technology to the processing of reactive metals. Int Mater Rev 36(1):221–252. https://doi.org/10.1179/imr.1991.36.1.221

25. Cremasco A, Messias A, Esposito A et al (2011) Effects of alloying elements on the cytotoxic response of titanium alloys. Mater Sci Eng C 31:833–839. https://doi.org/10.1016/j.msec.2010.12.013
26. Oliveira NTC, Aleixo G, Caram R et al (2007) Development of Ti–Mo alloys for biomedical applications: microstructure and electrochemical characterization. Mater Sci Eng A 452–453:727–731. https://doi.org/10.1016/j.msea.2006.11.061
27. Bertrand E, Gloriant T, Gordin DM et al (2010) Synthesis and characterization of a new superelastic Ti–25Ta–25Nb biomedical alloy. J Mech Behav Biomed Mater 3(8):559–564. https://doi.org/10.1016/j.jmbbm.2010.06.007
28. Hagihara K, Nakano T (2017) Experimental clarification of the cyclic deformation mechanisms of β-type Ti–Nb–Ta–Zr-alloy single crystals developed for the single-crystalline implant. Int J Plast 98:27–44. https://doi.org/10.1016/j.ijplas.2017.06.006
29. Karre R, Niranjan MK, Dey SR (2015) First-principles theoretical investigations of low Young's modulus beta Ti–Nb and Ti–Nb–Zr alloys compositions for biomedical applications. Mater Sci Eng C 50:52–58. https://doi.org/10.1016/j.msec.2015.01.061
30. Liu J, Chang L, Liu H et al (2017) Microstructure, mechanical behavior and biocompatibility of powder metallurgy Nb–Ti–Ta alloys as a biomedical material. Mater Sci Eng C 71:512–519. https://doi.org/10.1016/j.msec.2016.10.043
31. Liu J, Ruan J, Chang L et al (2017) Porous Nb–Ti–Ta alloy scaffolds for bone tissue engineering: Fabrication, mechanical properties and in vitro/vivo biocompatibility. Mater Sci Eng C 78:503–512. https://doi.org/10.1016/j.msec.2017.04.088
32. Li WL, Lu K, Walz JY (2012) Freeze casting of porous materials: review of critical factors in microstructure evolution. Int Mater Rev 57(1):37–60. https://doi.org/10.1179/1743280411Y.0000000011
33. Warnke PH, Douglas T, Wollny P et al (2008) Rapid prototyping: porous titanium alloy scaffolds produced by selective laser melting for bone tissue engineering. Tissue Eng Part C Methods 15(2):115–124. https://doi.org/10.1089/ten.tec.2008.0288
34. Pogrebnjak AD, Ponomarev AG, Shpak AP et al (2012) Application of micro- and nanoprobes to the analysis of small-sized 3D materials, nanosystems, and nanoobjects. Phys Usp 55(3):270–300. https://doi.org/10.3367/UFNe.0182.201203d.0287
35. Pogrebnjak AD (2013) Structure and properties of nanostructured (Ti–Hf–Zr–V–Nb)N coatings. J Nanomater 2013:1–12. https://doi.org/10.1155/2013/780125
36. Pogrebnjak AD, Beresnev VM, Smyrnova KV et al (2018) The influence of nitrogen pressure on the fabrication of the two-phase superhard nanocomposite (TiZrNbAlYCr)N coatings. Mater Lett 211:316–318. https://doi.org/10.1016/j.matlet.2017.09.121
37. Pogrebnjak AD, Ivashchenko VI, Skrynskyy PL et al (2018) Experimental and theoretical studies of the physicochemical and mechanical properties of multi-layered TiN/SiC films: Temperature effects on the nanocomposite structure. Compos Part B Eng 142:85–94. https://doi.org/10.1016/j.compositesb.2018.01.004
38. Pogrebnjak AD, Bondar OV, Abadias G et al (2016) Structural and mechanical properties of NbN and Nb–Si–N films: Experiment and molecular dynamics simulations. Ceram Int 42(10):11743–11756. https://doi.org/10.1016/j.ceramint.2016.04.095
39. Alliston T, Chang J (2011) TGFβ and Runx2 calibration of bone extracellular matrix quality for tissue-specific function. IBMS Bonekey 8(8):370–380. https://doi.org/10.1138/20110525
40. Li Y, Yang C, Zhao H et al (2014) New developments of Ti-based alloys for biomedical applications. Materials 7(3):1709–1800. https://doi.org/10.3390/ma7031709
41. Lai Y-S, Chen W-C, Huang C-H et al (2015) The effect of graft strength on knee laxity and graft in-situ forces after posterior cruciate ligament reconstruction. PLoS One 10(5):e0127293. https://doi.org/10.1371/journal.pone.0127293
42. Dumitrescu C, Răducanu D, Cojocaru VD et al (2011) Investigation of mechanical properties for A Ti–Ta–Nb alloy. UPB Sci Bull Ser B Chem Mater Sci 73:221–230
43. Tang X, Ahmed T, Rack HJ (2000) Phase transformations in Ti–Nb–Ta and Ti–Nb–Ta–Zr alloys. J Mater Sci 25:1805–1811. https://doi.org/10.1023/A:1004792922155

44. Mohammed MT, Khan ZA, Siddiquee AN (2014) Beta titanium alloys: the lowest elastic modulus for biomedical applications: a review. Int J Mater Metall Eng 8(8):822–827
45. Stráský J, Harcuba P, Václavová K et al (2017) Increasing strength of a biomedical Ti–Nb–Ta–Zr alloy by alloying with Fe, Si and O. J Mech Behav Biomed Mater 71:329–336. https://doi.org/10.1016/j.jmbbm.2017.03.026
46. He G, Hagiwara M (2006) Ti alloy design strategy for biomedical applications. Mater Sci Eng C 26(1):14–19. https://doi.org/10.1016/j.msec.2005.03.007
47. Ozan S, Lin J, Li Y et al (2017) New Ti–Ta–Zr–Nb alloys with ultrahigh strength for potential orthopedic implant applications. J Mech Behav Biomed Mater 75:119–127. https://doi.org/10.1016/j.jmbbm.2017.07.011
48. Williams D (2008) On the mechanisms of biocompatibility. Biomaterials 29:2941–2953. https://doi.org/10.1016/j.biomaterials.2008.04.023
49. Stenlund P, Omar O, Brohede U et al (2015) Bone response to a novel Ti–Ta–Nb–Zr alloy. Acta Biomater 20:165–175. https://doi.org/10.1016/j.actbio.2015.03.038
50. Schwalfenberg GK (2012) The alkaline diet: Is there evidence that an alkaline pH diet benefits health? J Environ Public Health 2012:727630. https://doi.org/10.1155/2012/727630
51. Pogrebnjak AD, Beresnev VM, Bondar OV et al (2018) Superhard CrN/MoN coatings with multilayer architecture. Mater Des 153:47–59. https://doi.org/10.1016/j.matdes.2018.05.001
52. Maksakova O, Simoẽs S, Pogrebnjak A et al (2018) The influence of deposition conditions and bilayer thickness on physical-mechanical properties of CA-PVD multilayer ZrN/CrN coatings. Mater Charact 140:189–196. https://doi.org/10.1016/j.matchar.2018.03.048
53. Pogrebnyak AD, Shpak A, Azarenkov NA et al (2009) Structures and properties of hard and superhard nanocomposite coatings. Phys Usp 52(1):29–54. https://doi.org/10.3367/UFNe.0179.200901b.0035
54. Kapoor K, Kain V, Gopalkrishna T et al (2003) High corrosion resistant Ti-5% Ta-1.8% Nb alloy for fuel reprocessing application. J Nucl Mater 322:36–44. https://doi.org/10.1016/S0022-3115(03)00302-7

Strain Effect on the Properties of Polar Dielectric Thin Films

Alexander Tkach, Olena Okhay, André Santos, Sebastian Zlotnik, Ricardo Serrazina, Paula M. Vilarinho and M. Elisabete Costa

Abstract Low cost scalable processing and substrates are critical for optimized polar dielectric performance of functional oxide thin films if they are to achieve commercialization. Here, we present a comprehensive investigation of the role low-cost MgO, Al_2O_3, $SrTiO_3$ and Si substrates on the structural and electrical properties of sol-gel derived $SrTiO_3$ (ST) and $K_{0.5}Na_{0.5}NbO_3$ (KNN) thin films. The substrate is found to have a strong effect on the stress/stain state and, consequently, on the dielectric and ferroelectric response of the films. A tensile stress induced in-plane by the thermal expansion mismatch between the substrates and the films observed for ST and KNN films deposited on platinized Al_2O_3 and Si substrates, respectively, lowers the relative permittivity and remanent polarization values in the parallel plate capacitor geometry. In contrast, a compressive stress/strain observed for ST films deposited on MgO/Pt and KNN films on $SrTiO_3$/Pt substrates result in superior polarization and dielectric permittivity, corresponding to enhanced out-of-plane displacement of Ti^{4+} ions in ST films and Nb^{5+} ions in KNN films. It is thus demonstrated that for polycrystalline polar dielectric thin films the relative permittivity and polarization may be optimized through an induced compressive stress state.

Keywords Thin films · Sol-gel · Thermal expansion · Stress/strain · Ferroelectric hysteresis · Dielectric properties

1 Introduction

Strontium titanate ($SrTiO_3$, ST) as a member of the perovskite structure family of materials has been extensively studied over the years. Undoped ST is a quantum

The original version of this chapter was revised: The chapter title has been updated from "Stain Effect on the Properties of Polar Dielectric Thin Films" to "Strain Effect on the Properties of Polar Dielectric Thin Films". The correction to this chapter is available at https://doi.org/10.1007/978-981-13-6133-3_38

A. Tkach (✉) · O. Okhay · A. Santos · S. Zlotnik · R. Serrazina · P. M. Vilarinho · M. E. Costa
Department of Materials and Ceramic Engineering, CICECO—Aveiro Institute of Materials, University of Aveiro, Aveiro 3810-193, Portugal
e-mail: atkach@ua.pt

© Springer Nature Singapore Pte Ltd. 2019, corrected publication 2023
A. D. Pogrebnjak and V. Novosad (eds.), *Advances in Thin Films, Nanostructured Materials, and Coatings*, Lecture Notes in Mechanical Engineering,
https://doi.org/10.1007/978-981-13-6133-3_32

paraelectric, where zero-point fluctuations preclude the condensation of the polar lattice soft mode, saturating the relative dielectric permittivity (ε_r) at low temperatures $T < 4$ K [1]. Hence, no ferroelectric but only an improper ferroelastic phase transition from cubic (Pm3m) to a centrosymmetric tetragonal (I4/mcm) phase with a doubled primitive unit cell, associated with tilts of the O-octahedra in antiphase around [001] direction, occurs at cooling below the transition temperature $T_a \sim 110$ K [2–4]. Therefore, high and electric-field tunable ε_r can be obtained in ST together with very low dielectric losses. Hence ST based thin films are excellent candidates for tunable device applications, in which low losses are required [5, 6].

The electrical properties of thin films are known to be highly dependent on the preparation method, fabrication conditions (temperature, atmosphere, heating cycles) and substrate or underneath layers. Reports prove that dielectric properties of ST films can be engineered by the mechanical misfit strain between the film and the substrate [6–9]. Thus, ST films deposited by reactive molecular-beam epitaxy on (110) $DyScO_3$ substrates reveal a ferroelectric anomaly accompanied by soft mode condensation near to room temperature (RT) due to in-plane stresses from the substrates [10, 11] and according to the lattice mismatch theoretical predictions [8]. As a result, high ε_r (nearly 7000 at 10 GHz) and its sharp dependence on the electric field E (with relative tunability $n_r = [\varepsilon_r(0) - \varepsilon_r(E)]/\varepsilon_r(0)$ of 82%) observed in these films at RT using interdigitated electrodes are promising for device applications, despite of the elevated losses (reflected in a dissipation factor $\tan\delta = 0.2$). However, film deposition by molecular-beam epitaxy is much more expansive than by sol-gel processing that, being a chemical solution deposition method, lowers the processing temperature, facilitates compositional control, provides good homogeneity within the films and may be used for the fabrication of large area devices at low cost. Then instead of $DyScO_3$ substrates, which are too costly and rare to be used in commercial applications, more common substrates such as Si, MgO, Al_2O_3, and $SrTiO_3$ single crystals should be considered. Finally, in contrast to interdigitated electrodes a more scalable parallel plate capacitor geometry should be used.

On the other hand, driven by the need to substitute lead based electroceramics, alkali niobates are considered as one of the most promising family of lead-free compounds [12]. Particularly, potassium sodium niobate ($K_{0.5}Na_{0.5}NbO_3$, KNN) has become one of the most extensively investigated piezoelectrics due to its high Currie temperature (T_C) up to 420 °C and owing to its high longitudinal piezoelectric coefficient (d_{33}) reported by Saito et al. in 2004 [13]. Since that time most of the literature has been focused on KNN-based bulk materials [12, 14, 15] as one of the promising alternatives in various applications, such as sensors, actuators, energy harvesting and microelectromechanical systems [16], while KNN based films have received some attention more recently.

Some KNN-based thin films with high remanent polarization (P_r) and d_{33} coefficient were synthesized by sol-gel methods [17–26] that being industrially appealing is promising for homogeneous thin film production. During the KNN thermal processing an additional problem of alkali volatilization appears that tends to induce point defects and non-stoichiometry in the resulting films, hence decreasing significantly both ferroelectric and piezoelectric performance [26]. Therefore, a rapid thermal annealing (RTA) should be used, while a solution containing 5% excess of potassium is found to be optimum for the volatilization compensation during KNN film fabrication at 750 °C for 5 min [21]. Resulting 250 nm thick KNN films presented a room-temperature P_r of ~8 $\mu C/cm^2$ and coercive field (E_c) of 80 kV/cm at an applied field of 700 kV/cm together with ε_r of 614 and tanδ of 0.015, at 1 kHz [21]. In spite of the numerous studies mainly focused on the preparation conditions of sol-gel derived KNN thin films, including alkali excess varying from 0 [19] to 40% [26] and annealing conditions varying between insertion into furnace [18, 26] and RTA [17, 19, 21], to the best of our knowledge, only platinized Si was reported as substrate for undoped sol-gel derived KNN film [17–21, 24–26]. Few works report also doped KNN films deposited on conductive niobium-doped $SrTiO_3$ [22, 23]. However, sol-gel derived KNN films deposited on platinized $SrTiO_3$ have not yet been studied. A comparative study of KNN thin films identically deposited on platinized Si and $SrTiO_3$ substrates to clearly identify the role of the substrate on their dielectric and ferroelectric properties is also missing.

As shown in Table 1, the sequence of the thermal expansion coefficients (TEC, α) is $\alpha_{MgO} > \alpha_{ST} > \alpha_{Al2O3} > \alpha_{KNN} > \alpha_{Si}$. Therefore, different strains on cooling from annealing to room temperature are anticipated, thus predicting different dielectric and ferroelectric properties for ST films on platinized Al_2O_3 and MgO substrates and KNN films on platinized Si and $SrTiO_3$ substrates. In this work and within this context, dense, highly crystalline and crack free ST films are deposited on Al_2O_3/Pt and MgO/Pt substrates and conventionally furnace annealed at 900 °C for 60 min. KNN films are also prepared by chemical solution deposition but from solution with 5% of potassium excess on $Si/SiO_2/TiO_2$/Pt and $SrTiO_3$/Pt substrates, being then annealed using RTA at 750 °C for 5 min. Using dielectric and ferroelectric characterization in combination with X-ray diffraction (XRD) analysis, a correlation between the electrical properties in the parallel plate capacitor geometry and the in-plane stress induced by thermal expansion mismatch in ST films deposited on MgO/Pt and KNN films deposited on $SrTiO_3$/Pt is found.

Table 1 Thermal expansion coefficients for materials of the films and substrates and annealing temperatures of 750 and 900 °C

Material	α ($\times 10^{-6}$ K^{-1})	
	For T_{ann} = 750 °C	For T_{ann} = 900 °C
Si	2.3[a]	–
$K_{0.5}Na_{0.5}NbO_3$	4.72[b]	–
Al_2O_3	–	7.7[c]
$SrTiO_3$	9[a]	10.3[c]
MgO	–	14.3[c]

[a]From supplier information
[b]From Ref. [27]
[c]From our dilatometry measurements [28]

2 Experimental Details

2.1 *Preparation of ST Films*

$SrTiO_3$ solution was prepared using strontium acetate (98%, ABCR), tetra-n-butyl orthotitanate (98%, Merck) as starting precursors. Acetic acid (99.8%, Merck), 1,2-propanediol (99.5%, Riedel-de Haën) and absolute ethanol (99.8%, Merck) were used as solvents. Strontium acetate was initially dissolved into heated acetic acid under constant stirring to form a transparent solution. After cooling to RT the former solution was diluted with 1,2-propanediol and then titanium isopropoxide was added. The resultant clear solution was continuously stirred during 12 h in a closed flask and ethanol was added as a final step to keep the concentration at 0.2 M. Using this solution, layers of ST were deposited on Al_2O_3/Pt and MgO/Pt substrates by spin-coating (Chemat Technology spin-coater KW-4A) at 4000 rpm for 30 s. Al_2O_3 and MgO substrates were purchased from Crystal GmbH, Germany with subsequent platinization in Inostek Inc., Korea, and cleaned in boiling ethanol before the film deposition. Each layer of the films was heated on a hot plate at 300 °C for 1 min to ensure complete removal of volatile species between the layers. After a multicycle deposition of 10 layers, ST thin films were conventionally furnace annealed at T_{ann} = 900 °C for 60 min to obtain high crystallinity [28, 29].

2.2 *Preparation of KNN Films*

For the synthesis of KNN films, a precursor solution was prepared using potassium acetate (≥99%, ChemPur, GmbH), sodium acetate (99%, Alfa Aeasar) niobium pentaethoxide (99%, Stark) and 2-methoxyethanol (99%, Sigma Aldrich). Initially, 2-methoxyethanol was placed in a closed flask and left for 30 min in nitrogen (N_2) under constant stirring. Using a glove box with argon (Ar) atmosphere, potassium acetate, sodium acetate and niobium pentaethoxide were weighted according to the requested ratio and dissolved in 2-methoxyethanol. 5% excess of potassium was used in this study to compensate alkali volatilization during the heat treatment [21]. After mixing the reagents and stirring in N_2 during ~30 min, the solution was refluxed at temperature of ~106 °C for 4 h and distilled at 124 °C. After cooling down, the KNN precursor solution was transferred to a flask and 2-methoxyethanol was added to keep the concentration at 0.2 M. Prepared KNN solution passed through a 0.2 μm filter was spin-coated on Si/SiO_2/TiO_2/Pt and $SrTiO_3$/Pt substrates at 3000 rpm for 30 s, forming layered films. Si/SiO_2/TiO_2/Pt substrates were supplied from Inostek Inc. $SrTiO_3$ substrates were purchased from Crystal GmbH, Germany with subsequent platinization in Inostek Inc. Pyrolysis of each as-deposited layers was performed at 350 °C for 2 min on a hot-plate in air. After a multicycle deposition of 10 layers, KNN films were annealed in air at 750 °C for 5 min with a heating/cooling rate of 30 °C/s, using a RTA system (Qualiflow, Jipelec, JetFirst 200) [30].

2.3 Characterization of the Films

Analysis of the thin films crystal phase evolution was performed by XRD with a Philips X'Pert MPD X-ray diffractometer, equipped with a mobile arm and X'Pert Data Collector controller software, using Cu-Kα radiation. The strain/stress analysis consisted of sequential $\theta/2\theta$ measurements with a 0.02° step width and 4 s of counting time per step in 29.5–33° range for KNN and 55–62° for ST films, for ψ values ranging from −70 to 70°, in a Philips X'Pert MRD equipment. The surface and cross-sectional morphologies as well as the thickness of KNN thin films were analyzed by scanning electron microscopy (SEM, Hitachi, SU-70 and S4100) and scanning force microscopy (SFM, Multimode Digital Instrument, Nanoscope IIIa). Dielectric and ferroelectric measurements were performed using Au, sputtered through a mask onto the films, as top electrode and the substrate Pt layer as the bottom one. The dielectric permittivity was obtained by impedance spectroscopy measurements under an oscillation level of the applied voltage of 50 mV and various biases at frequency of 10 kHz, using a precision LCR-meter (Agilent E4980A and HP 4284A). The electric field dependence of the polarization was evaluated using the ferroelectric analyzer (aixACCT, TF Analyzer 1000 and TF Analyzer 2000). Low-temperature measurements of ST film were performed in a He closed cycle cryogenic system (Displex ADP-Cryostat HC-2), controlled by a digital temperature controller (Scientific Instrument Model 9650) with silicon diode thermometers.

3 Results

3.1 Structure and Microstructure

From the cross-section SEM micrographs (not shown), the average thickness of ST films was found to be of ~425 nm and that of KNN films of ~335 nm as listed in Table 2. From top-view SEM and SFM images [28, 30], both kinds of the films reveal a dense and crack free microstructure with a smooth surface and ~200 nm grain size for ST films and ~100 nm grain size for KNN films as also shown in Table 2. Thus, the microstructure of the films does not significantly depend on the substrates, confirming the high reproducibility of the films synthesized by the sol-gel method in this work.

XRD analysis of ST and KNN thin films have revealed that all the films have well crystallized perovskite phase, according to the diffraction peaks of the standard powder samples of $SrTiO_3$ (JCPDS#35-0734) and $Na_{0.5}K_{0.5}NbO_3$ (JCPDS #01-079-7690), respectively, and no other phases were detected. Strain measurements were conducted by the XRD $\sin^2\psi$ method, in which the lattice spacings d of a specific *(hkl)* plane were measured at different tilt angles ψ (inclined exposure), resulting in d_ψ values normalized to the spacing at zero angle d_0. The variation of the corresponding strains induced in ST films and by the Al_2O_3/Pt and MgO/Pt substrates and

Table 2 Microstructural, structural and electrical parameters of ST films on Al_2O_3/Pt and MgO/Pt substrates and KNN films on $Si/SiO_2/TiO_2$/Pt and $SrTiO_3$/Pt substrates

Properties of films on platinized substrates	ST films		KNN films	
	Al_2O_3/Pt	MgO/Pt	Si//Pt	$SrTiO_3$/Pt
Average film thickness (nm)	400	450	340	330
Average grain size (μm)	175	244	99	106
Residual stress (MPa)	529	−531	234	−450
Theoretical thermal stress (MPa)	911	−1093	252	−445
Remanent polarization (μC/cm^2)	0.32[d]	0.47[d]	2.65	4.56
Relative dielectric permittivity	345[d]	613[d]	161	184

[d]At 30 K

KNN films by the $Si/SiO_2/TiO_2$/Pt and $SrTiO_3$/Pt substrates are presented in Fig. 1a and b, respectively. From the slope, ST films deposited on MgO/Pt and KNN films on $SrTiO_3$/Pt have a compressive strain, while the strain of ST films deposited on Al_2O_3/Pt and KNN films on $Si/SiO_2/TiO_2$/Pt is tensile. Using the normalized values of the obtained strain, the in-plane stress values can be calculated as:

$$\sigma = m \times Y/(1 + \nu), \tag{1}$$

where Y and ν corresponds to the Young's modulus and Poisson's ratio of the film, respectively, and m is the gradient of the $(d_\psi - d_0)/d_0$ versus $\sin^2\psi$ dependence. For ST, $Y = 265$ GPa and $\nu = 0.236$ [31]. Then a positive (tensile) residual stress value of 529 MPa for ST films deposited on Al_2O_3/Pt and a negative (compressive) one of −531 MPa for the films on MgO/Pt are obtained, as summarized in Table 2. The values of KNN Young's modulus of 104 GPa and Poisson's ratio of 0.27 were used [32, 33], resulting in a tensile stress value of 234 MPa for KNN films deposited on $Si/SiO_2/TiO_2$/Pt and a compressive one of −450 MPa for the films on $SrTiO_3$/Pt, as also listed in Table 2.

Thermal stress that originates between the film and substrate during cooling from the annealing to room temperature due to the differences in their TECs have been shown theoretically to be the main component of the residual stress in non-epitaxial films [34]. It can be calculated using the following equation

$$\sigma_{th} = \int_{RT}^{T_{\text{ann.}}} \frac{Y}{1 - \nu} \times (\alpha_f(T) - \alpha_{\text{sub}}(T))dT \tag{2}$$

where α_f (T) and α_{sub} (T) are the temperature dependence of the film TECs and corresponding substrate.

The calculated values of the thermal stress that should occur at RT in ST films deposited at 900 °C are shown in Table 2. The tensile stress of ~911 MPa was found

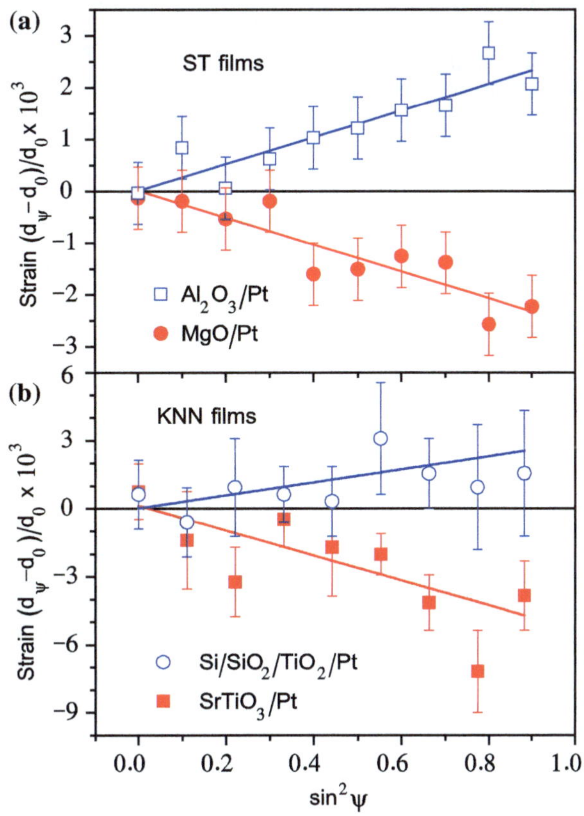

Fig. 1 Strain $(d_\psi - d_0)/d_0$ as a function of $\sin^2\psi$ for ST films on Al_2O_3/Pt and MgO/Pt substrates (**a**) and for KNN films on Si/SiO_2/TiO_2/Pt and $SrTiO_3$/Pt substrates (**b**) at RT together with linear fits used for stress calculations

for ST films deposited on Al_2O_3 substrates with smaller TEC and the compressive stress of ~−1093 MPa was determined for ST films deposited on MgO substrates with higher TEC in agreement with the measured residual stress values. According to the Table 1, TEC of KNN [27] is higher than that of Si substrates but lower than that of $SrTiO_3$. Therefore, a tensile thermal stress of 252 MPa is expected to occur at RT in KNN films deposited on Si substrates at 750 °C, while a compressive thermal stress of −445 MPa should be induced on $SrTiO_3$ substrates. These values are also shown in Table 2, revealing a complete agreement with the measured residual stress values. Thus, thermal stress is indeed the main component of the residual stress of the films obtained in this work.

3.2 Electrical Properties

Variation of the polarization under the *ac* electric field measured at 100 Hz and 30 K on ST films deposited on Al_2O_3/Pt and MgO/Pt is shown in Fig. 2a. At this

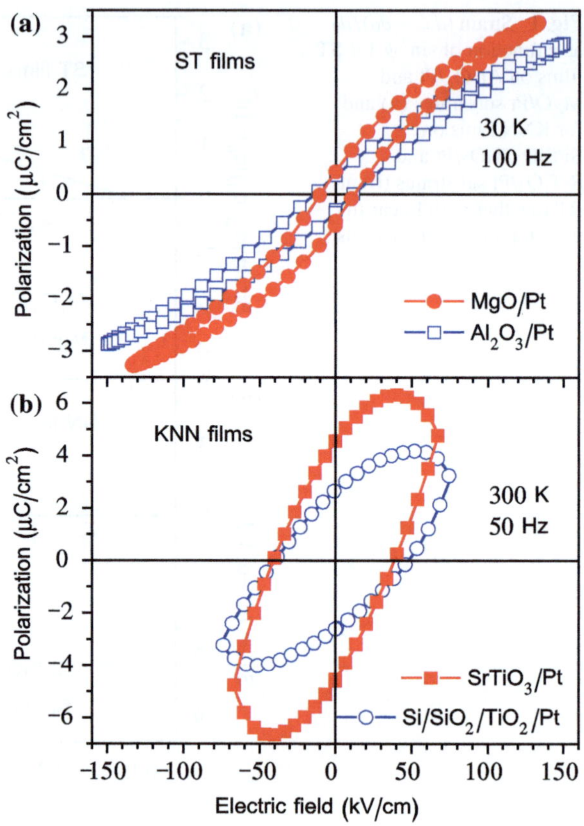

Fig. 2 Polarization *P*—electric field *E* loops for ST films on Al_2O_3/Pt and MgO/Pt substrates at 30 K (**a**) and for KNN films on $Si/SiO_2/TiO_2$/Pt and $SrTiO_3$/Pt substrates at RT (**b**)

temperature, both ST films show an *S*-shape hysteretic *P-E* dependence with quite similar E_c of 12–15 kV cm^{-1}. However, there is a stronger variation of the P_r from ~0.32 μC cm^{-2} for Al_2O_3/Pt to ~0.47 μC cm^{-2} for MgO/Pt substrate (see Table 2). Moreover, the nonlinearity for ST films on MgO/Pt is higher than that of the films on Al_2O_3/Pt substrates.

Comparing to the P_r of ~8 μC/cm^2 reported in literature for KNN films deposited on platinized Si [18, 21], P_r obtained of our KNN films deposited on $SrTiO_3$/Pt looks twice smaller. However, one should take into account that our maximum applied electric field is 3 to 10 times lower than that used in [18, 21].

Electric field dependence of the relative dielectric permittivity of ST films on all the substrates measured at 10 kHz and 30 K is presented in Fig. 3a. ST films on MgO/Pt reveal higher ε_r values and stronger dependence of ε_r on *dc* electric field comparing with those for the films on Al_2O_3/Pt substrates. As shown in Fig. 3b, the relative permittivity of KNN films is higher in the whole field range for those deposited on $SrTiO_3$/Pt, revealing zero-field ε_r of 184 at 10 kHz in comparison to ε_r of 161 at the same frequency for KNN films on $Si/SiO_2/TiO_2$/Pt, as summarized in Table 2 as well.

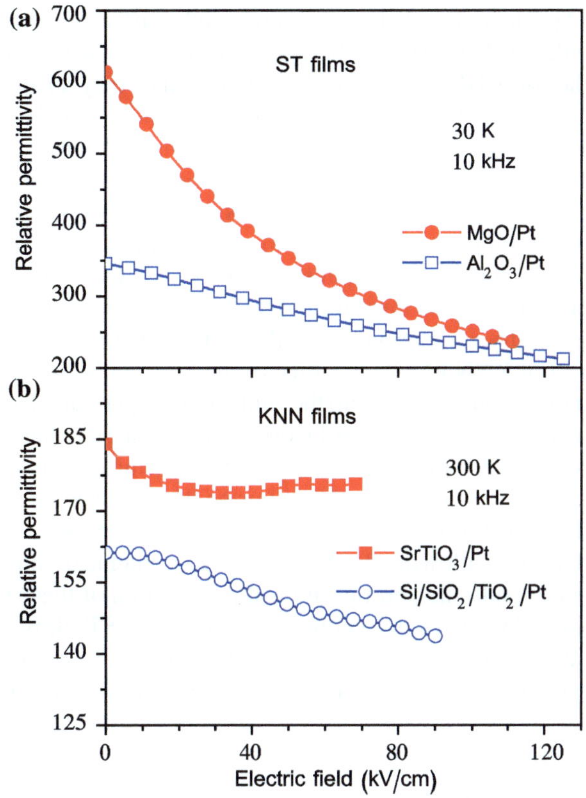

Fig. 3 Electric field dependence of the relative dielectric permittivity ε_r for ST films on Al_2O_3/Pt and MgO/Pt substrates at 30 K (**a**) and for KNN films on $Si/SiO_2/TiO_2$/Pt and $SrTiO_3$/Pt substrates at RT (**b**)

4 Discussion

Summarizing the results presented above, ST films grown on Al_2O_3/Pt substrates are found to show less defined *S*-shape *P*-*E* hysteresis loops (with lower value of P_r), inferior value of ε_r, and lower dependency of ε_r on *dc* electric field comparing to those for ST films deposited on MgO/Pt. Lower P_r and ε_r values are also obtained for KNN films deposited on $Si/SiO_2/TiO_2$/Pt substrates in comparison to those grown on $SrTiO_3$/Pt. Impurities, difference in oxygen vacancy concentration, grain size distribution, crystallographic texture, etc. could have an influence on the observed variation of polarization and permittivity. However, for the equally deposited and processed films, all these factors should be rather the same.

The residual stress/strain is another factor that can affect the ferroelectric and dielectric properties of the sol-gel derived polar dielectric films deposited on different substrates, since these properties are closely related to the ionic motions in the film crystal structures. As predicted from thermodynamic theory and demonstrated experimentally, ferroelectricity can be induced by lattice mismatch strain in epitaxial ST thin films deposited on the substrates with very different lattice parameters [8, 10,

11]. Thus, their behavior is drastically different from that of quantum paraelectric bulk ST. However, the lattice mismatch strain is not so strong, being also similar for the sol-gel derived films of this work, since they all are polycrystalline and grown on top of the Pt electrode layer. The differential TEC between substrate and film plays the major role due to the fact that ~1 mm thicknesses of the substrates is much larger than that of 150 nm for Pt layer deposited on them. Thus, a tetragonal distortion, allowing displacements of Ti ions dominantly out-of-plane, is induced in the lattice of ST films on MgO/Pt compressed in-plane on cooling from annealing temperature by the thermal expansion mismatch ($\alpha_{MgO} > \alpha_{ST}$). As a result, the polarization, permittivity and tunability of these films are enhanced in the parallel plate capacitor geometry comparing to the films on Al_2O_3/Pt.

For KNN films, the lattice of the films on $Si/SiO_2/TiO_2$/Pt is under tensile stress induced on cooling from the annealing temperature due to the thermal expansion mismatch ($\alpha_{Si} < \alpha_{KNN}$), thus favoring the in-plane displacement of Nb^{5+} ions, responsible for ferroelectric properties [35], and lowering the out-of-plane permittivity and polarization, similarly to $SrTiO_3$ films on Al_2O_3 substrates. For KNN films on $SrTiO_3$/Pt, the lattice of the films is compressed in plane ($\alpha_{SrTiO3} > \alpha_{KNN}$). Therefore, a lattice distortion allowing the Nb ionic displacements dominantly out-of-plane is induced. As a result, the permittivity and polarization of these films are enhanced in the parallel plate capacitor geometry, similarly to ST films on MgO substrates. Thus, we have demonstrated that TEC of the substrate can contribute to the final electrical properties of the ST and KNN polar dielectric thin films.

5 Conclusions

Sol-gel derived ~425 nm thick ST thin films were deposited on Al_2O_3/Pt and MgO/Pt substrates under identical processing conditions. A compressive stress/strain induced by the difference in TEC between the MgO substrate and ST film provokes a larger out-of-plane displacement of Ti^{4+} ions and hence STO films prepared on MgO/Pt show maximum values of polarization, higher values of ε_r with stronger dependency on *dc* electric field in comparison with ST films under tensile strain on Al_2O_3/Pt substrates.

The influence of stress/strain built during KNN films processing on their electrical properties is studied on dense and crack free ~335 nm thick KNN films produced from precursor solutions with 5% of potassium excess by RTA on platinised Si and $SrTiO_3$ substrates. As a result of the thermal expansion mismatch induced lattice distortions, determining the Nb ionic displacements, the room-temperature ferroelectric and dielectric measurements in the parallel plate capacitor geometry have also indicated that KNN films with lower remanent polarization and dielectric permittivity values are the films under tensile strain deposited on $Si/SiO_2/TiO_2$/Pt, whereas superior ε_r and P_r values were obtained for the films under compressive strain deposited on $SrTiO_3$/Pt substrates.

Thus, it is shown that variation of the electrical properties can be achieved by controlling the stain/stress level on the films via the choice of the substrate. Substrates inducing compressive strain/stress in polycrystalline thin films improve the dielectric permittivity and polarization in the parallel plate capacitor geometry. The importance of these results lies in the demonstration of the possibility to design the electrical performance of polar dielectric films for applications such as tunable devices and actuators, via the use of a simple cost effective process as sol-gel and commercially available substrates, without resorting to complex and expensive techniques and substrates.

Acknowledgements This work was developed within the scope of the project CICECO-Aveiro Institute of Materials, POCI-01-0145-FEDER-007679 (FCT Ref. UID/CTM/50011/2013), financed by national funds through the FCT/MEC and when appropriate co-financed by FEDER under the PT2020 Partnership Agreement as well as within FCT independent researcher grant IF/00602/2013. M. R. Soares is acknowledged for XRD strain measurements.

References

1. Müller KA, Burkard H (1979) $SrTiO_3$: an intrinsic quantum paraelectric below 4 K. Phys Rev B 19:3593–3602
2. Müller KA (1959) Electron paramagnetic resonance of manganese IV in $SrTiO_3$. Phys Rev Lett 2:341–343
3. Unoki H, Sakudo T (1967) Electron spin resonance of Fe^{3+} in $SrTiO_3$ with special reference to the 110°K phase transition. J Phys Soc Jpn 23:546–552
4. Shirane G, Yamada Y (1969) Lattice-dynamical study of the 110°K phase transition in $SrTiO_3$. Phys Rev 177:858–863
5. Vendik OG, Hollmann EK, Kozyrev AB, Prudan AM (1999) Ferroelectric tuning of planar and bulk microwave devices. J Supercond 12:325–338
6. Tagantsev AK, Sherman VO, Astafiev KF, Venkatesh J, Setter N (2003) Ferroelectric materials for microwave tunable applications. J Electroceram 11:5–66
7. Sirenko AA, Akimov IA, Fox JR, Clark AM, Li H-C, Si W, Xi XX (1999) Observation of the first-order Raman scattering in $SrTiO_3$ thin films. Phys Rev Lett 82:4500–4503
8. Pertsev NA, Tagantsev AK, Setter N (2000) Phase transitions and strain-induced ferroelectricity in $SrTiO_3$ epitaxial thin films. Phys Rev B 61:R825–R829
9. Astafiev K, Sherman V, Tagantsev A et al (2003) Shift of phase transition temperature in strontium titanate thin films. Integr Ferroelectr 59:1371–1379
10. Haeni JH, Irvin P, Chang W et al (2004) Room-temperature ferroelectricity in strained $SrTiO_3$. Nature 430:758–761
11. Nuzhnyy D, Petzelt J, Kamba S et al (2009) Soft mode behavior in $SrTiO_3/DyScO_3$ thin films: evidence of ferroelectric and antiferrodistortive phase transitions. Appl Phys Lett 95:232902
12. Li JF, Wang K, Zhu FY, Cheng LQ, Yao FZ (2013) $(K,Na)NbO_3$-based lead-free piezoceramics: fundamental aspects, processing technologies, and remaining challenges. J Am Ceram Soc 96:3677–3696
13. Saito Y, Takao H, Tani T et al (2004) Lead-free piezoceramics. Nature 432:84–87
14. Rafiq MA, Costa ME, Tkach A, Vilarinho PM (2015) Impedance analysis and conduction mechanisms of lead free potassium sodium niobate (KNN) single crystals and polycrystals: a comparison study. Cryst Growth Des 15:1289–1294
15. Rafiq MA, Tkach A, Costa ME, Vilarinho PM (2015) Defects and charge transport in Mn-doped $K_{0.5}Na_{0.5}NbO_3$ ceramics. Phys Chem Phys 17:24403–24411

16. Rodel J, Webber KG, Dittmer R, Jo W, Kimura M, Damjanovic D (2015) Transferring lead-free piezoelectric ceramics into application. J Eur Ceram Soc 35:1659–1681
17. Tanaka K, Hayashi H, Kakimoto KI, Ohsato H, Iijima T (2007) Effect of (Na,K)—excess precursor solutions on alkoxy-derived (Na,K)NbO_3 powders and thin films. Jpn J Appl Phys 46:6964–6970
18. Ahn CW, Lee SY, Lee HJ, Bae JS, Jeong ED, Choi JS (2009) The effect of K and Na excess on the ferroelectric and piezoelectric properties of $K_{0.5}Na_{0.5}NbO_3$ thin film. J Phys D Appl Phys 42:215304
19. Yan X, Ren W, Wu X, Shi P, Yao X (2010) Lead-free (K,Na)NbO_3 ferroelectric thin films: Preparation, structure and electrical properties. J Alloy Compd 508:129–132
20. Kang C, Park JH, Shen D, Ahn H, Park M, Kim D-J (2010) Growth and characterization of ($K_{0.5}Na_{0.5}$)NbO_3 thin films by a sol-gel method. J Sol-Gel Sci Technol 58:85–90
21. Kupec A, Malic B, Tellier J, Tchernychova E, Glinsek S, Kosec M (2012) Lead-free ferroelectric potassium sodium niobate thin films from solution: composition and structure. J Am Ceram Soc 95:515–523
22. Yu Q, Li J-F, Sun W, Zhou Z, Xu Y, Xie Z-K, Lai F-P, Wang Q-M (2013) Electrical properties of $K_{0.5}Na_{0.5}NbO_3$ thin films grown on Nb:$SrTiO_3$ single-crystalline substrates with different crystallographic orientations. J Appl Phys 113:024101
23. Vendrell X, Raymond O, Ochoa DA, García JE, Mestres L (2015) Growth and physical properties of highly oriented La-doped (K,Na)NbO_3 ferroelectric thin films. Thin Solid Films 577:35–41
24. Deng Q, Zhang J, Huang T, Xu L, Jiang K, Li Y, Hu Z, Chu J (2015) Optoelectronic properties and polar nano-domain behavior of sol-gel derived $K_{0.5}Na_{0.5}Nb_{1-x}Mn_xO_{3-\delta}$ nanocrystalline films with enhanced ferroelectricity. J Mater Chem C 3:8225–8234
25. Won SS, Lee J, Venugopal V et al (2016) Lead-free Mn-doped ($K_{0.5}$,$Na_{0.5}$)NbO_3 piezoelectric thin films for MEMS-based vibrational energy harvester applications. Appl Phys Lett 108:232908
26. Weng C, Tsai C, Hong C, Lin C (216) Effects of non-stoichiometry on the microstructure, oxygen vacancies, and electrical properties of KNN-based thin films. ECS J. Solid State Sci Technol 5:49–56
27. Malic B, Razpotnik H, Koruza J, Kokalj S, Cilensek J, Kosec M (2011) Linear thermal expansion of lead-free piezoelectric $K_{0.5}Na_{0.5}NbO_3$ ceramics in a wide temperature range. J Am Ceram Soc 94:2273–2275
28. Tkach A, Okhay O, Reaney IM, Vilarinho PM (2018) Mechanical strain engineering of dielectric tunability in polycrystalline $SrTiO_3$ thin films. J Mater Chem C 6:2467–2475
29. Okhay O, Tkach A, Wu A, Vilarinho PM (2013) Manipulation of dielectric permittivity of sol-gel $SrTiO_3$ films by deposition conditions. J Phys D Appl Phys 46:505315
30. Tkach A, Santos A, Zlotnik S, Serrazina R, Okhay O, Bdikin I, Costa ME, Vilarinho PM (2018) Strain-mediated substrate effect on the dielectric and ferroelectric response of potassium sodium niobate thin films. Coatings 8:449
31. Wachtman JB Jr, Wheat ML, Marzullo S (1963) A method for determining the elastic constants of a cubic crystal from velocity measurements in a single arbitrary direction; application to $SrTiO_3$. J Res NBS 67:A193–A209
32. Jeager RE, Egerton L (1962) Hot pressing of potassium sodium niobates. J Am Ceram Soc 45:209–213
33. Egerton L, Dillon DM (1959) Piezoelectric and dielectric properties of ceramics in the system potassium sodium niobate. J Am Ceram Soc 42:438–442
34. Zhang J, Weiss CV, Alpay SP (2011) Effect of thermal stresses on the dielectric properties of strontium titanate thin films. Appl Phys Lett 99:042902
35. Levin I, Krayzman V, Cibin G, Tucker MG, Eremenko M, Chapman K, Paul RL (2017) Coupling of emergent octahedral rotations to polarization in (K,Na)NbO_3 ferroelectrics. Sci Rep 7:15620

Determination of the Percolation Threshold for $(FeCoZr)_x(SiO_2)_{(100-x)}$ Nanocomposite

V. Bondariev

Abstract This work presents an explanation of the effect of metallic phase content x on AC electrical properties of nanocomposites $(FeCoZr)_x(SiO_2)_{(100-x)}$ and determination of the percolation threshold for series of nanocomposite samples with metallic phase content $x = 48.3$–82.8 at.%. Frequency and temperature dependencies of conductivity $\sigma(f, T)$ and phase shift angle $\theta(f, T)$ for series of nanocomposite produced by ion-beam sputtering in pure argon atmosphere were determined. Samples with metallic phase content $x < 79.8$ at.% exhibit dielectric type of conductivity $d\sigma/dT > 0$. Increase in concentration of metallic phase $x \geq 79.8$ at.% causes a change of conductivity character from dielectric to metallic $d\sigma/dT < 0$. Comparison of the conductivity at $T_p = 20$ K and conductivity at room temperature allowed to determine the percolation threshold for a series of samples of nanocomposite $(FeCoZr)_x(SiO_2)_{(100-x)}$. Metallic phase particles form the "endless cluster" that crosses though sample from the electrode to electrode in nanocomposites with $x \geq 80$ at.%.

Keywords Nanocomposites · Conductivity · Percolation · Metallic phase

1 Introduction

Heterogeneous systems (HS) including composite materials are widely used in different industries [1–8]. Unique properties of nanocomposites have found applications in mechanical engineering [9, 10], optics [11, 12], electronics [13, 14]. One of the most important factors determining the properties of nanocomposites is their structure and composition. For example, the shape, and size of metallic phase particles, and disperse phase distribution. Also, when metallic phase particles with a size in the range nano- are separated by insulating material, the quantum-mechanical phenomena [15] can occur in the nanocomposite [16–20].

V. Bondariev (✉)
Department of Electrical Devices and High Voltage Technology,
Lublin University of Technology, Lublin, Poland
e-mail: v.bondariev@pollub.pl

© Springer Nature Singapore Pte Ltd. 2019

A. D. Pogrebnjak and V. Novosad (eds.), *Advances in Thin Films, Nanostructured Materials, and Coatings*, Lecture Notes in Mechanical Engineering,
https://doi.org/10.1007/978-981-13-6133-3_33

To understand the nature of the origin of the unique electrical properties of nanocomposites, it is worth examining the mechanism of transferring electric charges in it [21]. The effective electrical conductivity σ is the main characteristic describing charge transfer in a randomly inhomogeneous heterogeneous system. For example, for binary metal-dielectric nanocomposites the effective electrical conductivity σ is primarily determined by the electrical conductivity of the σ_i components and their volume content V. A binary system is considered highly inhomogeneous if the electrical conductivity of the dispersion (dielectric matrix) phase σ_d is very different (much less or much greater) of the electrical conductivity of the dispersed (metal) phase σ_m. Effective electrical conductivity σ of metal-dielectric nanocomposite can change from dielectric σ_d to metallic σ_m depending on the ratio of phases contents, and this change can reach up to 10 orders of magnitude. The greatest increase in σ is observed at a certain metallic phase concentration $x \approx x_c$, determined by the percolation theory as the percolation threshold [22, 23]. Metallic phase particles form the "endless cluster" that crosses though sample from the electrode to electrode in nanocomposites. In real heterogeneous systems, phase layers are always present between the phase particles, which are characterized by their local conductivity values, and therefore $\sigma(r)$ can vary continuously over a wide range. And the transfer of charge between nanoparticles isolated between them is carried out by hopping mechanism of charge transfer [24, 25].

The use of impedance spectroscopy of heterogeneous systems (for example nanocomposites) based on the measurement and analysis of the dependence of the impedance on the frequency of alternating current makes it possible to determine the structural features, microscopic parameters and other characteristics of the HS [26–30]. The frequency dependencies of the impedance of the metal-dielectric nanocomposites are determined by many physical causes, some of which appear at the same frequencies.

1. Electronic relaxation processes at the interface boundaries.
2. Dispersion of electrical characteristics (specific conductance and permittivity) of the phases entering into the nanocomposite.
3. Relaxation processes at the interface boundaries, associated with ion transport and electrochemical reactions.
4. The presence of structural (having a certain geometry) elements with reactive impedance.

All these physical causes are those factors that determine the unique properties of nanocomposites and the phenomena that occur in them.

Most often the original properties of nanocomposites are shown when the concentration of the metal phase x is about the percolation threshold x_c. Determination of percolation threshold is an important task in the study of nanocomposites and their properties.

In previous works, the percolation threshold for nanocomposite $(FeCoZr)_x(CaF_2)_{(100-x)}$ samples sputtered in the pure argon [31] and mixed (argon and oxygen) [32] atmosphere was determined. The percolation threshold

x_c for the series of oxygen samples, compared to argon samples ($x_c \approx (65.8 \pm 2.0)$ at.%) is shifted to the value (82 ± 2.0) at.%.

Annealing of the nanocomposite $(FeCoZr)_x(CaF_2)_{(100-x)}$ sample with metallic phase content $x = 62.7$ at.% (x near x_c) in the temperature of 398 K and higher intensifies the coilless-like inductance phenomenon, which results in frequency change causing an increase of the phase angle to above ± 90 [33].

This work aims to the explain the effect of metallic phase content x on AC electrical properties of nanocomposites $(FeCoZr)_x(SiO_2)_{(100-x)}$ and determination of the percolation threshold for series of nanocomposite samples with metallic phase content $x = 48.3$–82.8 at.%.

2 Experimental Details

Nanocomposite $(FeCoZr)_x(SiO_2)_{(100-x)}$ has been produced by ion-beam sputtering in a pure argon atmosphere. As a result, a series of samples with a variable content of metallic phase concentration $x = 48.3$–82.8 at.% were obtained. Determination of their chemical composition was performed by X-ray microanalysis in the scanning electron microscope. Immediately after obtaining the nanocomposite samples, their AC electrical properties were tested by using the test station for frequency-domain dielectric spectroscopy of nanocomposites and semiconductors [34]. The station allows measuring a wider measurement temperature range $T_p = 20$–373 K and in the frequency range from 50 Hz to 5 MHz. Frequency and temperature dependencies on conductivity $\sigma(f, T)$ and phase shift angle $\theta(f, T)$ for series of nanocomposite were determined.

3 Results and Discussion

Figure 1 shows the frequency-temperature dependence of conductivity for the nanocomposite $(FeCoZr)_x(SiO_2)_{(100-x)}$ with metallic phase content $x = 58.15$ at.% for some selected measurement temperatures. As can be seen from Fig. 1, in the low frequencies area conductivity hardly depends on the frequency. In the high-frequency area (started from ~2000 Hz) the conductivity value is almost constant. The conductivity increased by almost one order of magnitude with increasing measurement temperature T_p from 34 to 347 K. This means that a dielectric conductivity type occurs in the material, for which the derivative of conductivity after temperature $d\sigma/dT_p$ takes positive values.

The frequency-temperature dependencies of phase angle θ for the same nanocomposite sample is shown on Fig. 2. From the figure, it can be seen that the frequency dependence of the phase shift angle behaves according to the model of impedance in nanocomposites with a hopping mechanism of charge transfer. Namely, in the area of low frequencies the phase shift angle has a value close to zero. As the frequency

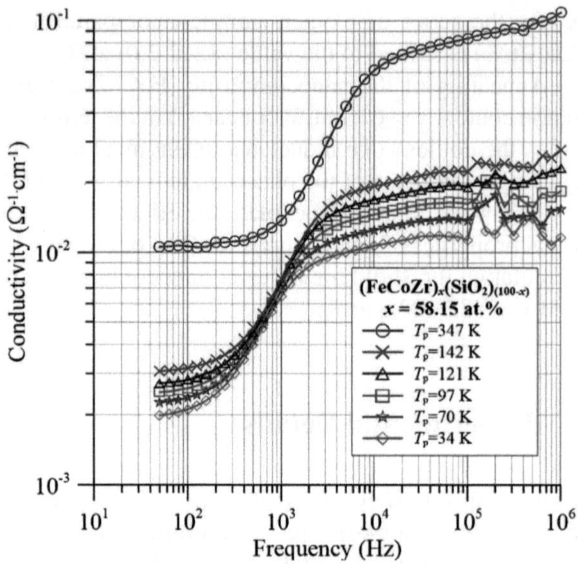

Fig. 1 Frequency dependences of conductivity σ for nanocomposite $(CoFeZr)_x(SiO_2)_{(100-x)}$ with the metallic-phase content of $x = 58.15$ at.%

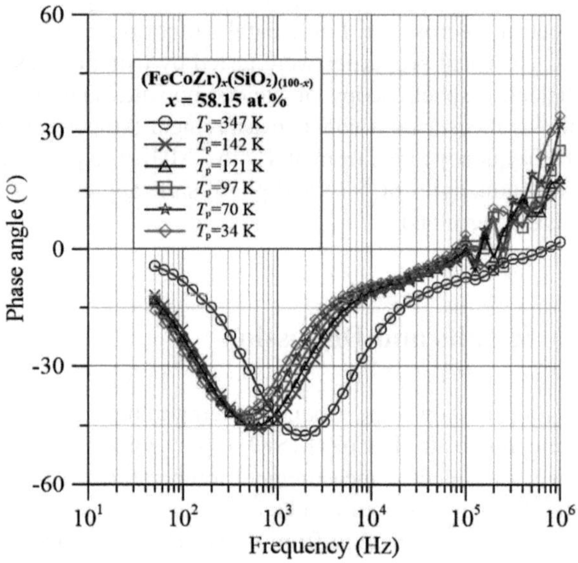

Fig. 2 Frequency dependences of phase-shift angle θ for nanocomposite $(CoFeZr)_x(SiO_2)_{(100-x)}$ with the metallic-phase content of $x = 58.15$ at.%

increases, it assumes ever greater negative values until a minimum is reached. Further increase in frequency causes the phase shift angle to approach zero and then go into the area of positive values. That is, in the area of low frequencies conductivity is of the resistive type. In the intermediate frequency range conductivity of the capacitive type is dominant. In yet higher frequencies the inductive type conductivity prevails.

In samples of a nanocomposite from the same series but with a higher concentration of the metal phase, a different type of conductivity was observed. For example, for a sample with a metallic phase content of $x = 79.77$ at.% with the rise of the measurement temperature T_p, the conductivity decreases (Fig. 3), which indicates a metallic type of conduction $d\sigma/dT < 0$. In addition, conductivity in almost the entire range does not depend on the frequency, only at 0.4 MHz a slight increase in the conductivity is visible.

And for the sample with $x = 82.81$ at.% concentration no temperature and frequency dependencies were detected (Fig. 4). On the frequency dependence of the phase shift angle for the same sample (Fig. 5) just a weak decrease of θ to ~−5° can be observed. Therefore, the following method was chosen to determine the percolation threshold.

To determine the percolation threshold for series of samples $(FeCoZr)_x(SiO_2)_{(100-x)}$ sputtering in the pure argon atmosphere the dependence of the conductivity on metallic phase content for measurement temperature $T_p = 20$ K and room temperature $T_p = 298$ K for frequency 100 Hz was determinate (Fig. 6). Both of these waveforms have been approximated, and the intersection point of the approximating lines was determined. As can be seen from Fig. 6 conductivity in the range of metallic phase content from 48.3 to 82.8 at.% increases by almost six orders of magnitude. The value at which the approximation line intersects must be accepted as a percolation threshold x_c of series of nanocomposite $(FeCoZr)_x(SiO_2)_{(100-x)}$ samples, and this value located at the metallic phase content about (79.8 ± 2) at.%.

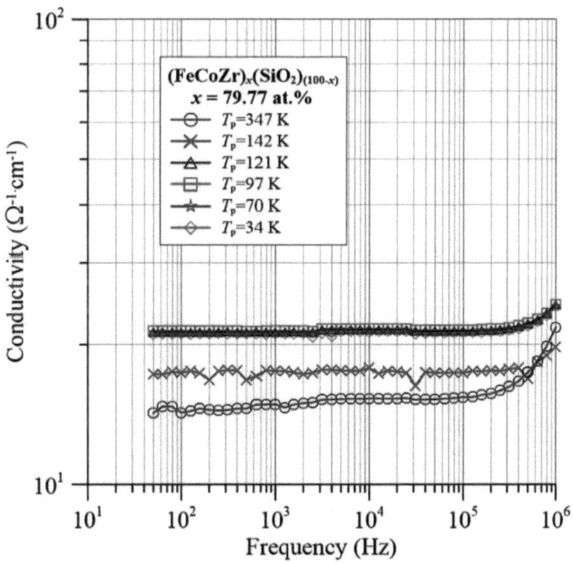

Fig. 3 Frequency dependences of conductivity σ for nanocomposite $(CoFeZr)_x(SiO_2)_{(100-x)}$ with the metallic-phase content of $x = 79.77$ at.%

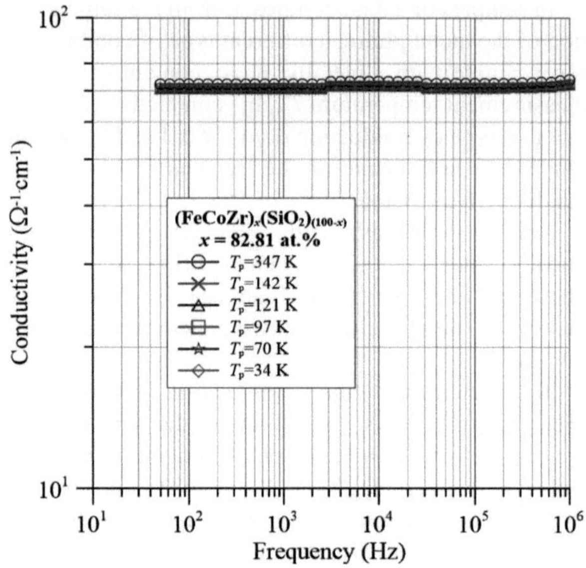

Fig. 4 Frequency dependences of conductivity σ for nanocomposite $(CoFeZr)_x(SiO_2)_{(100-x)}$ with the metallic-phase content of $x = 82.81$ at.%

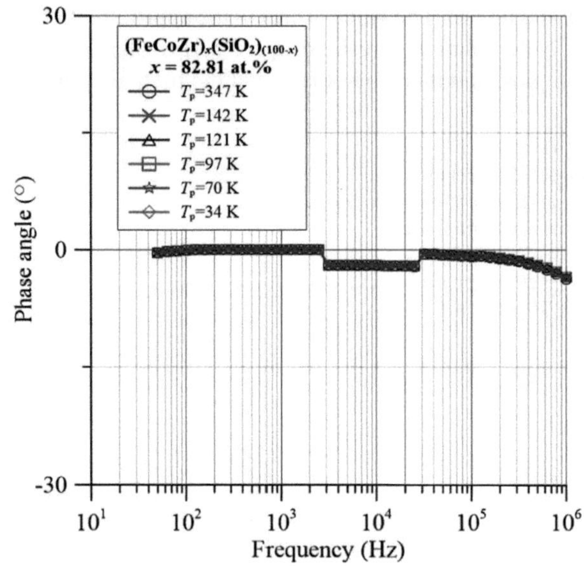

Fig. 5 Frequency dependences of phase-shift angle θ for nanocomposite $(CoFeZr)_x(SiO_2)_{(100-x)}$ with the metallic-phase content of $x = 82.81$ at.%

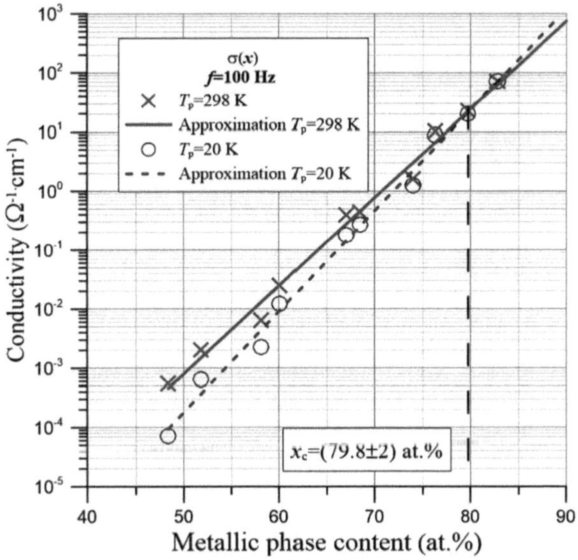

Fig. 6 Dependence of the conductivity of nanocomposite $(FeCoZr)_x(SiO_2)_{(100-x)}$ on metallic phase content for measurement temperatures T_p 20 and 298 K for frequency 100 Hz

4 Conclusions

This study examined the frequency and temperature dependences of the AC parameters of $(FeCoZr)_x(SiO_2)_{(100-x)}$ nanocomposites with the metal phase content within the scope of 48.3 at.% < x < 82.8 at.%, produced by sputtering a beam of pure argon. Frequency and temperature dependences of conductivity $\sigma(f, T)$ show that for the content of the metallic phase under ~79.8 at.% temperature dependences of the dielectric type occur, i.e., the conductivity increases with increasing temperature. From the frequency dependences of phase shift angle $\theta(f, T)$ can be seen that in nanocomposite the transfer of charge between nanoparticles is carried out by hopping mechanism of charge transfer. Increase in concentration of metallic phase $x \geq$ 79.8 at.% causes a change of conductivity character from dielectric to metallic.

Based on the analysis of the conductivity dependence on the content of the metallic phase, the percolation threshold was determined for the $(FeCoZr)_x(SiO_2)_{(100-x)}$ nanocomposite, obtained by ion sputtering of a pure argon beam, the value of which is $x_c \approx (79.8 \pm 2)$ at.%.

References

1. Komarov FF, Zukowski P, Kryvasheyeu RM et al (2015) Effects of surfactant and fabrication procedure on the electrical conductivity and electromagnetic shielding of single-walled carbon nanotube films. Phys Status Solidi Appl Mater Sci 212:425–432. https://doi.org/10.1002/pssa.201431493

2. Pogrebnjak AD, Berestok TO, Opanasyuk AS et al (2014) Structural properties and elemental composition of Au^{+} implanted ZnO films, obtained by sol-gel method. J Nano- Electron Phys 6:1–5
3. Astashynskaya MV, Astashynski VV, Kvasov NT et al (2016) Microstructure of amorphous copper-carbon thin films formed by plasma-enhanced chemical vapor deposition. High Temp Mater Process Int Q High-Technol Plasma Process 20:115–126. https://doi.org/10.1615/HighTempMatProc.2016017797
4. Boiko O (2017) Ferromagnetic alloy–ferroelectric ceramic nanocomposites for nanoelectronics: the influence of heat treatment on electrical properties. High Temp Mater Process Int Q High-Technol Plasma Process 21:251–259. https://doi.org/10.1615/HighTempMatProc.2018025554
5. Fedotova J, Kasiuk J, Przewoznik J et al (2011) Effect of oxide shells on the magnetic and magnetotransport characteristics of oxidized FeCoZr nanogranules in Al_2O_3. J Alloys Compd 509:9869–9875
6. Kopia A, Nowakowski P, Suliga I et al (2006) Porównanie własności katalitycznych nanokrystalicznych tlenków ceru CeO2 i rutenu RuO2. Inżynieria Mater 27:1069–1072
7. Judy JH (2005) Advancements in PMR thin-film media. J Magn Magn Mater 287:16–26. https://doi.org/10.1016/j.jmmm.2004.10.004
8. Timopheev AA, Ryabchenko SM, Kalita VM et al (2011) Growth-induced perpendicular anisotropy of grains in Co-Al-O nanogranular ferromagnetic films. Phys Solid State 53:494–503. https://doi.org/10.1134/S1063783411030309
9. Yatsun EI, Ageev EV, Ugrimov AS (2017) Preparation of protective nanocomposite coatings by ultra-fast gas-dynamic formation using hard alloy electro-erosion powders. Chem Pet Eng 52:774–778. https://doi.org/10.1007/s10556-017-0269-y
10. Beresnev VM, Sobol' OV, Pogrebnjak AD et al (2017) Structure and properties of vacuum arc single-layer and multiperiod two-layer nitride coatings based on Ti(Al): Si layers. J Nano-Electron Phys 9:1033–1033. https://doi.org/10.21272/jnep.9(1).01033
11. Wang TL, Yang CH, Shieh YT et al (2009) Synthesis of CdSe-poly(N-vinyl carbazole) nanocomposite by atom transfer radical polymerization for potential optoelectronic applications. Macromol Rapid Commun 30:1679–1683. https://doi.org/10.1002/marc.200900349
12. Pantsialeyeu KU, Krautsevich AU, Rovba IA et al (2017) Analysis of the electrophysical and photoelectric properties of nanocomposite polymers by the modified Kelvin probe. Devices Methods Meas 8:386–397. https://doi.org/10.21122/2220-9506-2017-8-4-55-62
13. Haiduk YS, Reutskaya OG, Savitsky AA et al (2016) Micropower gas sensor based on the composition tungsten oxide and multiwall carbon nanotubes. Devices Methods Meas 7:41–49. https://doi.org/10.21122/2220-9506-2016-7-1-60-69
14. Shklovskii BI, Efros AL (1985) Electronic properties of doped semiconductors. Springer, Berlin, Heidelberg, New York, Tokyo
15. Jönsson BJ, Turkki T, Ström V et al (1996) Oxidation states and magnetism of Fe nanoparticles prepared by a laser evaporation technique. J Appl Phys 79:5063–5065. https://doi.org/10.1063/1.361572
16. Kravchenko YO, Coy LE, Peplińska B et al (2018) Nano-multilayered coatings of (TiAlSiY)N/MeN (Me = Mo, Cr and Zr): influence of composition of the alternating layer on their structural and mechanical properties. J Alloy Compd 767:483–495
17. Pogrebnjak AD, Beresnev VM, Bondar OV et al (2018) Superhard CrN/MoN coatings with multilayer architecture. Mater Des 153:47–59
18. Smyrnova KV, Pogrebnjak AD, Beresnev VM et al (2018) Microstructure and physical-mechanical properties of (TiAlSiY)N nanostructured coatings under different energy conditions. Met Mater Int 24(5):1024–1035
19. Pogrebnjak AD, Ivashchenko VI, Skrynskyy PL et al (2018) Experimental and theoretical studies of the physicochemical and mechanical properties of multi-layered TiN/SiC films: temperature effects on the nanocomposite structure. Compos B Eng 142:85–94
20. Pogrebnjak AD, Beresnev VM, Smyrnova KV et al (2018) The influence of nitrogen pressure on the fabrication of the two-phase superhard nanocomposite (TiZrNbAlYCr)N coatings. Mater Lett 211:316–318

21. Stauffer D, Aharony A (1994) Introduction to percolation theory. Taylor and Francis, London
22. Efros AL, Shklovskii BI (1976) Critical behavior of conductivity and dielectric constant near the metal-nonmetal transition threshold. Phys Status Solidi 76:475–485. https://doi.org/10.1002/pssb.2220760205
23. Grimmett G (1999) Percolation. Springer, Berlin
24. Zukowski P, Kołtunowicz T, Partyka J et al (2009) Electrical properties of nanostructures $(CoFeZr)_x + (Al2O3)_{1-x}$ with use of alternating current. Vacuum 83:S275–S279. https://doi.org/10.1016/j.vacuum.2009.01.081
25. Zukowski P, Kołtunowicz T, Partyka J et al (2007) Dielectric properties and model of hopping conductivity of GaAs irradiated by H^+ ions. Vacuum 81:1137–1140. https://doi.org/10.1016/j.vacuum.2007.01.070
26. Chockalingam R, Basu S (2011) Impedance spectroscopy studies of Gd-CeO_2-$(LiNa)CO_3$ nano composite electrolytes for low temperature SOFC applications. Int J Hydrogen Energy 36:14977–14983. https://doi.org/10.1016/j.ijhydene.2011.03.165
27. Svito I, Fedotova JA, Milosavljević M et al (2015) Influence of sputtering atmosphere on hopping conductance in granular nanocomposite $(FeCoZr)_x(Al_2O_3)_{1-x}$ films. J Alloys Compd 615:S344–S347. https://doi.org/10.1016/j.jallcom.2013.12.061
28. Koltunowicz TN, Zukowski P, Boiko O et al (2016) Capacitive properties of nanocomposite $(FeCoZr)_x(PZT)_{(100-x)}$ produced by sputtering with the use of argon and oxygen ions beam. J Mater Sci Mater Electron 27:1171–1176. https://doi.org/10.1007/s10854-015-3868-4
29. Koltunowicz TN, Zukowski P, Bondariev V et al (2015) Voltage and current resonance in nanocomposite $(FeCoZr)_x(CaF_2)_{100-x}$ produced by ion-beam sputtering in pure argon atmosphere. Acta Phys Pol A 128:897–900
30. Boiko O, Koltunowicz TN, Zukowski P et al (2017) The effect of sputtering atmosphere parameters on dielectric properties of the ferromagnetic alloy—ferroelectric ceramics nanocomposite $(FeCoZr)_x(PbZrTiO_3)_{(100-x)}$. Ceram Int 43:2511–2516. https://doi.org/10.1016/j.ceramint.2016.11.052
31. Zukowski P, Koltunowicz TN, Bondariev V et al (2016) Determining the percolation threshold for $(FeCoZr)_x(CaF_2)_{(100-x)}$ nanocomposites produced by pure argon ion-beam sputtering. J Alloys Compd 683:62–66. https://doi.org/10.1016/j.jallcom.2016.05.070
32. Bondariev V (2017) Percolation threshold of granular metal-dielectric nanocomposites $(FeCoZr)_x(CaF_2)_{1-x}$ produced in atmosphere of argon and oxygen. In: Romaniuk RS, Linczuk M (eds) International society for optics and photonics, p 104455A
33. Koltunowicz TN, Zukowski P, Bondariev V et al (2015) The effect of annealing on the impedance of $(FeCoZr)_x(CaF_2)_{(100-x)}$ nanocomposite films produced by the ion-beam sputtering in vacuum. Vacuum 120:44–50. https://doi.org/10.1016/j.vacuum.2015.01.030
34. Koltunowicz TN (2015) Test station for frequency-domain dielectric spectroscopy of nanocomposites and semiconductors. J Appl Spectrosc 82:653–658. https://doi.org/10.1007/s10812-015-0158-0

Mechanisms of Surface State Formation at Si/SiO$_2$ Interface in the Nanosized MOS Transistors

A. N. Volkov, D. V. Andreev and V. M. Maslovsky

Abstract The paper demonstrates the main physical mechanisms of the surface state generation at the interface Si/SiO$_2$, which are typical for nanosized MOS transistors. We demonstrate the most common model of the surface state generation. One analyzes graphs showing a dependence of lifetime on substrate current (I_{sub}), which are obtained from the literature. We demonstrate a method to determine charge carrier energy participating in the process of surface state generation and a method to ascertain the mechanism of surface state generation in nanosized MOS transistors. We ascertain main parameters of MOS transistors affecting the process of surface state generation.

Keywords Surface states · Si/SiO$_2$ interface · MOSFET · Si–H bonds · MOS transistors lifetime

1 Introduction

The interface between bulk silicon and its dioxide (Si/SiO$_2$) has a large concentration of the Si–H bonds formed after the formation of thin films of silicon dioxide during annealing in a hydrogen-containing atmosphere. The breakdown of Si–H bonds can lead to the process of formation of surface states at Si/SiO$_2$ interface of the nanosized metal-oxide-semiconductor-field-effect-transistors (MOSFETs). This process can proceed according to several mechanisms [1, 2], depending on the energy of the charge carriers initiating the dissociation reaction of the Si–H bond. The surface

A. N. Volkov
Research Institute of Physical Problems, Zelenograd, Russia

D. V. Andreev (✉)
Bauman Moscow State Technical University, The Kaluga Branch, Kaluga, Russia
e-mail: dandreeveic@gmail.com

V. M. Maslovsky
Moscow Institute of Physics and Technology (State University), Dolgoprudnyi, Moscow Region, Russia

© Springer Nature Singapore Pte Ltd. 2019
A. D. Pogrebnjak and V. Novosad (eds.), *Advances in Thin Films, Nanostructured Materials, and Coatings*, Lecture Notes in Mechanical Engineering,
https://doi.org/10.1007/978-981-13-6133-3_34

states can significantly affect the lifetime of the MOSFETs and devices based on them.

2 Generation of Surface States in Nanosized MOS Transistors

2.1 *Physical Mechanisms of Surface State Generation*

A well-known fact is that one of the main processes of surface state generation in MOS transistors is the dissociative reaction of Si–H bonds at the Si/SiO_2 interface. The mechanism, caused by the influence of charge carriers with energies sufficient to initialize the dissociative reaction of Si–H bonds, is the catalyst of this reaction for the MOS transistors having the long channel. This mechanics is well described by a concept named "lucky electron model" (LEM) [1]. In accordance with this concept, the charge carriers obtain sufficient for Si–H bond cutting energy by the influence of lateral electric field in the MOS transistor channel caused by the relatively high supply voltage. Thus, for nanosized MOS transistors with the relatively low strength of the lateral electric field in the channel, by LEM, the generation of surface states should come to a minimum. However, as practice shows, the generation of surface states in nanosized MOS transistors as before is one of the main reasons of failure of electric devices based on them.

By papers [1, 2], mechanisms, which are responsible for the dissociative process of Si–H bonds, and, as a consequence, responsible for the surface state generation, do not depend on the amount of the lateral electric field in the channel. The basis of these mechanisms is physical aspect of the distribution of charge carriers by energies featured for every mechanism. Papers [1, 2] describe three main mechanisms of the surface state generation as a result of Si–H bond cutting at the Si/SiO_2 interface, depending on charge carrier energy, featured for MOS transistors:

1. The first one is a mechanism of single excitation (SE)—bond cutting by one electron having enough energy (energy of electrons ~3.7 eV);
2. The second one is electron-electron scattering (EES)—bond cutting by an electron that has received sufficient energy due to scattering (energy of electrons $\sim 3.6 < E < 2.5$ eV);
3. The third one is a mechanism of multi-vibrational excitation (MVE)—bond cutting due to the influence of electrons group having low energy (energy of electrons $\sim E < 2.5$ eV).

Independently on mechanism of surface state generation, increasing of their concentration at Si/SiO2 interface leads to the degradation of electrical characteristics of MOS transistors (drain current, slope, threshold voltage, etc.), what in its turn leads to parametric failure of devices based on nanosized MOS transistors, and, as a consequence, to reduction of their lifetime [3–5].

2.2 Empirical Model of Surface State Formation

To model processes of surface state generation, and, as a consequence, to predict the lifetime of nanosized MOS transistors, there is a significant amount of empirical models [1, 6]. One of the most known models is the model of the dependence of MOS transistors lifetime on substrate current (I_{sub}) [6]:

$$t_{\text{L}} = A \cdot (I_{\text{sub}})^{-m}, \tag{1}$$

where A and m are empirical coefficients.

By paper [1], $m = \varphi_{\text{it}}/\varphi_{\text{i}}$, where φ_{it}—the energy of charge carriers taking part in the process of surface state generation; φ_{i}—1.3–1.4 eV—threshold (critical) energy to create the impact ionization.

The basis of this model is an assumption that the charge carriers, taking part in the process of creation of surface states by Si–H bond cutting, drain into the substrate creating substrate current due to the impact ionization. Observing the amount of the substrate current and the level of degradation of the electrical parameter of MOS transistor (drain current, slope, etc.) allow concluding about the intensity of the process of creation of surface states in MOS transistor.

Plotting of graphs, showing dependence of lifetime (t_{L}) on substrate current (I_{sub}) using results of accelerated tests, at which the operating conditions are selected, characterized by maximal amount of the substrate current, allows to clearly demonstrate an influence of the process of surface state generation on the degradation of MOS transistor characteristics.

These graphs give an estimate not only about the influence of the process of surface state generation on the lifetime but also can allow to determine:

1. The energy of the charge carriers taking part in the process of surface state generation.
2. Physical mechanism featured for the given process of generation of surface states.
3. Parameters of MOS transistors influencing the process of surface state generation.

To verify these assumptions, we analyzed the graphs of the dependence of lifetime (t_{L}) on substrate current (I_{sub}) borrowed from the papers [7–9].

3 Description of Samples

We analyzed the following graphs:

1. Graph of the dependence of the lifetime (t_{L}) on substrate current (I_{sub}) of n-channel MOS transistors made by 0.35 μm CMOS technology with lightly doped drain (LDD) and large angle tilt implanted drain (LATID) architectures. Samples had different length of the channel $L_{\text{m}} = 0.4\text{–}1$ μm (channel mask length), channel width $W = 40$ μm, thickness of oxide film $T_{ox} = 8$ nm, and length of

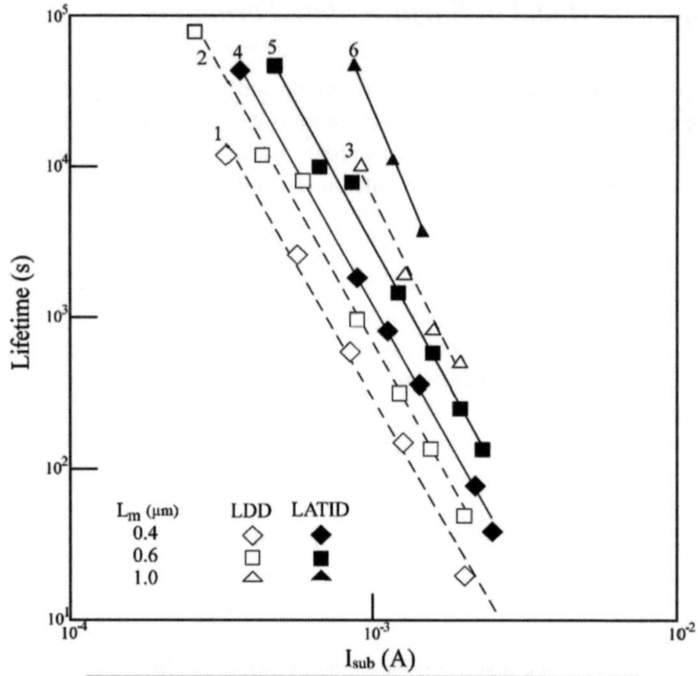

1) $Lifetime_1 = 1{,}04 \cdot 10^{-8} \cdot (I_{sub})^{-3{,}4}$	4) $Lifetime_4 = 1{,}79 \cdot 10^{-8} \cdot (I_{sub})^{-3{,}6}$
2) $Lifetime_2 = 1{,}13 \cdot 10^{-8} \cdot (I_{sub})^{-3{,}6}$	5) $Lifetime_5 = 1{,}16 \cdot 10^{-8} \cdot (I_{sub})^{-3{,}8}$
3) $Lifetime_3 = 5{,}39 \cdot 10^{-8} \cdot (I_{sub})^{-3{,}6}$	6) $Lifetime_6 = 0{,}02 \cdot 10^{-8} \cdot (I_{sub})^{-4{,}6}$

Fig. 1 Dependence of lifetime on the substrate current for samples made by CMOS 0.35 μm technology with different architectures (LDD, LATID)

spacer $Ls = 0.12$ μm. The only difference between LATID and conventional LDD devices consisted in performing the LDD implant at $\theta = 42°$ angle implant for LATID compared to the usual $\theta = 7°$ LDD angle implant [7]. As a criterion of loss of operating capacity, one chose the degradation of the maximum value of slope $\Delta G_m = 10\%$ (Fig. 1).

2. Graph of the dependence of the lifetime (t_L) on substrate current (I_{sub}) of n-channel MOS transistors made by 0.18 μm CMOS technology with different length of the channel $L_m = 0.5$, 0.25, and 0.18 μm, and channel width $W = 10$ μm [8]. As a criterion of loss of operating capacity, one chose the degradation of the maximum value of slope $\Delta G_m = 10\%$ (Fig. 2).
3. Graph dependence of the lifetime (t_L) on reduced substrate current (I_{sub}/W) of n- and p-channel MOS transistors made by 40 nm CMOS technology with different length of the channel $L_m = 35$, 50, 60 nm and 0.14 and 0.16 μm for n-channel MOS transistors and $L_m = 0.14$ and 0.16 μm for p-channel MOS transistors, and channel width $W = 10$ μm [9]. As a criterion of loss of operating capacity, one chose the degradation of drain current $\Delta I_{ds} = 10\%$ (Fig. 3).

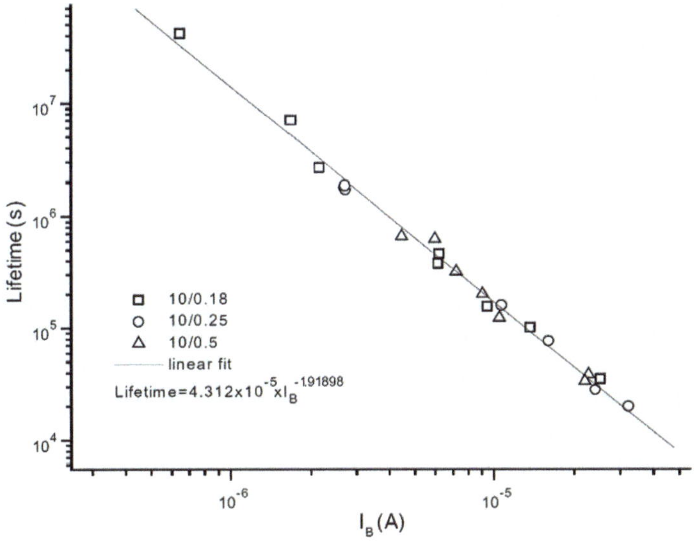

Fig. 2 Dependence of the lifetime on substrate current for the samples made by CMOS 0.18 μm with different channel length

Fig. 3 Dependence of the lifetime (t_L) on reduced substrate current (I_{sub}/W) of n- and p-channel MOS transistors with different lengths of channel and different thicknesses of oxide film

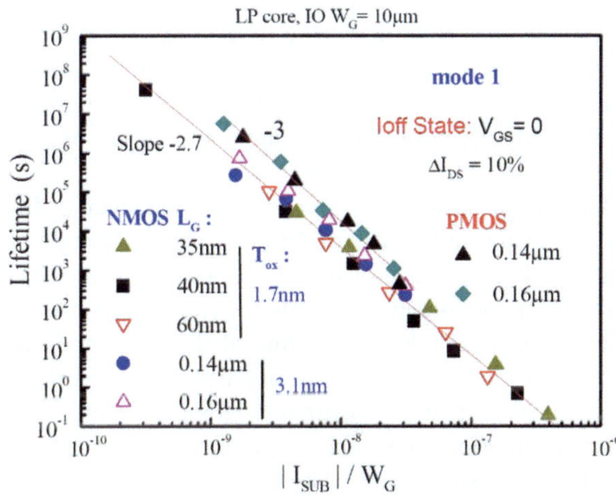

4 Analysis Results

Figure 1 shows graphs of the dependence of the lifetime (t_L) on substrate current (I_{sub}) for the samples with different channel length and architecture (LDD and LATID). Figure 1 demonstrates that the inclinations of lines, representing given dependencies, are almost the same. Proceeding from the expression (1), the empirical coefficient m, causing the line inclinations, is the ratio of the energy of charge carriers, taking part in the process of generation of surface states (φ_{it}), to the threshold (critical) energy of creation of the impact ionization ($\varphi_i = 1.3$ eV in accordance with the paper [1]). By determining from the graphs the dependencies of a lifetime (t_L) on the substrate current, we can calculate the energy of charge carriers taking part in the process of generation of surface states (φ_{it}) by using the following expression:

$$\varphi_{it} = \mathrm{m} \cdot \varphi_i. \quad (2)$$

Thus, using the expression (2) and values of the m coefficient derived from the graphs of Fig. 1, we obtain values of the energies of charge carriers, taking part in the process of surface state generation, for the samples presented in Fig. 1 (4.42 eV for sample 1; 4.68 eV for samples 2, 3, and 4; 4.94 eV for sample 5; 5.98 eV for sample 6). Here and below, one calculates the energies of charge carriers, taking part in the process of generation of surface states (φ_{it}), at $\varphi_i = 1.3$ eV.

Figure 1 demonstrates that the energy of charge carriers for all the samples is higher than 3.7 eV. This can mean that for the case SE mechanism of surface state generation takes place.

It is important to notice that for this experiment with increasing of the length of the channel the empirical coefficient "A" increases, and, correspondingly, the lifetime of samples at the same values of the substrate current raises. This demonstrates that the decreasing of channel length leads to raising the intensity of the process of surface state generation and lowering of the lifetime. Besides, comparing the samples with the same length of the channel but different architecture (LDD and LATID), one can observe insignificant increasing of the "A" coefficient in case of LATID samples. This indirectly points at the change of intensity of process of surface state generation at altering of the implantation angle.

Figure 2 shows the graph of the dependence of lifetime (t_L) on substrate current (I_{sub}) for the samples with different channel length made by CMOS 0.18 μm technology. Figure 2 demonstrates that the experimental data for all the samples lay in frames of one line in contrast to the experiment presented in Fig. 1 including the participation of samples with different channel length. Proceeding from (2), the energy of charge carriers, taking part in the process of surface state generation (φ_{it}), is equal to 2.47 eV for this experiment. This can signify that the MVE process of surface state generation takes place for this case ($\varphi_{it} < 2.5$ eV).

Figure 3 shows the graphs of the dependence of the lifetime (t_L) on reduced substrate current (I_{sub}/W) of n- and p-channel MOS transistors made by 40 nm CMOS technology with different lengths of the channel and different thicknesses of the oxide film.

Figure 3 demonstrates that n- and p-channel samples have the different implantation angle. However, as it is for the second experiment, for this case the experimental data of the samples with different channel length and thickness of oxide lay in frames of one line. Thus, we can assume that for this case the length of channel and thickness of oxide do not influence the intensity of the process of surface state generation.

In accordance with (2), the energy of charge carriers, taking part in the process of surface state generation (φ_{it}), is equal to 3.51 eV for the n-channel samples and 3.9 eV for the p-channel samples for this experiment. Proceeding from this, one can assume that for the n-channel samples the EES mechanism ($2.5 < \varphi_{it} < 3.6$ eV) of surface state generation takes place whereas for the p-channel samples the SE mechanism ($\varphi_{it} > 3.6$ eV) of surface state generation takes place. Thus, using the expression (2) and analyzing the graphs obtained from the experimental data of the accelerated tests, one can determine the energy of charge carriers, taking part in the process of the surface state generation, and the process itself (SE, EES, MVE).

Analysis of the presented above experiments shows that the influence of geometrical sizes of the samples (length of the channel) on the process of surface state generation takes place only in the case of samples made by CMOS 0.35 μm technology. For the samples made by CMOS 0.18 μm and 40 nm technologies (which are more considered as nanosized MOS transistors) this dependence is not observed.

One should notice that all the experiments were realized in modes corresponding to the maximal substrate current. Evidently, for the different technologies, these modes will mean different electrical stresses. Thus, one can conclude that the main influence on the intensity of the process of surface state generation will be affected by the mode of accelerated tests (electrical parameters of the mode). At the same time, to accelerate each of the three mechanisms of surface state generation (ES, EES, and MVE) one needs to choose a certain mode of tests. The papers [1, 10–13] suggest the following to choose electrical parameters to accelerate ES, EES, and MVE mechanisms:

1. Use a low value of the drain current and low value of the voltage at the gate (ES mechanism).
2. Use the mean value of the drain current; mean value of the voltage at the drain and voltage at the gate (EES mechanism).
3. Use the high value of the drain current and low value of the voltage at the gate (MVE mechanism).

5 Conclusions

Modern nanosized MOS transistors and devices based on them, despite the LEM, are subjected by the influence of the process of surface state generation caused by Si–H bond cutting. The basis of this process for nanosized MOS transistors is the three physical mechanisms, such as ES, EES, and MVE. These mechanisms are characterized by the energy of charge carriers taking part in the process of surface state generation. A simple analysis of the graphs of the dependence of the lifetime (t_L) on reduced substrate current (I_{sub}/W), obtained from results of the accelerated tests, allows to approximately determine the energy of charge carriers taking part in the process of surface state generation. Besides, this analysis allows to assume the mechanism of the process. In addition, the analysis provides the parameters of MOS transistors influencing on the process of surface state generation. We ascertained that the length of the channel, the angle of the implantation, and modes of the accelerated tests could make an impact on the process of surface state generation.

References

Grasser T (2015) Hot carrier degradation in semiconductor devices. Springer, Switzerland

Guerin C, Huard V, Braviax A (2009) General framework about defect creation at the Si/SiO2 interface. J Appl Phys 105:114513

Andreev VV, Bondarenko GG, Maslovsky DV et al (2015) Modification and reduction of defects in thin gate dielectric of MIS devices by injection-thermal and irradiation treatments. Phys Status Solidi C 12(1–2):126–130

Andreev VV, Bondarenko GG, Maslovsky VM et al (2015) Modification of MOS devices by high-field electron injection and arc plasma jet treatment. Acta Phys Pol A 128(5):887–890

Andreev VV, Maslovsky VM, Andreev DV et al (2016) Method of stress and measurement modes for research of thin dielectric films of MIS structures. In: International conference on micro- and nano-electronics, vol 10224. SPIE, Dec 2016, p 1022429

Prabhakar M (2002) Characterization and modeling of hot carrier degradation in sub-micron n-MOSFETs. Nashville, Tennessee

Rafi JM, Campabadal F (2001) Hot-carrier degradation in deep-submicrometer nMOSFET's: lightly doped drain vs. large angle tilt implanted drain. Solid State Electron 45:1391

Cui Z, Liou JJ, Yue Y, Vinson J (2003) Empirical reliability modeling for 0.18-μm MOS devices. Solid State Electron 47:1515–1522

Bravaix A, Guerin C, Goguenheim D et al (2010) Off state incorporation into the 3 energy mode device lifetime modeling for advanced 40 nm CMOS node. In: Proceedings of the IEEE International Reliability Physics Symposium 2010, p 55

Guérin C, Huard V, Bravaix A (2007) The energy driven hot-carrier degradation modes in NMOSFETs. IEEE Trans Device Mater Reliability 7(2):225–235

Ivashchenko VI, Pogrebnjak AD, Sobol' OV et al (2015) The effect of Al target current on the structure and properties of (Nb2Al) N films with an amorphous AlN phase. Tech Phys Lett 41(7):697–700

Pogrebnjak AD, Bondar OV, Abadias G et al (2016) Structural and mechanical properties of NbN and Nb-Si-N films: experiment and molecular dynamics simulations. Ceram Int 42(10):11743–11756

Pogrebnjak AD, Bagdasaryan AA, Yakushchenko IV et al (2014) The structure and properties of high-entropy alloys and nitride coatings based on them. Russ Chem Rev 83(11):1027–1061

ZnO Doped Nanosized Composite Material Based on Hydroxyapatite and Sodium Alginate Matrix

Alexander D. Pogrebnjak, L. F. Sukhodub, L. Sukhodub, O. V. Bondar and A. Turlybekuly

Abstract The presented work shows new results of a nanostructured composite material investigation. The fabricated and studied material was in the form of beads based on hydroxyapatite (HA), sodium alginate (Alg) and ZnO particles. The HA was produced under the influence of microwave radiation. The beads of HA–ZnO–Alg were fabricated by dropping slurry into the calcium chloride solution. The main purpose of the research was to study the physical and chemical properties of HA–ZnO–Alg composite and approve its biocompatibility in vitro and in vivo. The morphology and elemental composition investigations were conducted by transmission electron microscopy with diffraction (TEM) and scanning electron microscopy with energy-dispersive analysis (SEM with EDX) methods. It was shown that synthesized HA consists of crystallites with a size of 40 nm. The ZnO inclusion appeared in the form of nanosized crystallites 25 nm in size. The average Ca/P ratio was 2.15, which is close to stoichiometric one.

Keywords Bioapatite · Biocomposite materials with nanoscale architecture · Hydroxyapatite · Sodium alginate · Zinc oxide

1 Introduction

The human bone tissue is a natural biocomposite material and demonstrates quite complicated properties. The bone composition (in general) consists of collagen fibrils, hydroxyapatite, and metallic microparticles. It is known that biological apatite includes doping elements' ions, such as Ca^{2+}, Mg^{2+}, Sr^{2+}, Ba^{2+}, Pb^{2+}, Zn^{2+}, Cu^{2+}, Na^{+}, K^{+}, Fe^{3+}. Understanding the own role of every consisting element and their

A. D. Pogrebnjak · L. F. Sukhodub · L. Sukhodub · O. V. Bondar
Department of Nanoelectronics, Sumy State University, Sumy, Ukraine

A. Turlybekuly (✉)
East-Kazakhstan State Technical University, Oskemen, Kazakhstan
e-mail: aturlybekuly@gmail.com

© Springer Nature Singapore Pte Ltd. 2019

A. D. Pogrebnjak and V. Novosad (eds.), *Advances in Thin Films, Nanostructured Materials, and Coatings*, Lecture Notes in Mechanical Engineering,
https://doi.org/10.1007/978-981-13-6133-3_35

influence on properties of the biological apatite has priority in the field of the development of synthetic composite materials for orthopedic application [1–15].

Nowadays, titanium-based materials are also widely used for biomedical applications, in particular—for prostheses construction, due to its extremely high biocompatibility with natural tissues [16–29].

Zinc ions stimulate the osteoinduction process. As a result, the density of bone tissue and the adhesion of proteins increase, the bone mass becomes constant [30–33] and surges the antibacterial activity [32]. Today, the problem of microorganisms' resistance to antibiotics becomes more and more notable, and as a result, the development of new materials with antibacterial properties is extremely actual. Antibacterial properties of HA doped by ions Zn, Cu, Au were studied in [33–41]. Thus, in [35] it was demonstrated that zinc sulfide with sodium alginate (ZnS + Alginate) has high antibacterial properties concerning some Gram-positive and Gram-negative bacteria. Sodium alginate is known to be biocompatible, non-toxic natural polysaccharide consisting of monomers of β-D-mannuronic and α-L-guluronic acids. It can bind and remove radionuclides from the body, decrease cholesterol in the blood and do not cause allergies. It is also known that materials based on ZnO are characterized by pronounced bioactivity, high tensile strength, ability to withstand extreme operating conditions.

This work is a pioneer in the proposed series of works, dedicated to the fabrication and further complex studies of a composite material, which will demonstrate properties quite close to natural bone tissue and consist of the hydroxyapatite, nanostructured ZnO particles, and sodium alginate. The studied material is promising in terms of use for various biomedical applications.

2 Experimental Details

The analytically pure bulk materials: calcium nitrate tetrahydrate $Ca(NO_3)_2{\cdot}4H_2O$, ammonium hydrophosphate $(NH_4)_2HPO_4$, 25% hydrous ammonia NH_4OH, zinc nitrate hexahydrate Zn $(NO_3)_2$ $6H_2O$, calcium chloride $CaCl_2$ were obtained from "Merck" firm; sodium alginate (Alg) (E401) with a molecular mass of 15 kDa (manufactured by Shanghai Chemical Company Ltd, China).

To fabricate the HA–ZnO–Alg beads, a 2 wt% aqueous solution of sodium alginate was added to the HA–ZnO slurry in a 1:1 weight ratio. The mixture was sonicated in a USD device (power of 75 W, 12 min duration) until a homogeneous mass was formed. By dropping the resulting slurry into 0.125 M calcium chloride solution, beads of the HA–ZnO–Alg composite material were obtained (further HA–ZnO–Alg sample). The exposure time in the $CaCl_2$ crosslinking solution was 25 min and 18 h accordingly. The resulting materials were filtered, washed with distilled water and dried. The preparation methods were described in [42, 43].

The synthesized composite material based on HA–Alg–ZnO was analyzed by scanning electron microscopy (SEM) using the JSM-6390LV microscope with the energy dispersive microanalysis system INCA Energy Penta FET X3, transmission electron microscopy (TEM) on JEM-2100 electron microscope. All these analytical investigations were conducted at university laboratories of East Kazakhstan State Technical University named after D. Serikbayev.

3 Results and Discussion

Investigation of the microstructure of the synthesized material showed that the composite material has a porous structure, Fig. 1. Microcrystals of ZnO, consisting of individual microcrystallites, are localized in the pores, Fig. 2. Localization of microcrystallites in porous can result in the material antibacterial properties increasing, due to the availability of active substances. Also, the addition of zinc oxide increases the swelling properties, as the osmotic pressure increases, which contribute to the absorption of the material.

The chemical composition of the synthesized material is shown in Table 1. From the spectra data, obtained from different sites, it can be concluded that the HA is distributed in the Alg matrix since the sodium concentration is almost unchanged. In general, we can conclude that the ratio of Ca/P is close to stoichiometric and equal to 2.02.

Figure 3 shows a TEM image of a HA–Zn sample, where represented crystallites of a needle-shape form, which is typical for bone apatite.

For a more detailed determination of the elements' distribution, we carried out the mapping on TEM, Fig. 4. As can be seen, the distribution of zinc is clearly limited by the region of the ZnO microparticle, while a small amount of phosphorus and calcium are present in the region of the ZnO microparticle. This fact may indicate that the HA molecules enter into the ZnO microparticle.

The swelling properties were measured at different pH values 4.5, 7.3 taking into account the natural conditions of a living organism where start the formation of new

Fig. 1 SEM image of the synthesized composite material HA–Alg–ZnO

Fig. 2 SEM image of ZnO microcrystalline

Table 1 Composition of the synthesized compozied material HA–Alg–ZnO (in wt%)

Spectrum	O	Na	Mg	Al	Si	P	Cl	Ca	Zn	Sum
Spec. 1	40.7	4.69	0.85	0.49	9.98	5.95	1.91	11.8	23.7	100.0
Spec. 2	38.8	3.99	0.72	0.40	10.7	6.61	1.83	12.7	24.3	100.0
Spec. 3	48.4	4.09	0.87	0.55	11.1	8.01	2.52	17.4	7.1	100.0
Average	42.6	4.25	0.81	0.48	10.6	6.86	2.09	14.0	18.3	100.0
Stand. deviation	5.08	0.38	0.08	0.08	0.56	1.05	0.38	3.0	9.8	
Max	48.4	4.69	0.87	0.55	11.1	8.01	2.52	17.4	24.3	
Min	38.8	3.99	0.72	0.40	9.98	5.95	1.83	11.8	7.1	

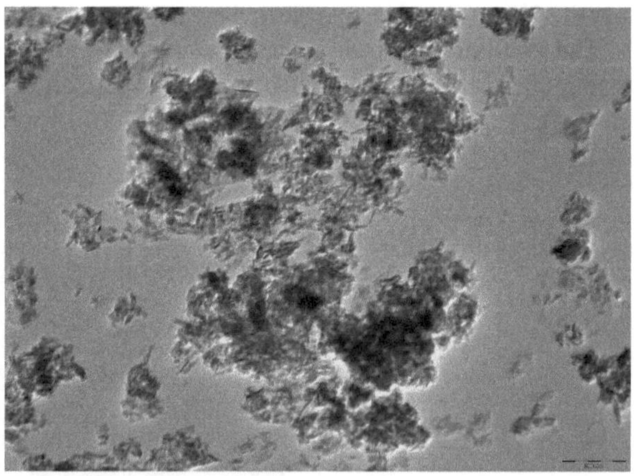

Fig. 3 TEM image of composite material

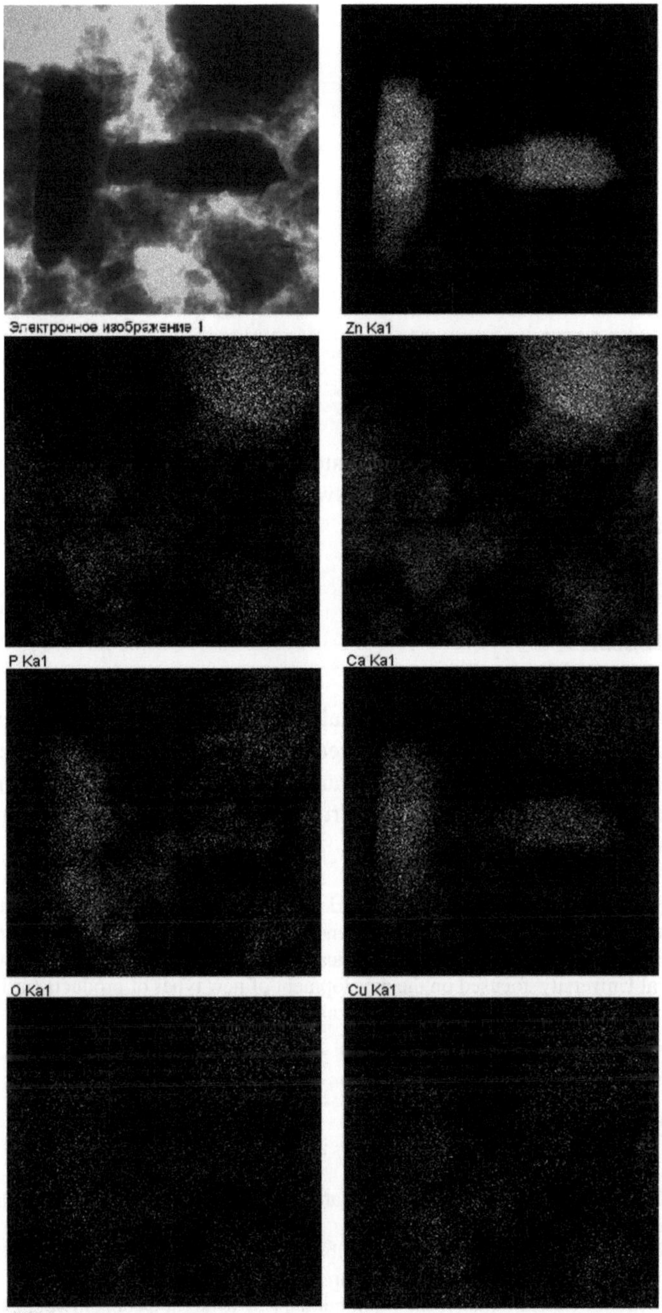

Fig. 4 TEM image (map) of composite material

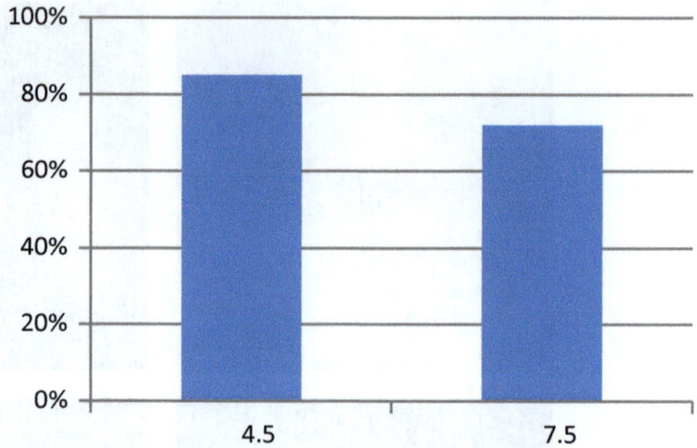

Fig. 5 The swelling diagram of HA–Alg–ZnO at different pH

bone tissue caused by osteoblasts and osteoclasts within the indicated pH values. Samples are slightly less susceptible to swelling in a neutral medium at pH = 7.3 than in acidic pH 4.5 (Fig. 5).

4 Conclusion

Synthesized ZnO doped composite material was studied. The samples have nanoscale architecture. The swelling properties decrease with pH level growing. The Ca/P ratio corresponds to the generally accepted parameters. In the study, it was found that HA molecules are embedded in the crystal structure of ZnO and form a chemical bond with the Ca atom substitution.

Acknowledgements The present work was carried out under the financial support of the Science Committee of the Ministry of Education and Science of the Republic of Kazakhstan for the targeted financing program "Targeted scientific and technical program of the D. Serikbayev East Kazakhstan State Technical University, focused on the development of new types of products for fabrication at the leading industrial enterprises of the East Kazakhstan region" for 2017–2019 years.

References

1. Ong K, Yun M, White J (2015) New biomaterials for orthopedic implants. Orthop Res Rev 7:107–130
2. Emami-Karvani Z, Chehrazi P (2011) Antibacterial activity of ZnO nanoparticle on Gram-positive and Gram-negative bacteria. African J Microbiol Res 5(12):1368–1373
3. Elliott JC (2002) Calcium phosphate biominerals. Rev Mineral Geochem 48(1):427–453
4. Elliott JC (1994) General chemistry of the calcium orthophosphates. In: Studies in organic chemistry
5. Dorozhkin SV (2015) Calcium orthophosphate deposits: preparation, properties, and biomedical applications. Mater Sci Eng C 55:272–326

6. Choudhury P, Agrawal DC (2012) Hydroxyapatite (HA) coatings for biomaterials. In: Nanomedicine, pp 84–127
7. Atri F, Rasaie V, Abbas A, Nikzad S, Fazel Anvari-Yazdi A (2017) Deposition of hydroxyapatite nano-particle on zirconia to improve its bonding. In: 41st annual conference of European Prosthodontic Association PA OP26, 2017, pp 28–29
8. Wahab R, Mishra A, Yun S-I et al (2010) Antibacterial activity of ZnO nanoparticles prepared via non-hydrolytic solution route. Appl Microbiol Biotechnol 87(5):1917–1925
9. Raghupathi KR, Koodali RT, Manna AC (2011) Size-dependent bacterial growth inhibition and mechanism of antibacterial activity of zinc oxide nanoparticles. Langmuir 27(7):4020–4028
10. Grigolo M, Dolzani B, Giannetti P, Tenucci C (2016) Use of a fully-resorbable, biomimetic composite hydroxyapatite as a bone graft substitute for posterolateral spine fusion: a case report. Int J Clin Exp Med 9(11):22458
11. Sukhodub LB (2014) Metal ions doped chitosan nanoparticles. J Nano Electron Phys 6:04034–04040
12. Dorozhkin SV (2016) Calcium orthophosphates ($CaPO_4$): occurrence and properties. Prog Biomater 5(1):9–70
13. Dorozhkin SV (2012) Amorphous calcium orthophosphates: nature, chemistry and biomedical applications. Int J Mater Chem 2(1):19–46
14. Ślósarczyk A, Paszkiewicz Z, Paluszkiewicz C (2005) FTIR and XRD evaluation of carbonated hydroxyapatite powders synthesized by wet methods. J Mol Struct 744–747:657–661
15. Tripathi A, Saravanan S, Pattnaik S et al (2012) Bio-composite scaffolds containing chitosan/nano-hydroxyapatite/nano-copper–zinc for bone tissue engineering. Int J Biol Macromol 50(1):294–299
16. Rokosz K, Hryniewicz T, Pietrzak K, Dudek Ł (2017) Development and SEM/EDS characterization of porous coatings enriched in magnesium and copper obtained on titanium by PEO with ramp voltage. World Sci News 80:29–42
17. Popovich A, Sufiiarov V, Polozov I et al (2016) Producing hip implants of titanium alloys by additive manufacturing. Int J Bioprinting 2(2):187–193
18. Pogrebnjak AD, Rogoz VM, Bondar OV et al (2016) Structure and physicomechanical properties of NbN-based protective nanocomposite coatings: a review. Prot Met Phys Chem Surfaces 52(5):802–813
19. Popovich VA et al (2017) Functionally graded inconel 718 processed by additive manufacturing: crystallographic texture, the anisotropy of microstructure and mechanical properties. Mater Des 114:441–449
20. Salahshoor M, Guo Y (2012) Biodegradable orthopedic magnesium-calcium (MgCa) alloys, processing, and corrosion performance. Materials (Basel) 5(12):35–155
21. Popovich AA et al (2016) Use of additive techniques for preparing individual components of titanium alloy joint endoprostheses. Biomed Eng (NY) 50(3):202–205
22. Pogrebnjak AD, Bratushka SN, Beresnev VM, Levintant-Zayonts N (2013) Shape memory effect and superelasticity of titanium nickelide alloys implanted with high ion doses. Russ Chem Rev 82(12):1135–1159
23. Pogrebnjak AD, Beresnev VM (2012) Nanocoatings nanosystems nanotechnologies. Bentham Science Publishers
24. Li X et al (2017) Novel bio-functional magnesium coating on porous Ti_6Al_4V orthopaedic implants: in vitro and in vivo study. Sci Rep 7(1):40755
25. Ahmed T, Rack HJ (1999) Low modulus biocompatible titanium base alloys for medical devices. U.S. Patent 5,871,595, issued 16 Feb 1999
26. Stráský J et al (2017) Increasing the strength of a biomedical Ti-Nb-Ta-Zr alloy by alloying with Fe, Si and O. J Mech Behav Biomed Mater 71:329–336
27. Kopova I, Stráský J, Harcuba P (2016) Newly developed Ti-Nb-Zr-Ta-Si-Fe biomedical beta titanium alloys with increased strength and enhanced biocompatibility. Mater Sci Eng C 60:230–238
28. El-Wassefy NA, Reicha FM, Aref NS (2017) Electro-chemical deposition of nano hydroxyapatite-zinc coating on titanium metal substrate. Int J Implant Dent 3(1):39

29. Hasan A, Waibhaw G, Saxena V et al (2018) Nano-biocomposite scaffolds of chitosan, carboxymethyl cellulose and silver nanoparticle modified cellulose nanowhiskers for bone tissue engineering applications. Int J Biol Macromol 111:923–934
30. Ito A et al (2005) Zinc-containing tricalcium phosphate and related materials for promoting bone formation. Curr Appl Phys 5(5):402–406
31. Ito A et al (2002) Zinc-releasing calcium phosphate for stimulating bone formation. Mater Sci Eng C 22(1):21–25
32. Moseke C, Gelinsky M, Groll J et al (2013) Chemical characterization of hydroxyapatite obtained by wet chemistry in the presence of V, Co, and Cu ions. Mater Sci Eng C 33(3):1654–1661
33. Gunalan S, Sivaraj R, Rajendran V (2012) Green synthesized ZnO nanoparticles against bacterial and fungal pathogens. Prog Nat Sci Mater Int 22(6):693–700
34. Kanhed S, Awasthi S, Midha S et al (2018) Microporous hydroxyapatite ceramic composites as tissue engineering scaffolds: an experimental and computational study. Adv Eng Mater 1701062:1–11
35. Mieshkov AV, Grebenik LI, Ivahnuk TV et al (2016) Antibacterial properties of the nanoparticles with the zinc sulfide quantum dots. IFMBE Proc 55:267–270
36. Pogrebjak AD, Krivets AS, Dyadyura KA et al (2016) Research of the relief and element composition of the surface coatings based on hydroxyapatite implants from titanium alloys. In: Proceedings of the international conference on Nanomaterials: Application & Properties (NAP), 2016, p 02NABM06
37. Shypylenko A et al (2016) Effect of ion implantation on the physical and mechanical properties of Ti-Si-N multifunctional coatings for biomedical applications. Mater Des 110:821–829
38. Saxena V, Hasan A, Pandey LM (2018) Effect of Zn/ZnO integration with hydroxyapatite: a review. Mater Technol 33(2):79–92
39. Dorozhkin SV, Epple M (2002) Biological and medical significance of calcium phosphates. Angew Chemie Int Ed 41(17):3130–3146
40. Bhowmick A et al (2015) Development of porous and antimicrobial CTS-PEG-HAP-ZnO nano-composites for bone tissue engineering. RSC Adv 120
41. Livitska O et al (2016) Copper (II), zinc (II) and copper (II)/zinc (II)-containing carbonate-substituted hydroxyapatite: synthesis, characterization and thermal behavior. Materwiss Werksttech 47(2–3):85–91
42. Sukhodub S et al (2014) Nanocomposite apatite-biopolymer materials and coatings for biomedical applications. J Nano-Electron Phys 6(2):1–5
43. Martynyuk OO, Sukhodub LF, Sukhodub LB et al (2016) Structural properties of apatite-biopolymer nanocomposites, doped ZnO. J Nano-Electron Phys 8(4):1–6

AC Dependence of Electrical Properties of SiO_x/ZrO_2 Multilayer Nanocomposites with Si Nanocrystals

T. N. Koltunowicz, K. Czarnacka and A. K. Fedotov

Abstract The following paper presents the method of obtaining silicon nanocrystals in a matrix of zirconium dioxide and the results of measurements of electrical properties. The tested material was produced by alternating vacuum evaporation with SiO_x and ZrO_2 and then annealed to obtain silicon nanocrystals. The measurement parameters in the function of temperature and frequency were: capacitance, resistance, the angle of phase shift and tangent of dielectric losses. On this basis, and referring to the dimensions of the sample, conductivity was determined as a function of temperature and frequency. Thanks to this, the mechanism of charge transfer and the nature of the material have been proposed.

Keywords Nanocomposites · Nanocrystals · Conductivity · Electrical properties

1 Introduction

The variety of properties of nanocomposites depends to a large extent on the method of their production. The process of nanocomposites producing is quite complicated and depends on many parameters. The main parameters that influence the formation process of nanostructures are the crystalline orientation of the substrate and the quality of its preparation, the temperature of the substrate, the time of deposition of the thin layer or the environment in which the layer is formed [1–4]. We can

T. N. Koltunowicz (✉)
Department of Electrical Devices and High Voltage Technology,
Lublin University of Technology, Lublin, Poland
e-mail: t.koltunowicz@pollub.pl

K. Czarnacka
University of Life Sciences in Lublin, 20-950 Lublin, Poland
e-mail: karolina.czarnacka@up.lublin.pl

A. K. Fedotov
Belarusian State University, 220030 Minsk, Belarus
e-mail: fedotov@bsu.by

© Springer Nature Singapore Pte Ltd. 2019

A. D. Pogrebnjak and V. Novosad (eds.), *Advances in Thin Films, Nanostructured Materials, and Coatings*, Lecture Notes in Mechanical Engineering,
https://doi.org/10.1007/978-981-13-6133-3_36

distinguish thin-layer nanocomposites, the structure of which may be layered [5, 6] or granular [7–9].

Existing technologies of producing thin nanostructure layers can be divided into physical and chemical ones, namely: laser evaporation [10], ion-beam sputtering [11], ion sputtering [12, 13], magnetron sputtering [14, 15], reactive vapor deposition [16], deposition from the gas phase [17], sol-gel [18], or atomic layers deposition ALD [19]. One of the tried and tested methods for controlling the size of silicon nanoparticles is the deposition of a silicon dioxide layer enriched with silicon between SiO_2 layers to form a multilayer system. During annealing, the silica layer acts as a diffusion barrier, thus limiting the growth of crystals [20, 21].

The purpose of the following work was an investigation of AC electrical parameters of multilayered SiO_x/ZrO_2 nanomaterials with Si nanocrystals, which were subjected to annealing to obtain silicon nanocrystals in the oxide matrix. On this basis, dielectric properties were determined, and AC charge transfer mechanism was proposed in this type of systems.

Impedance spectroscopy is a widely used method for testing nanocomposite materials [22]. The AC method gives the possibility to determine the basic electrical parameters as a function of frequency, which in turn, when measuring properties in a wide frequency range, gives the possibility to obtain information on the charge transfer mechanism in the material [23–25]. In addition, nanostructured materials can exhibit a number of optical properties, which is why extensive luminescence tests are carried out [26, 27].

2 Experimental Details

The SiO_x/ZrO_2 layered nanocomposite was deposited on a silicon substrate of orientation (100) by alternating vacuum evaporation of the suitable materials from two separate sources. In this way, 22 layers of alternately arranged SiO_x and ZrO_2 were deposited. Thicknesses of individual layers were obtained at the level of approx. 2 nm for ZrO_2 and approx. 8 nm for SiO_x. Then the obtained material was subjected to two-hour annealing at 1110 °C under a nitrogen atmosphere. As a result, structures with silicon nanoparticles were obtained. Silver contacts were applied to the surface of the samples to minimize the influence of point contact during electrical measurements.

Studies of AC electrical properties of manufactured and annealed layers were carried out using measurement stand described in the paper [28]. The measurement was carried out in the range of measuring temperatures T_p from 20 to 375 K. The measured parameters were resistance R_p, capacity C_p, phase shift angle θ and the tangent of the dielectric loss angle tgδ using alternating current in the frequency range 50 Hz–1 MHz. On this basis, the frequency dependence of the conductivity σ was determined.

3 Results and Discussion

Figure 1 shows the frequency-temperature dependence of capacity C_p of the material SiO_x/ZrO_2 for several selected measuring temperatures. It can be noticed that the capacity increases slightly as the measurement temperature increases, and decreases significantly with increasing frequency—by more than an order of magnitude. Figure 2 shows the frequency and temperature dependence of the phase shift angle. In the entire measuring temperature and frequency range, the angle is negative and does not exceed −90°. This shows the capacitive nature of the structure under study.

Figure 3 shows the frequency-temperature dependence of the tangent of the dielectric loss angle tgδ, and Fig. 4 presents the frequency and temperature dependence of conductivity σ of the material a-SiO_x/ZrO_2 for several selected measuring temperatures. The analysis of the graphs suggests that the material exhibits a dielectric conductivity. The conductivity value increases with the increase of the measurement frequency in the full temperature range. We can determine that in the studied material there is tunneling or hopping of electrons between silicon nanoparticles through the dielectric ZrO_2 barrier, so the AC model of the hopping conductivity proposed in the works [29, 30] can be used.

Figure 5 shows Arrhenius's dependence of conductivity determined at 10^5 Hz. It is possible to distinguish three change intervals with temperature. For low temperatures, conductivity practically does not depend on it. The conductivity activation energy is very low and amounts to $\Delta E_1 \approx 2.3 \times 10^{-4}$ eV. It is only in the higher temperature range; the conductivity starts to grow by almost an order of magnitude. And at this moment we have two sections of activation energy of the conductivity: at temperatures from approx. 100 to 200 K, where energy is $\Delta E_2 \approx 5.8 \times 10^{-3}$

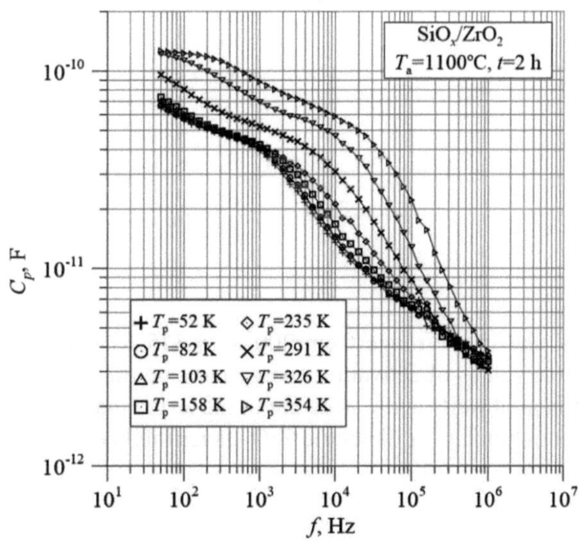

Fig. 1 Frequency-temperature dependence on capacity C_p of the nanocomposite a-SiO_x/ZrO_2

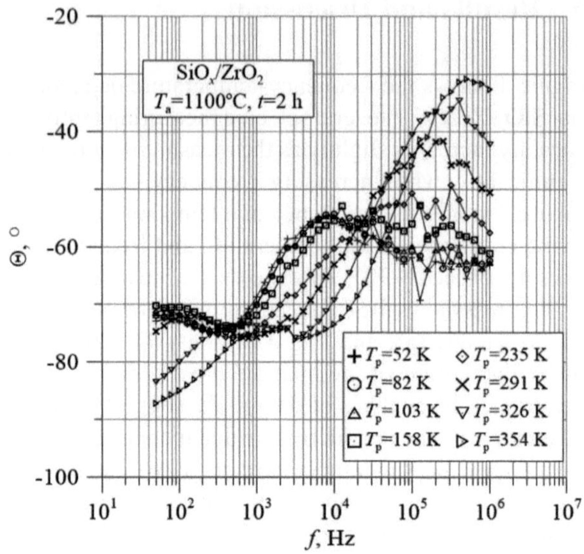

Fig. 2 Frequency-temperature dependence on the phase shift angle θ of the nanocomposite a-SiO_x/ZrO_2

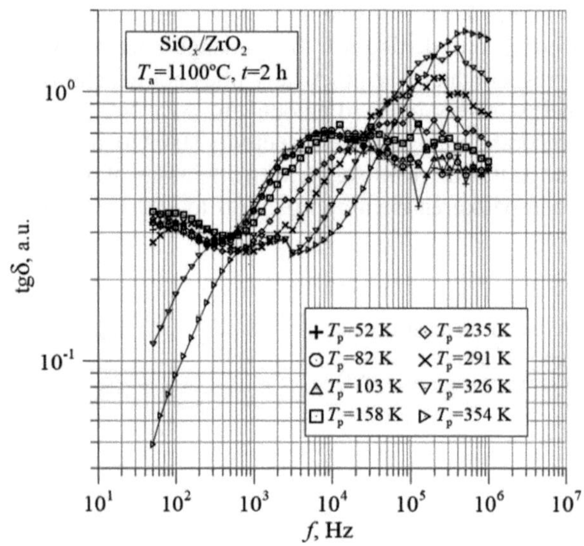

Fig. 3 Frequency-temperature dependence of the tangent of dielectric losses tgδ of the nanocomposite a-SiO_x/ZrO_2

eV and in the range from 200 K to 375 K, where $\Delta E_3 \approx 9.2 \times 10^{-2}$ eV. These three values of activation energy indicate that there is a minimum of three types of potential wells in the structure of the tested material, depending on the size of silicon nanogranules, or dielectric ZrO_2 layer thicknesses.

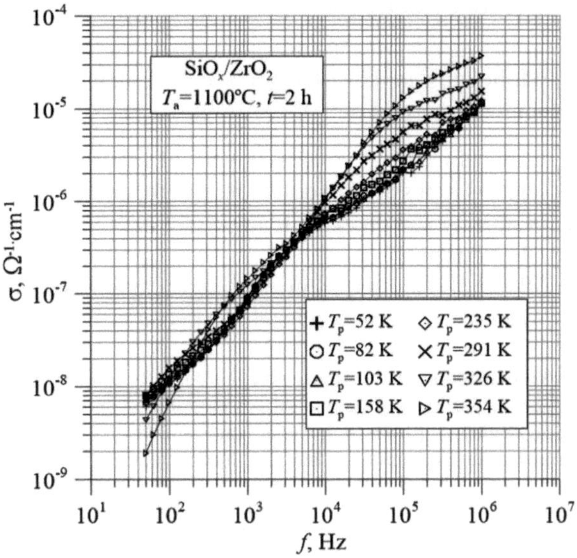

Fig. 4 Frequency-temperature dependence of conductivity σ of the nanocomposite a-SiO_x/ZrO_2

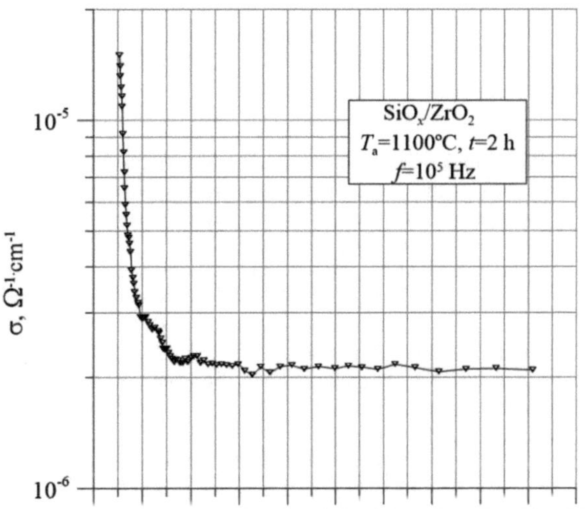

Fig. 5 Graph of Arrhenius conductivity at the measuring frequency 10^5 Hz of the nanocomposite a-SiO_x/ZrO_2

4 Conclusions

The following conclusions can be drawn on the basis of the electrical measurements which were carried out:

1. Annealing of alternating layers of amorphous silicon oxide and zinc dioxide makes it possible to obtain nanocomposites with silicon nanogranules;
2. Capacity and phase shift angle measurements determine the capacitive nature of the material;
3. By analyzing the frequency and temperature dependence of conductivity, the dielectric conduction mechanism can be confirmed.

Acknowledgements This research was partially supported by the Polish Ministry of Science and Higher Education as a statute tasks of the Lublin University of Technology, at the Faculty of Electrical Engineering and Computer Science, 8620/E-361/S/2018 (S-28/E/2018), entitled "Researches of electrical, magnetic, thermal and mechanical properties of modern electrotechnical and electronic materials, including nanomaterials and electrical devices and their components, in order to determination of suitability for use in electrical engineering and to increase the efficiency of energy management".

References

1. Pogrebnjak AD, Ivashchenko VI, Skrynskyy PL et al (2018) Experimental and theoretical studies of the physicochemical and mechanical properties of multi-layered TiN/SiC films: temperature effects on the nanocomposite structure. Compos B Eng 142:85–94
2. Kravchenko YO, Coy LE, Peplińska B et al (2018) Nano-multilayered coatings of (TiAlSiY)N/MeN (Me = Mo, Cr and Zr): influence of composition of the alternating layer on their structural and mechanical properties. J Alloy Compd 767:483–495
3. Pogrebnjak AD, Beresnev VM, Bondar OV et al (2018) Superhard CrN/MoN coatings with multilayer architecture. Mater Des 153:47–59
4. Smyrnova KV, Pogrebnjak AD, Beresnev VM et al (2018) Microstructure and physical-mechanical properties of (TiAlSiY)N nanostructured coatings under different energy conditions. Met Mater Int 24(5):1024–1035
5. Pogrebnjak AD, Beresnev VM (2012) Nanocoatings nanosystems nanotechnologies. Bentham Science, Oak Park. https://doi.org/10.2174/97816080541691120101
6. Rizal C, Ued Y, Karki BR (2012) Magnetic properties of Fe/Cu Multilayers prepared using pulsed-current electrodeposition. J Nano-Electron Phys 4(1):01001
7. Bondariev V, Zukowski P, Luhin V et al (2017) Thermo-gravimetric analysis of the nanocomposite (FeCoZr)x(CaF2)(100-x). High Temp Mater Process 21(4):289–298. https://doi.org/10.1615/HighTempMatProc.2018025484
8. Boiko O (2017) Ferromagnetic alloy—ferroelectric ceramic nanocomposites for nanoelectronics: the influence of a heat treatment on electrical properties. High Temp Mater Process 21(3):251–259. https://doi.org/10.1615/HighTempMatProc.2018025554
9. Koltunowicz TN, Zukowski P, Boiko O et al (2015) AC hopping conductance in nanocomposite films with ferromagnetic alloy nanoparticles in a PbZrTiO3 matrix. J Electron Mater 44(7):2260–2268. https://doi.org/10.1007/s11664-015-3685-9

10. Yu X, Shen MR (2008) Fabrication of anatase-type TiO2 films by reactive pulsed laser deposition for photocatalyst application. J Mater Process Technol 202(1–3):301–306. https://doi.org/10.1016/j.jmatprotec.2007.09.015
11. Kalinin YE, Ponomarenko AT, Sitnikov AV et al (2001) Granular metal-insulator nanocomposites with an amorphous structure. Fiz Khim Obrab Mater 5:14–20
12. Canham LT (1990) Silicon quantum wire array fabrication by electrochemical and chemical dissolution of wafers. Appl Phys Lett 57(10):1046–1048. https://doi.org/10.1063/1.103561
13. Choi SH, Elliman RG, Cheylan S et al (2000) Intrinsic defect-related blue-violet and ultraviolet photoluminescence from Si+-implanted fused silica. Appl Phys Lett 76(15):2062–2064. https://doi.org/10.1063/1.126255
14. Meng LJ, Dossantos MP (1993) Investigations of titanium-oxide films deposited by DC reactive magnetron sputtering in different sputtering pressures. Thin Solid Films 226(1):22–29. https://doi.org/10.1016/0040-6090(93)90200-9
15. Pshyk AV, Coy LE, Yate L et al (2016) Combined reactive/non-reactive DC magnetron sputtering of high temperature composite AlN-TiB2-TiSi2. Mater Des 94:230–239. https://doi.org/10.1016/j.matdes.2015.12.174
16. Ershov AV, Tetelbaum DI, Chugrov IA et al (2011) Annealing-induced evolution of optical properties of the multilayered nanoperiodic SiO_x/ZrO_2 system containing Si nanoclusters. Semiconductors 45(6):731–737. https://doi.org/10.1134/S1063782611060108
17. Kusdianto K, Jiang DP, Kubo M et al (2017) Fabrication of TiO2-Ag nanocomposite thin films via one-step gas-phase deposition. Ceram Int 43(6):5351–5355. https://doi.org/10.1016/j.ceramint.2017.01.009
18. Rashidi MJ, Mashkani MA, Khoshrou S et al (2017) Light harvesting applications of ZnFe2O4/NiTiO3 nanocomposite through the two-step sol-gel method. J Mater Sci-Mater Electron 28(15):11393–11400. https://doi.org/10.1007/s10854-017-6933-3
19. Mahajan AM, Khairnar AG, Thibeault BJ (2011) Pt-Ti/ALD-Al2O3/p-Si MOS capacitors for future ULSI technology. J Nano-Electron Phys 3(1):647–650
20. Gourbilleau F, Ternon C, Maestre D et al (2009) Silicon-rich SiO2/SiO2 multilayers: a promising material for the third generation of solar cell. J Appl Phys 106(1):013501. https://doi.org/10.1063/1.3156730
21. Ershov AV, Chugrov IA, Tetelbaum DI et al (2013) Thermal evolution of the morphology, structure, and optical properties of multilayer nanoperiodic systems produced by the vacuum evaporation of SiO and SiO2. Semiconductors 47(4):481–486. https://doi.org/10.1134/S1063782613040064
22. Adamchuck DV, Ksenevich VK, Gorbachuk NI et al (2016) Impedance spectroscopy of polycrystalline tin dioxide films. Devices Methods of Meas 7(3):312–321
23. Koltunowicz TN, Zukowski P, Czarnacka K et al (2015) Percolation phenomena in nanofilms Cux(SiOy)100-x produced by ion beam-sputtering. Acta Phys Pol A 128(5):908–911. https://doi.org/10.12693/APhysPolA.128.908
24. Svito I, Fedotov AK, Koltunowicz TN et al (2014) Hopping of electron transport in granular Cux(SiO2)1-x nanocomposite films deposited by ion-beam sputtering. J Alloy Compd 615(Suppl. 1):S371–S374. https://doi.org/10.1016/j.jallcom.2014.01.136
25. Koltunowicz TN, Zukowski P, Czarnacka K et al (2015) Dielectric properties of nanocomposite (Cu)x(SiO2)(100-x) produced by ion-beam sputtering. J Alloy Compd 652:444–449. https://doi.org/10.1016/j.jallcom.2015.08.240
26. Mudryi AV, Mofidnahai F, Karotki AV et al (2012) Silicon-germanium nanostructures with germanium quantum dots for optoelectronic applications. Devices Methods Meas 1:44–50
27. Czarnacka K, Komarov FF (2016) The influence of annealing on the electrical and optical properties of silicon-rich silicon nitride films. In: Proceedings of SPIE photonics applications in astronomy, communications, industry, and high-energy physics experiments 2016, Spie-Int Soc Optical Engineering, 2016, p 10031. https://doi.org/10.1117/12.2249111
28. Koltunowicz TN (2015) Measurement station for frequency dielectric spectroscopy of nanocomposites and semiconductors. J Appl Spectrosc 82(4):653–658. https://doi.org/10.1007/s10812-015-0158-0

29. Kołtunowicz TN, Fedotova JA, Zhukowski P et al (2013) Negative capacitance in (FeCoZr)-(PZT) nanocomposite films. J Phys D-Appl Phys 46(12):125304. https://doi.org/10.1088/0022-3727/46/12/125304
30. Koltunowicz TN, Zukowski P, Bondariev V et al (2015) Study of dielectric function of (FeCoZr)x(CaF2)(100-x) nanocomposites produced with a beam of argon ions. J Alloy Compd 650:262–267. https://doi.org/10.1016/j.jallcom.2015.07.276

Effect of Implantation Temperature and Annealing on Synthesis of ZnSe Nanocrystals in Silica by Ion Implantation

M. A. Makhavikou, F. F. Komarov, O. V. Milchanin, L. A. Vlasukova, I. N. Parkhomenko, E. Wendler and J. Żuk

Abstract ZnSe nanocrystals have been synthesized in the silicon dioxide matrix by ion implantation of Zn^+ and Se^+ ions at 25 and 550 °C with subsequent rapid thermal annealing at 1000 °C for 3 min. Structural and optical properties of implanted films were analyzed using Rutherford backscattering spectrometry, transmission electron microscopy, Raman spectroscopy, and photoluminescence. It was shown that the temperature of implantation, as well as thermal treatment, affects the structural and optical properties of implanted films. In the case of high-temperature implantation, ZnSe nanocrystals have been formed already during the implantation process. In the case of room-temperature implanted samples, ZnSe nanocrystals have been synthesized only after subsequent rapid thermal annealing. It was found that implanted silica layers exhibit photoluminescence in wide visible spectral range. The origin of photoluminescence of the as-implanted and annealed silica samples is discussed.

Keywords Silica layer · Ion implantation · Zinc selenide nanocrystals · Structure · Photoluminescence

1 Introduction

Up to now, there is still a problem of absent effective light source built-in silicon-based microelectronic chip. Creation of direct-band gap nanocrystals (NCs) in the Si-based matrix is one of the promising approaches to the solution of this problem.

M. A. Makhavikou (✉) · F. F. Komarov · O. V. Milchanin
A.N. Sevchenko Institute of Applied Physical Problems, Minsk, Belarus
e-mail: m.makhavikou@gmail.com

L. A. Vlasukova · I. N. Parkhomenko
Belarusian State University, Minsk, Belarus

E. Wendler
Institute Für Festkörperphysik, Friedrich-Schiller-Universität, Jena, Germany

J. Żuk
Institute of Physics, Maria Curie-Skłodowska University, Lublin, Poland

© Springer Nature Singapore Pte Ltd. 2019
A. D. Pogrebnjak and V. Novosad (eds.), *Advances in Thin Films, Nanostructured Materials, and Coatings*, Lecture Notes in Mechanical Engineering,
https://doi.org/10.1007/978-981-13-6133-3_37

The ideal matrix to synthesize such light-emitting quantum dots is the amorphous silicon dioxide. SiO_2 is the most important dielectric in the current planar silicon technology and the main material in the optical fiber. ZnSe is attracting much interest owing to its unique optical and electronic properties. Thin films of silicon dioxide with ZnSe inclusions can be used in the creation of long-lived light-emitting devices [1], solar cells [2], sensors [3] and optical recording materials [4]. Following the demonstration of possible applications of zinc selenide in photoelectronic devices, there has been increasing scientific attention to ZnSe as an alternative to the more toxic Cd-based materials.

Various synthesis methods for ZnSe nanocrystals, such as thermochemical processes [5], molecular precursor decomposition [6], reverse micelle synthesis [7], sol-gel [8] and other wet chemical processes [9] were reported. For practical applications, however, ion beam technique is becoming increasingly important due to its compatibility with the common Si technology. Indeed, the ion beam implantation is a widely applied technique not only to modify the electronic and optical properties of semiconducting and insulating materials. Moreover, the ion implantation is a suitable technique for the synthesis of nanostructured films and NCs embedded in different dielectric and semiconductor matrixes.

In this work, we demonstrate the creation of ZnSe NCs in silica by this technique. In our previous work [10], we studied the effect of implanted species order on structural properties of synthesized ZnSe NCs. It was shown that Zn^+ implantation at first results in a perfect crystalline quality of the synthesized nanocrystals. The present study is focused on the effect of implantation temperature and post-implantation treatment on the formation of ZnSe clusters.

2 Experimental Details

The samples of SiO_2 (600 nm)/Si were sequentially implanted with Zn and Se ions at 25 and 550 °C. Based on the TRIM simulation, the implantation energies (150 keV for Zn^+ ions and 170 keV for Se^+ ions) were chosen to obtain overlapping concentration profiles of Zn and Se species. Both types of ions were implanted with the same fluence of 4×10^{16} ions/cm^2. For a part of the implanted samples, rapid thermal annealing (RTA) at 1000 °C for 3 min in Ar atmosphere was carried out. Rutherford backscattering spectrometry (RBS) using 2.5 meV He^+ ions was employed to study the elemental composition of the implanted silicon oxide films. The structural composition was investigated by the cross-section as well as plan-view techniques of transmission electron microscopy (TEM) with a Hitachi H-800 electron microscope operating at 200 keV. Phase composition of the samples was investigated using Raman scattering (RS). RS measurements were performed in backscattering geometry with a Nanofinder High-End micro-Raman spectrometer (LOTIS TII) using a 532-nm laser beam as the excitation source. Photoluminescence was studied using the He-Cd laser beam at the wavelength $\lambda = 325$ nm as the excitation source at room temperature.

3 Results and Discussion

The depth profiles of the impurity concentration of the as-implanted and annealed SiO_2 films calculated from the experimental RBS spectra are shown in Fig. 1. The maximum of the concentration profiles for Zn and Se ions calculated with the computer simulation (TRIM) is located at a depth of ~113 nm. In the case of room temperature implantation, the concentration profile of Se atoms almost coincides with a simulated one. However, in the case of Zn profile, concentration in maximum is lower (4.9%), and profile is broader in comparison with the simulated Zn profile. The implantation at high-temperature results in the broadening of concentration profiles for both species and shift of their maximum positions to the surface (~59 and 95 nm for Zn and Se atoms, respectively). It is caused by non-equilibrium radiation-enhanced diffusion during high-temperature implantation. The post-implantation RTA results in further impurity redistribution. Namely, the profiles become slightly narrower, and diffusion tails stretched up to the SiO_2/Si interface appear after treatment.

Figure 2 shows the cross-section and plan-view TEM images of the as-implanted samples. In the case of the sample implanted at room temperature, the small clusters

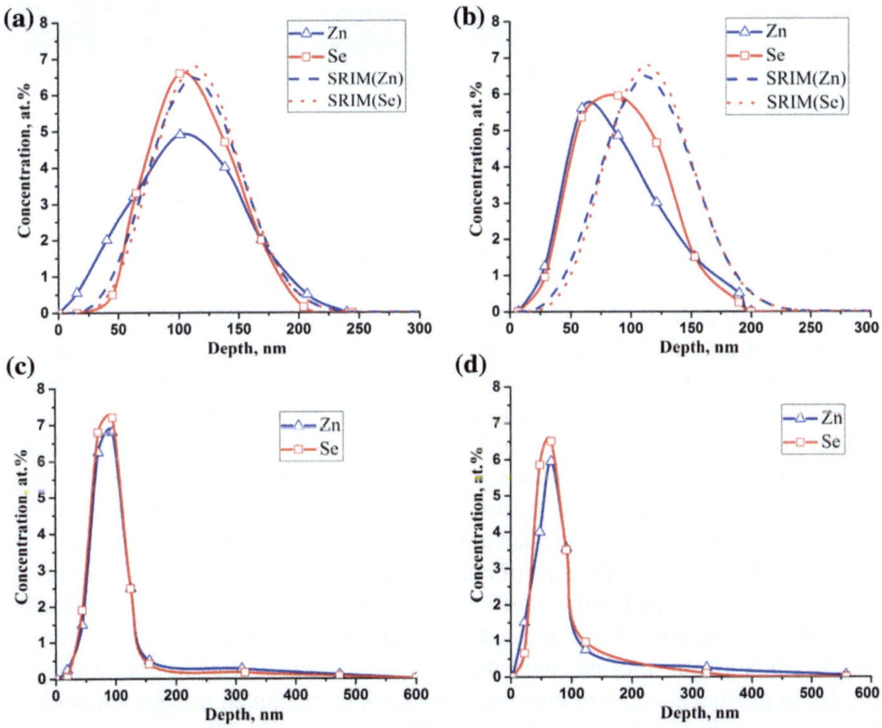

Fig. 1 Depth distributions of impurities in SiO_2 films after room temperature implantation (**a**, **c**) and "hot" implantation (**b**, **d**) for the as-implanted (**a**, **b**) and annealed (**c**, **d**) samples

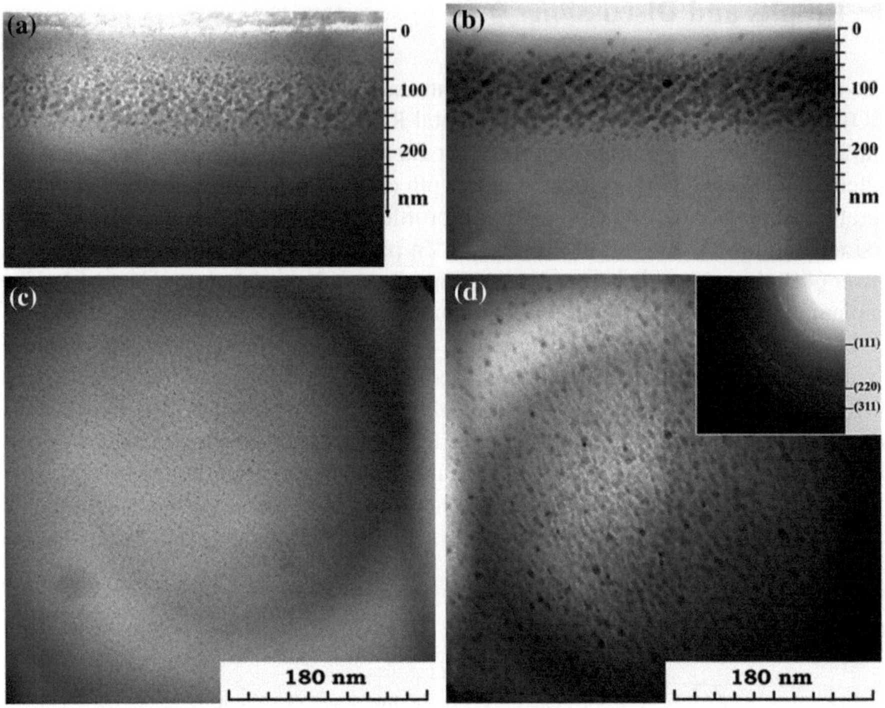

Fig. 2 Cross-section (**a**, **b**) and plan-view (**c**, **d**) TEM images of the as-implanted at 25 °C (**a**, **c**) and 550 °C (**b**, **d**) samples of SiO_2 films. Insert shows electron diffraction pattern

with the sizes less than 4–5 nm are observed at the layer located from the surface to the depth of 190 nm. Similar cluster distribution is also registered for the film implanted at 550 °C. However, in this case, the sizes of the cluster are bigger (2–10 nm). Besides, electron diffraction pattern shows that synthesized during implantation at high-temperature clusters are the nanocrystalline ZnSe with cubic sphalerite lattice.

Figure 3 presents the cross-section and plan-view TEM images of the implanted samples after RTA at 1000 °C. As can be seen, annealing results in the depth cluster redistribution and an increase of their sizes. The three spatial separated layers can be designed in the implanted at 25 °C film after RTA. The first layer contained clusters with sizes from 4 to 20 nm is located at the depth 0–120 nm from the surface.

It should be noted that the size of clusters in this layer increases with the depth. The second layer located at the depth 120–240 nm contains big clusters with sizes of 55–60 nm. The third layer located at the depth from 240 nm to the interface SiO_2/Si contains rare clusters with the average size of 15 nm. In contrast to the samples implanted at room temperature, the sample implanted at high temperature characterized with a lower concentration and bigger sizes (50–60 nm) of synthesized precipitates. Besides, the clusters have regular faceted shape, and secondary defects (twin boundaries) are registered inside them. These defects can be formed inside

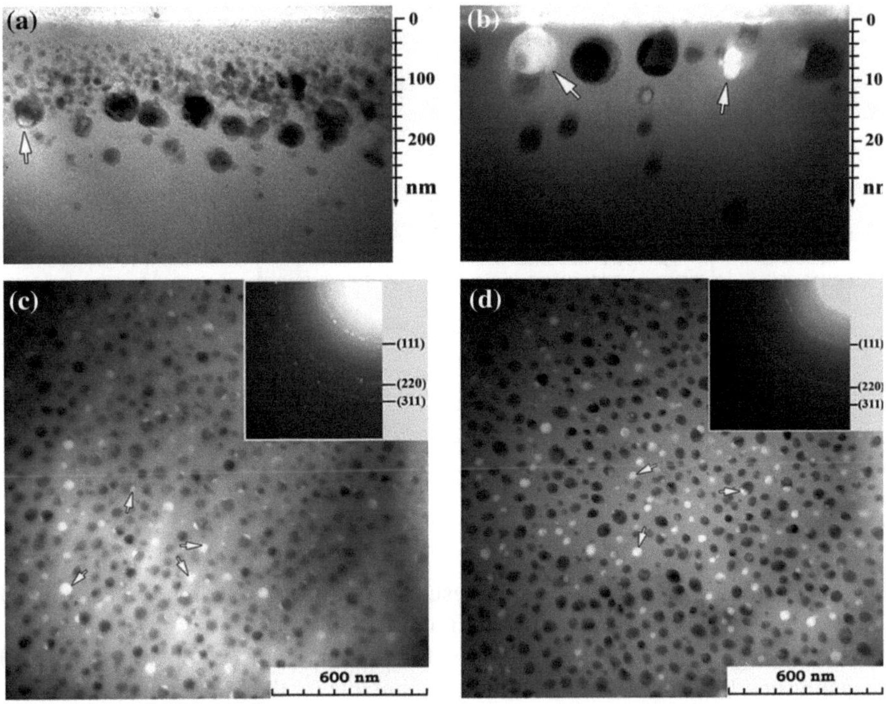

Fig. 3 Cross-section (**a**, **b**) and plan-view (**c**, **d**) TEM images of annealed samples of SiO_2 films implanted at 25 °C (**a**, **c**) and 550 °C (**b**, **d**). Inserts show electron diffraction patterns

clusters during their growth and via stress relaxation in crystal structure at thermal annealing.

It should be noted that the layer with big clusters is located at the depth of 60–70 nm from the surface what is in agreement with a maximum of depth concentration impurity profiles. The rare small clusters are registered in more deep layers. The electron diffraction pattern shows that silica films implanted at 25 °C as well as at 550 °C and undergone RTA at 1000 °C contain ZnSe nanocrystals with sphalerite structure. It should be noted that many voids are registered at the interfaces "cluster/SiO_2" in plan-view images of the annealed samples (shown by arrows) for both implantation temperatures.

Figure 4 shows the Raman spectra of the as-implanted and annealed samples. The second order 2TA (transverse acoustical) phonon mode of Si substrate is exhibited at ~302 cm^{-1} in all spectra. In contrast to the sample implanted at room temperature, the Raman spectrum of the as-"hot" implanted samples are characterized with the band at ~252 cm^{-1} assigned to LO (longitudinal optical) phonon of crystalline ZnSe [5, 6, 11]. Thus, it can be concluded that ZnSe nanocrystals are already formed in SiO_2 layers during the high-temperature implantation. In the case of room-temperature implantation, there is no any signal from crystalline ZnSe in the spectrum of the as-

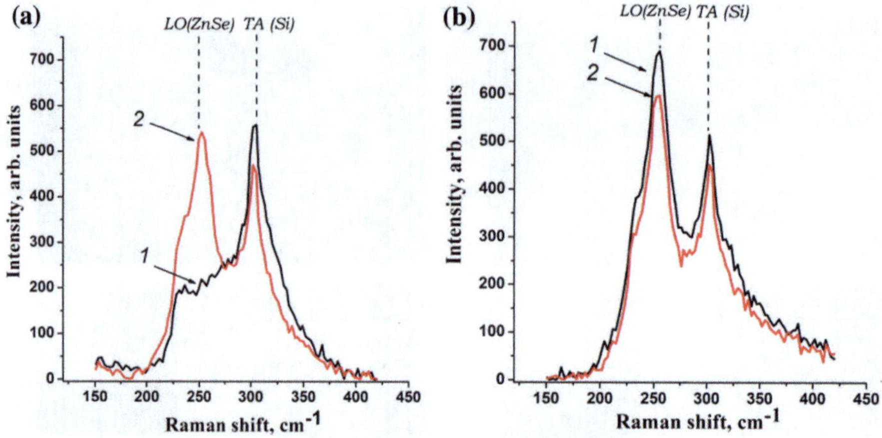

Fig. 4 Raman spectra of the as-implanted (curves 1) and annealed (curves 2) samples implanted at 25 °C (**a**) and 550 °C (**b**)

implanted sample. However, annealing results in an appearance of the band assigned to crystalline ZnSe phase in RS spectrum of the room-temperature implanted silica. These results are in agreement with the results of electron diffraction. So, the selected annealing conditions (1000 °C for 3 min) are suitable for the formation of ZnSe NCs after implantation at 25 °C. In the case of the high-temperature implanted samples, the RTA at corresponding conditions results in insignificant intensity decrease of ZnSe band. It should be noted that the intensity of the ZnSe band for the sample implanted at 550 °C is comparable with that one for the sample implanted at 25 °C and annealed at 1000 °C for 3 min.

Let us consider light-emitting properties of the ZnSe nanostructures. It is known that the room temperature PL spectrum of ZnSe nanostructures is typically dominated by two characteristic emission peaks. They are a near band edge (NBE) emission peak in the blue spectral range (for bulk ZnSe $E_g = 2.788–2.82$ eV) and a broad, deep defect (DD) emission band in the spectral range of 1.8–2.48 eV [12, 13]. The NBE emission is generally assigned to the excitons bound to neutral and charged acceptors and donors of different chemical nature and donor-acceptor pair recombination. The lines of acceptor-bound excitons (I_1) and donor-bound (I_2) in bulk ZnSe crystals are observed at low temperature in the energy range of (2.78–2.794 eV) and (2.794–2.802 eV), respectively [14–16]. In the case of pure 'not-intentionally' doped ZnSe, Zn interstitials and Se vacancies can play the role of donors, and Zn vacancies can play the role of acceptors [14].

On the other hand, the deep level emission is generally known to be mainly due to the intrinsic point defects such as vacancies, interstitials, stacking faults, and antisites [12]. Also, the greenish emission band was ascribed to deep radiative levels generated by strained crystal lattices or imperfections such as dislocations and vacancies [17, 18]. Additionally, the DD emission band can be assigned to the donor-acceptor pair recombination mechanism involving Zn vacancies as the acceptor species and Zn interstitials as the donor species with a large separation between them and surface emission [13, 19].

The SiO_2 samples implanted with Zn and Se ions also exhibit these bands with some peculiarities. Figure 5 shows room-temperature PL spectra of the as-implanted and annealed samples. The PL spectra of the samples implanted at 25 and 550 °C are composed of a strong blue band and less intensive green band. The intensity of PL spectra of the sample implanted at room temperature is significantly higher than for that one implanted at high-temperature. One can see a slight blue shift in the positions of blue and green PL bands for the sample implanted at 550 °C (2.85 and 2.32 eV) in comparison with ones for the sample implanted at 25 °C (2.75 and 2.23 eV.) Taking into account that ZnSe nanocrystals were synthesized during implantation process only in the sample implanted at 550 °C, the origin of the blue and green band in PL spectra of as-implanted samples could not be attributed to emission from ZnSe NCs. We supposed that the PL of as-implanted samples could be explained by radiative recombination via radiation defects in silica matrix generated during implantation (for example, Zn- or Se- related oxygen deficient centers) [20]. It is obviously that concentration of such defects is higher for the sample implanted at a lower temperature. Our assumption is supported by the fact of a more intensive luminescence which we observed for the sample implanted at room temperature than for the sample implanted at high temperature.

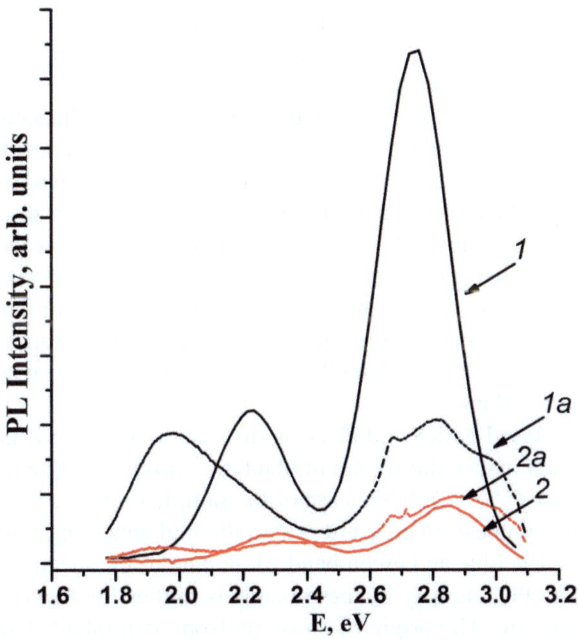

Fig. 5 Room-temperature PL spectra of SiO_2 (600 nm)/Si implanted by Zn and Se at 25 °C (1 and 1a) and 550 °C (2 and 2a): as-implanted (1 and 2) and after RTA (1a and 2a)

Annealing results in the change of intensity and spectral redistribution of PL spectra. PL intensity significantly decreases for room temperature implanted sample and slightly increases for the "hot" implanted one. It should be noted complicating of spectral shape in blue range for both types of the samples, and red shift of wide bands in spectral range (1.8–2.5 eV) after annealing. Since the ZnSe NCs were formed in both annealed samples, we supposed that in this case, the origin of luminescence could be attributed to emission from ZnSe NCs. Taking into account this assumption the complicated shape of the blue band can be explained by the different mechanism of NBE emission (mainly, excitons bound to neutral and charged acceptor and donor of different chemical nature, and donor-acceptor pair recombination). The bands in red spectral range can be attributed to deep defect emission from ZnSe NCs. It should also be noted that the PL intensity of the annealed sample implanted at room temperature is higher than for the sample implanted at high temperature. It can be explained by the larger concentration of small clusters with higher crystal quality in the annealed samples implanted at 25 °C.

4 Conclusions

ZnSe nanocrystals have been synthesized in the silicon dioxide matrix by ion implantation of Zn^+ and Se^+ ions at 25 and at 550 °C with subsequent rapid thermal annealing at 1000 °C for 3 min. The RBS data analysis shows that, in the case of room temperature implantation, a diffusion of Zn atoms in silica is more significant in comparison with Se atoms. In the case of "hot" implantation, the broadening of concentration profiles and their maximum shift towards the surface are observed for both implanted impurities due to non-equilibrium radiation-enhanced diffusion. Post-implantation rapid thermal annealing results in narrowing of concentration profiles with the simultaneous appearance of diffusion tails stretched up to the SiO_2/Si interface.

TEM images show the formation of a cluster at a depth of (0–190 nm) during implantation at 25 °C as well as at 550 °C. However, in the case of the sample implanted at high temperature, the sizes of clusters are bigger (2–10 nm) in comparison of the ones in the samples implanted at room temperature (less than 4–5 nm). Thermal annealing results in decreasing concentration of small clusters (especially for the sample implanted at 550 °C) and formation of large precipitates with the sizes up to 60 nm.

Based on RS and electron diffraction data, the ZnSe nanocrystals are created in silica matrix during the implantation process only at elevated implantation temperature (550 °C). In the case of the sample implanted at room temperature, ZnSe NCs are synthesized only after rapid thermal annealing at 1000 °C.

The blue and green bands dominate in PL spectra of both as-implanted samples. The PL intensity of these bands is higher for the sample implanted at room temperature. The origin of emission from as-implanted samples has been attributed to radiative recombination via defects in the silica matrix. Annealing results in spectral redistribution and intensity change of PL spectra for the samples implanted at the room temperature as well as for the ones implanted at the elevated temperature. In the case of annealed samples, one can conclude that the measured luminescence is due to emission from ZnSe NCs.

Acknowledgements This research was partly supported by the Belarusian Republican Foundation for Fundamental Research, grant no. F17M-053.

References

1. Mora-Sero I, Munoz V, Barbe M et al (1999) Study of the chemically activated sublimation of ZnSe. J Cryst Growth 197:497–503. https://doi.org/10.1016/S0022-0248(98)00796-9
2. Kale RB, Lokhande CD (2004) Room temperature deposition of ZnSe thin films by successive ionic layer adsorption and reaction (SILAR) method. Mater Res Bull 39:1829–1839. https://doi.org/10.1016/j.materresbull.2004.06.014
3. Li J, Li X, Yang R et al (2013) A sensitive electrochemical chlorophenols sensor based on nanocomposite of ZnSe quantum dots and cetyltrimethyl ammonium bromide. Anal Chim Acta 804:76–83. https://doi.org/10.1016/j.aca.2013.09.049
4. Krivoshlykov SG (2015) Holographic recording of infrared diffractive optics based on ZnSe material. Appl Optic 54(12):3569–3575. https://doi.org/10.1364/AO.54.003569
5. Pol SV, Pol VG, Calderon-Moreno JM et al (2008) Facile synthesis of photoluminescence ZnS and ZnSe nanopowders. Langmuir 24:10462–10466. https://doi.org/10.1021/la800921a
6. Norris DJ, Yao N, Chamock FT et al (2000) High-quality manganese-doped ZnSe nanocrystals. Nano Lett 1:3–7. https://doi.org/10.1021/nl005503h
7. Gong K, Kelley DF, Kelly AM (2016) Resonance raman spectroscopy and electron-phonon coupling in zinc selenide quantum dots. J Phys Chem C 120:29533–29539. https://doi.org/10.1021/acs.jpcc.6b12202
8. Y-l Duan, S-l Yao, Dai C et al (2014) Characterization of ZnSe microspheres synthesized under different hydrothermal conditions. Trans Nonferrous Met Soc China 24:2588–2597. https://doi.org/10.1016/S1003-6326(14)63387-2
9. Peng LL, Wang YH, Li CY (2010) Ultraviolet-blue photoluminescence of ZnSe quantum dots. J Nanosci Nanotechnol 10:2113–2118. https://doi.org/10.1166/jnn.2010.2086
10. Makhavikou M, Komarov F, Parkhomenko I et al (2018) Structure and optical properties of SiO_2 films with ZnSe nanocrystals formed by ion implantation. Surf Coat Techol 344:596–600. https://doi.org/10.1016/j.surfcoat.2018.03.017
11. Wang HI, Tang WT, Liao LW et al (2012) Femtosecond laser-induced formation of wurtzite phase ZnSe nanoparticles in air. J Nanomat 2012:278364. https://doi.org/10.1155/2012/278364
12. Zhang XB, Ha KL, Hark SK (2001) Selenium-related luminescent centers in metalorganic chemical-vapor-phase deposition grown ZnSe epilayers on GaAs. Appl Phys Lett 79:1127–1129. https://doi.org/10.1063/1.1394949
13. Philipose U, Xu T, Yang S et al (2006) Enhancement of band edge luminescence in ZnSe nanowires. J Appl Phys 100:084316. https://doi.org/10.1063/1.2362930
14. Kudlek GH, Pohl UW, Fricke Ch et al (1993) Electronic structure and dynamical behavior of different bound-exciton complexes in ZnSe bulk crystals. Phys B 185:325–331. https://doi.org/10.1016/0921-4526(93)90255-5
15. Kawakami Y, Ohnakado T, Tsuka M (1993) P-type ZnSe grew by molecular beam epitaxy with remote microwave plasma of N_2. J Vac Sci Technol, B 11:2057–2061. https://doi.org/10.1116/1.586542
16. Roppischer H, Jacobs J, Novikov BV (1975) The influence of Zn and Se heat treatment on the exciton spectra of ZnSe Single Crystals. Phys Stat sol (a) 27:123–127. https://doi.org/10.1002/pssa.2210270115

17. Chen HS, Wang SJJ, Lo CJ et al (2005) White-light emission from organics-capped ZnSe quantum dots and application in white-light-emitting diodes. Appl Phys Lett 86:131905. https://doi.org/10.1063/1.1886894
18. Deng Z, Lie FL, Shen S et al (2009) Water-based route to the ligand-selective synthesis of ZnSe and Cd-doped ZnSe quantum dots with tunable ultraviolet A to blue photoluminescence. Langmuir 25:434–442. https://doi.org/10.1021/la802294e
19. Wei J, Li K, Chen J et al (2012) Synthesis and photoluminescence of semiconductor ZnSe hollow microspheres by two–sourced evaporation. J Alloys Compd 531:86–90. https://doi.org/10.1016/j.jallcom.2012.03.112
20. Salh R (2011) Concentration and annealing effects on luminescence properties of ion-implanted silica layers. J At, Mol, Opt Phys 2011:1–7. https://doi.org/10.1155/2011/326368

Correction to: Strain Effect on the Properties of Polar Dielectric Thin Films

Alexander Tkach, Olena Okhay, André Santos, Sebastian Zlotnik, Ricardo Serrazina, Paula M. Vilarinho, and M. Elisabete Costa

Correction to:
Chapter "Strain Effect on the Properties of Polar Dielectric Thin Films" in: A. D. Pogrebnjak and V. Novosad (eds.), *Advances in Thin Films, Nanostructured Materials, and Coatings*, Lecture Notes in Mechanical Engineering, https://doi.org/10.1007/978-981-13-6133-3_32

In the original version of the book, the title of Chapter has been changed from "Stain Effect on the Properties of Polar Dielectric Thin Films" to "Strain Effect on the Properties of Polar Dielectric Thin Films".

The correction chapter and the book have been updated with the change.

The updated original version of this chapter can be found at
https://doi.org/10.1007/978-981-13-6133-3_32

© Springer Nature Singapore Pte Ltd. 2023

A. D. Pogrebnjak and V. Novosad (eds.), *Advances in Thin Films, Nanostructured Materials, and Coatings*, Lecture Notes in Mechanical Engineering,
https://doi.org/10.1007/978-981-13-6133-3_38

MIX
Papier aus verantwortungsvollen Quellen
Paper from responsible sources
FSC® C105338

If you have any concerns about our products,
you can contact us on
ProductSafety@springernature.com

In case Publisher is established outside the EU,
the EU authorized representative is:
Springer Nature Customer Service Center GmbH
Europaplatz 3, 69115 Heidelberg, Germany

Printed by Libri Plureos GmbH
in Hamburg, Germany